Merrill

Ecological Studies, Vol. 123

Analysis and Synthesis

Edited by

M.M. Caldwell, Logan, USA
G. Heldmaier, Marburg, Germany
O.L. Lange, Würzburg, Germany
H.A. Mooney, Stanford, USA
E.-D. Schulze, Bayreuth, Germany
U. Sommer, Kiel, Germany

Ecological Studies

Volumes published since 1990 are listed at the end of this book.

Springer

Berlin
Heidelberg
New York
Barcelona
Budapest
Hong Kong
London
Milan
Paris
Santa Clara
Singapore
Tokyo

W. Schramm   P.H. Nienhuis (Eds.)

Marine Benthic Vegetation
Recent Changes
and the Effects of Eutrophication

With 133 Figures and 55 Tables

 Springer

Dr. Winfrid Schramm
Institut für Meereskunde
Universität Kiel
Düsternbrooker Weg 20
24105 Kiel, Germany

Prof. Dr. Pieter H. Nienhuis
Department of Environmental Studies
University of Nijmegen
P.O. Box 9010
6500 Gl Nijmegen, The Netherlands
and
Netherlands Institute of Ecology
Centre for Limnology
Rijksstraatweg 6
3631 AC Niewersluis, The Netherlands

Front cover:
Locations in European coastal waters where changes in marine
benthic vegetation have been observed and documented

ISBN 3-540-58106-5 Springer-Verlag Berlin Heidelberg New York

Library of Congress Cataloging-in-Publication Data. Marine benthic vegetation: recent changes and
the effects of eutrophication Winfrid Schramm, Pieter H. Nienhuis (eds.). p. cm. – (Ecological
studies: (vol. 123) Includes bibliographical references and index. ISBN 3-540-58106-5 1. Benthic
plants – Europe. 2. Marine eutrophication – Europe. 3. Coasts – Europe. I. Schramm. Winfrid, 1937 – .
II. Nienhuis, P.H. III. Series: Ecological studies: v. 123. QK303.B46M37 1996 581.5'2638'094-dc20
96-3923

© Springer-Verlag Berlin Heidelberg 1996
Printed in Germany

The use of general descriptive names, registered names, trademarks, etc. in this publication does not
imply, even in the absence of a specific statement, that such names are exempt from the relevant
protective laws and regulations and therefore free for general use.

Cover design: Design & Production, Heidelberg
Typesetting: Thomson Press (India) Ltd., Madras

SPIN: 10104864   31/3137/SPS – 5 4 3 2 1 0 – Printed on acid-free paper

Preface

This book has resulted from joint activities of an international group of scientists associated with the EC-COST (acronym for "European Cooperation in the Field of Scientific and Technical Research") action programme 48 on Aquatic Primary Biomass, dealing particularly with marine macroalgae. In the framework of this cooperation, in an early phase the attention was directed to the effects of human impact on the marine macrophyte resources along the European coasts, in particular the ubiquitous and sometimes conspicuous changes in benthic macro-vegetation as a result of pollution and eutrophication. Consequently, one of the COST 48 working groups was established to study especially the macrophytic biomass production in relation to coastal eutrophication, the possibilities of removal of nutrients with the help of marine macrophytes, and ways of making use of the plant material produced.

In this context, at the beginning of 1994, a cooperative EC-project EUMAC (Eutrophication – Macrophytes) was launched to investigate the dynamics and mechanisms of the impact of eutrophication on marine benthic macrophytes in eight different locations along the Baltic, North Sea, Atlantic and Mediterranean coasts.

Although a lot of scattered information existed on algal blooms in Europe, this had never been collected in a readily available form. Therefore, during one of the working group meetings the idea was raised of compiling a synopsis of the existing knowledge related to the effects of eutrophication and pollution on the marine vegetation of the coasts of Europe. The result is this book to which 43 marine scientists from Europe and abroad have actively contributed.

The editors wish to thank the authors of the chapters for their efforts, the referees for their critical annotations, and the members of the COST 48 Management Committee for support of this book project.

One of the editors (W.S.) would like to express his special thanks to the University of San Carlos (USC) in Cebu City, Philippines, where most of the editing work was done. Special thanks are also due to the members of the USC Marine Research Section for their supportive interest in the work, to Dia Sotto-Alibo and Irene Caballes for secretarial assistance, as well as to Beate Wenskowski for the preparation of the index. Without all their assistance and patience the book project would not have advanced very far.

Spring 1996   W. Schramm
                                                   P. H. Nienhuis

Contents

Contributors

V. Aysel

Faculty of Science, Ege University, 35100 Bornova, Izmir, Turkey

V. Bombelli

KOBA Company (Biomass Provision and Conversion), via del Bollo 4, 20134 Milano, Italy

J. Buela

ENCE, Lourizan, Pontevedra, Spain

G. Cabeçadas

Portuguese Institute of Marine Research, AV. Brasilia, 1400 Lisboa, Portugal

J. Camp

Instituto de Ciencias del Mar, CSIC, Paseo Nacional s/n, Barcelona, Spain

R.H. Charlier

Free University of Brussels (Earth Sciences, VUB), 2 Avenue du Congo (Box 23), 1050 Brussels, Belgium

V. Clavero

Department of Ecology, University of Málága, Campus de Teatinos, 29071 Málaga, Spain

M.-L. De Casabianca

Station Méditerranéenne de l' Environment Littoral, Université Montpellier II, 1 Quai de la Daurade, 34200 Sète, France

E.G. De groodt

   Delft Hydraulics, P.O. Box 177, 2600 MH Delft, The Netherlands

I. De Vries

   Rijkswaterstaat, National Institute for Coastal and Marine
   Management, P.O. Box 20907, 2500 EX The Hague,
   The Netherlands

P. Dion

   CEVA (Centre d'Etude et de Valorisation des Algues), BP 3,
   22610 Pleubian, France

M. Espejo

   Conselleria de Pesca, Autonomic Government of Galicia,
   Santiago, Spain

C. Fernández

   Department of Ecology, University of Oviedo, Oviedo, Spain

J.A. Fernández

   Department of Plant Physiology, University of Málaga,
   29071 Málaga, Spain

F.G. Figueiras

   Instituto de Ciencias Marinas Vigo, Spain

F.L. Figueroa

   Department of Ecology, University of Málaga, Campus
   de Teatinos, 29071 Málaga, Spain

R.L. Fletcher

   The Marine Laboratory, School of Biological Sciences,
   University of Portsmouth, Ferry Road, Hayling Island,
   Hampshire PO11 ODG, UK

J.M. Fuentes

   Centro de Investigaciones Marisqueras de Galicia, Vilaxoan,
   Pontevedra, Spain

M.C. Garcia-Jiménez

Department of Ecology, University of Málaga, Campus
de Teatinos, 29071 Málaga, Spain

M.J. Garcia-Sánchez

Department of Plant Physiology, University of Málaga,
29071 Málaga, Spain

H. Güner

Faculty of Science, Ege University, 35100 Bornova, Izmir, Turkey

S. Haritonidis

Botanical Institute, Aristotle University, 54006 Thessaloniki,
Greece

I. Hernández

Department of Ecology, University of Cádiz, Cádiz, Spain

C. Jiménez

Department of Ecology, University of Málaga, Campus
de Teatinos, 29071 Málaga, Spain

S. Le Bozec

CEVA (Centre d'Etude et de Valorisation des Algues),
BP 3, 22610 Pleubian, France

M. Lenzi

Laboratorio die Ecologia Lagunare e Acquacoltura via
Leopardi 15, 58015 Orbetello (Gr.), Italy

Th. Lonhienne

State University of Liège, Liège, Belgium

A. Marcomini

Department of Environmental Science, University of Venice,
Calle Larga S. Marta 2137, 30123 Venice, Italy

A. Martínez-Arroyo

Centro de Ciencias de la Atmósfera, Universidad Nacional
Autónoma de México, CP 04510 México DF, México

J.E. Merrill

Department of Microbiology, 178 Giltner Hall, Michigan State
University, East Lansing, MI 48824, USA

I.M. Munda

Centre for Scientific Research, Slovene Academy of Science
and Arts, 6100 Ljubljana, Slovenia

F.X. Niell

Department of Ecology, University of Málaga, Campus
de Teatinos, 29071 Málaga, Spain

P.H. Nienhuis

Department of Environmental Studies, University of Nijmegen,
P.O. Box 9010, 6500 GL Nijmegen, The Netherlands
and Netherlands Institute of Ecology, Centre for Limnology,
Rijksstraatweg 6, 3631 AC Nieuwersluis, The Netherlands

J.C. Oliveira

Portuguese Institute of Marine Research, Av. Brasilia,
1400 Lisboa, Portugal

M. Pérez

Department of Ecology, University of Barcelona, Diagonal 645,
08028 Barcelona, Spain

J.L. Pérez-Llorens

Department of Ecology, University of Cádiz, Cádiz, Spain

C.J.M. Philippart

Institute of Forestry and Nature Research, P.O. Box 165,
1790 AD Den Burg, The Netherlands

J. Romero

Department of Ecology, University of Barcelona, Diagonal 645,
08028 Barcelona, Spain

W. Schramm

Institut für Meereskunde, Universität Kiel, Düsternbrooker
Weg 20, 24105 Kiel, Germany

A. Sfriso

Department of Environmental Science, University of Venice,
Calle Larga S. Marta 2137, 30123 Venice, Italy

M.W.M. van der Tol

Rijkswaterstaat, National Institute for Coastal and Marine
Management, P.O. Box 20907, 2500 EX The Hague,
The Netherlands

F. Vasiliu

Complexui Muzeal de Stiinte ale Naturii-Constanta,
Bul. Mamaia nr. 255, 8700 Constanta, Romania

I. Wallentinus

Dept. of Marine Botany and Göteborg Marine Research Centre,
University of Göteborg, Carl Skottbergs gata 22,
41319 Göteborg, Sweden

Introduction

W. SCHRAMM[1] and P.H. NIENHUIS[2]

There is a tendency towards an increasing burden to marine coastal systems as a direct result of growing human activities, which had already started at the end of the last century, but became especially manifest during the past three to four decades. Increasing amounts of various types of wastes and pollutants including nutrients enter the coastal waters via rivers, direct discharges from land drainage systems, diffuse land runoff, dumping and via the atmosphere. These wastes and pollutants are derived from industry, urbanization, traffic, agriculture and mariculture together with other forms of intensified use of land and sea.

As a result, in many shallow coastal areas which are the primary recipients of land runoff, a trend towards increasing levels of pollutants and nutrients has been observed. Particularly in enclosed or semi-enclosed water bodies such as estuaries, bays, harbours or fjords, the discharge of nutrient-carrying wastes has led to eutrophication and in extreme cases to hypertrophication.

Eutrophication is defined here as the process of natural or man-made enrichment with inorganic nutrient elements, mainly nitrogen and phosphorus, beyond the maximum critical level of the self-regulatory capacity of a given system for a balanced flow and cycling of nutrients. The critical levels may of course vary considerably between oligotrophic and eutrophic systems. Eutrophication may induce changes in the structure and in the flow of energy and matter through a system, although such changes are not necessarily harmful to the system. In fact, increased supply of nutrients per se into the sea, where nutrients are often limiting primary production, has generally been considered as beneficial in enhancing pelagic and benthic productivity and thus improving the harvest by man from the sea.

[1] Institut für Meereskunde, Universität Kiel, Düsternbrooker Weg 20, 24105 Kiel, Germany
[2] Netherlands Institute of Ecology, Centre for Limnology, Rijksstraatweg 6, 3631 AC Nieuwersluis, The Netherlands

Ecological Studies, Vol. 123
Schramm/Nienhuis (eds) Marine Benthic Vegetation
© Springer-Verlag Berlin Heidelberg 1996

In contrast, hypertrophication, also termed "nutrient pollution", means over-enrichment or excess supply of nutrients beyond the maximum critical self-regulatory level to an extent that detrimental processes cause irreversible changes in the aquatic communities, as long as nutrient loadings are not reduced.

The direct and indirect effects of landborne, seaborne and atmospheric pollution on coastal ecosystems have been documented in an ever growing number of observations on structural and functional changes leading to a general deterioration of the coastal environment. Ecological responses to pollution, either correlative or causal, comprise many examples such as enhanced plankton production, resulting in increased sedimentation and benthic production. In turn, this may lead to oxygen depletion events and die-back of the benthic system, and diseases and mass mortality among pelagic and benthic organisms.

Several of the most conspicuous effects which probably can be related to both pollution in general and to eutrophication and hypertrophication in particular, are the changes in the marine benthic vegetation, composed of microalgae, macrolgae (seaweeds) and seagrasses. Specific changes and secondary effects are typically conneted with increasing eutrophication:

1. Increasing nutrient levels.
2. Occurrence of phytoplankton blooms.
3. Decline or disappearance of certain perennial plant communities, often replaced by annual, fast growing forms (e.g. folious green algae or filamentous algae).
4. Reduced diversity of the flora and associated fauna.
5. Mass development (blooms) of short-lived annual forms which may become "nuisance algae", owing to the fact that they hinder fishing activities and navigation, or pollute amenity beaches when cast ashore.
6. Changing depth distribution of benthic algae owing to the reduced light transmittance through the water column.

As a result of these changes, secondary effects can be noted:

1. Increased plant biomass production necessarily leads to enrichment of the pool of particulate and dissolved organic material both in seawater and sediments.
2. Increasing numbers of benthic filter-feeders and detritus-feeders.
3. Enhancement of heterotrophic oxygen-consuming processes which lead to oxygen depletion and finally hydrogen sulphide development.
4. Mass mortalities of plants and animals, in particular in the benthic system.
5. Decline in zooplankton diversity and fish species.

Although all the above mentioned phenomena have been connected to eutrophication, the data are very variable, and hence it appears to be difficult to formulate causal relations between increasing nutrient loadings and biological responses.

Until recently, coastal eutrophication and the resulting effects on marine macrophytes were mainly treated as local short-term problems. In fact, these problems have been neglected in favour of eutrophication-related plankton blooms in the open sea. However, the local nearshore problems developed into overall coastal and inshore phenomena, and recently we are facing coastal eutrophication problems on a global scale (see Chap. 3).

Along the European coasts, in numerous places ranging from the innermost parts of the Baltic and the northernmost Atlantic shores to the eastern parts of the Mediterranean and into the Black Sea, changes of the macrophytobenthos, sometimes conspicuous and even dramatic, have been observed and documented (Fig. 1). Similar observations could probably be made in many more locations. The fact that changes in marine

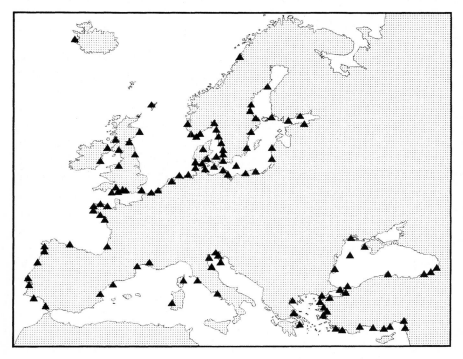

Fig. 1. Locations in European coastal waters where changes in marine benthic vegetation have been observed and documented

vegetation seem to occur mainly along northern European coasts (French and British Atlantic, North Sea, Baltic Sea) must certainly be related to higher industrialization and population density, but partly also just to the fact that more investigations have been carried out in the northern countries compared to the other European regions.

Up to now, no attempt has been made to approach the "marine eutrophication-macrophyte problem" on a European scale, and the long-term chronic effects of coastal eutrophication and "nutrient pollution", together with the ultimate offshore consequences have largely been ignored. In a few cases, these problems have been recognized and included in either regional, national or international projects or agree-ments (e.g. HELCOM, North Sea Convention); however, mostly on a descriptive, monitoring level.

Owing to sucessful abatement of the pollution problem in some Euro-pean countries, recent signals of de-eutrophication, decreasing levels of inorganic nutrients in coastal waters, can be observed. The alarming phenomenon is, however, that decreased loading has not yet led to de-creased marcophyte blooms.

Studies into the mechanisms of the impact of eutrophication and pollution on benthic macro-vegetation have recently been launched in Europe, for example in the framework of a cooperative EC-ENVIRON-MENT project EUMAC (Eutrophication and Macrophytes) of several Euro-pean universities and research centres. EUMAC has been initiated mainly because of the insight that effects of eutrophication on the coastal systems, in particular on phytobenthic communities, are very complex and cannot simply be explained as monofactorial effects. In other words, increased nutrient levels in most cases do not lead as a matter-of-course to mass development of algae or to the other observed changes in biological communities. On the way from the introduction of additional nutrients into a coastal system to the actual occurrence of typical eutrophication effects, other (combined) physical, chemical and biological factors are effective. More knowledge of the mechanisms of these interactions is a prerequisite for successful abatement and control, in order to protect and, hopefully, restore the coastal ecosystems.

Part A
General Aspects

1   The Occurrence of "Green Tides" – a Review

R.L. FLETCHER

1.1   Introduction

The eutrophication, or nutrient enrichment, of coastal waters as a result of man's activities is now widely recognized as a major, world-wide pollution threat. Essentially, the increased anthropogenic source of inorganic plant nutrients interferes with the natural annual nutrient cycles and can artificially enhance primary production during periods when activity is normally low. This can have quite considerable ecological consequences for both pelagic and benthic organisms. For example, phytoplankton activity will be increased (Hoogweg et al. 1991) and, although this can be generally beneficial by increasing fisheries (Raymont 1947; Fonselius 1978; Elmgren 1989), there is some evidence that it has resulted in the occurrence of some phytoplankton blooms, both toxic and non-toxic, which have had serious effects on local fisheries and leisure activities (Braarud 1945; Ruud 1968; O'Sullivan 1971; Zou and Dong 1983; Rosenberg 1985; Kimor 1991). One such bloom occurred in the northern Adriatic in 1988 when large quantities of mucilaginous material was washed up on many tourist beaches (Degobbis 1989; Vukadin 1991). A number of authors have similarly speculated on the possible relationship between eutrophication and the occurrence of toxic blooms of microalgae in the North Sea (Cole 1972).

An increasing number of world-wide reports are also highlighting the response of benthic algal communities to the impact of nutrients. Usually, these refer to the occurrence of excessive growths of macroalgae and the resultant detrimental ecological and environmental consequences. Notably, these excessive growths are largely comprised of green algae and, as deposits on the shore line, are commonly termed "green tides". The present paper examines the phenomenon of "green tides" and reviews aspects of

The Marine Laboratory, School of Biological Sciences, University of Portsmouth, Ferry Road, Hayling Island, Hampshire P01 ODG, UK

Ecological Studies, Vol. 123
Schramm/Nienhuis (eds) Marine Benthic Vegetation
© Springer-Verlag Berlin Heidelberg 1996

their world-wide distribution, ecological requirements, floristic composition, influence on the local ecology and methods of their control and utilization.

1.2   Eutrophication – The Background

Primary production in the sea is governed primarily by the natural cycles of nutrients. In general, nutrient levels are higher in the winter and lower in the summer (Nasr and Aleem 1948; Fonselius 1978) with the cycle largely controlled by the activity of phytoplankton which removes much of the available nitrogen and phosphorous from seawater during the spring bloom period (Wallentinus 1981, 1991). Nutrients can, therefore, be in short supply for benthic algae during the spring/summer period with nitrogen being the most likely nutrient limiting plant growth (Foster 1914; Alfimov 1959; Ryther and Dunstan 1971; Prince 1974; Wallentinus 1981; Sfriso et al. 1988a,b). Whilst some perennial algae are known to be capable of binding large quantities of N and P in their biomass [e.g. a bioaccumulation factor for nitrate of 28000 was reported by Chapman and Craigie (1977) for the large kelp alga *Laminaria longicruris*] which can be utilized during periods of low ambient levels, the growth and seasonal response of many algae largely revolves around the natural availability of the nutrients. Not surprisingly, therefore, any additional inputs of these nutrients will upset the natural seasonal cycles of growth and development.

Although there are a number of natural mechanisms outside the growth cycle of phytoplankton by which nutrients are released into the marine environment e.g. the release of nutrients from sediments disturbed by storms, from rotting benthic seaweed and as a result of upwelling systems (Fujita et al. 1989), many reports relate to man's activities. These include: the release of nutrients from "disturbed" sediments as a result of dredging activities, coastal construction works etc. (Perez-Ruzafa et al. 1991); nutrients derived from atmospheric fall-out, which usually occurs in coastal waters down-line from industrial areas (considered by Rosenberg 1985, 1990 and Wilson et al. 1990 to be a source of nutrients entering the Kattegat region of the Baltic and Dublin Bay, Ireland, respectively); nutrient-containing effluents discharged from various coastal refuge tips; nutrients originating from the growing number of fish farms which are in operation (Lumb and Fowler 1989; Tsutsumi et al. 1991); nutrients from river input (runoff) and percolating groundwater coming from agricultural land treated with fertilizers (Fitzgerald 1978; Rosenberg 1985; Wilson et al.

1990); nutrients associated with sewage effluents discharged from urban areas (Letts and Richards 1911; Wilkinson 1963; Sawyer 1965; Buttermore 1977; Wilson et al. 1990). Probably the last two systems are responsible for the majority of land-derived nutrients entering coastal waters and are, therefore, primarily responsible for eutrophication processes.

The relative contribution made by these two anthropogenic nutrient sources will, however, vary geographically. Not surprisingly, large catchment areas, such as the Po River Delta in Italy, are major sources of agriculturally-based nutrients (Justic 1987; Pugnetti et al. 1990), whilst nutrient contributions from urban areas are largely derived from sewage discharges (e.g. Letts and Richards 1911; Wilkinson 1963; Sawyer 1965; Bellan 1970; Buttermore 1977).

Although these sources represent quite a substantial input of nutrients into coastal waters, given suitable dilution factors and widespread dispersal, they would not give rise to eutrophication phenomena. Indeed, in some oligotrophic seas such as the Mediterranean, the overall effect of such added nutrients on primary production would probably be considered beneficial (Friligos 1987). It is only when the nutrients become entrapped in coastal systems and relatively quite high levels are reached that the recipient waters can be termed eutrophic and ecological imbalances ensue. Not surprisingly, therefore, reports of eutrophication phenomena are generally confined to largely enclosed water systems, both artificial and natural, where water exchange/movement is reduced. These include, for example, basins (Nasr and Aleem 1948), loughs (Letts and Richards 1911), bays (Fletcher 1974; Wilson et al. 1990; Park 1992), harbours (Nicholls et al. 1981) and lagoons (Buttermore 1977; Sfriso 1987; Wilson et al. 1990). However, given the right hydrodynamic conditions, eutrophication phenomena can also occur on apparently more open coasts, e.g. in the Bay of St. Brieuc, Brittany, as a result of nutrient entrapment by tidal currents (Piriou and Menesguen 1990). Similar hydrodynamic phenomena might explain reports of "green tides" in some bays on the relatively exposed south-east tip of England (Fletcher 1974). There have also been a number of reports of coastal eutrophication problems arising as a result of hydrographical alterations, construction work, sand movement etc. limiting tidal interchange (Sawyer 1965; Fahy et al. 1975; Buttermore 1977; Sfriso et al. 1987; Ruiz et al. 1990; Park 1992). In the Venice Lagoon, northern Adriatic, this was the result of channel excavations (Sfriso et al. 1987). In a shallow lagoon in Hobart, Tasmania, and in the Rogerstown Estuary, Ireland, it was the result of causeway constructions (Buttermore 1977 and Fahy et al. 1975 respectively), whilst in Boston Harbour it was the result of a channel becoming blocked and preventing fresh ocean water entering (Sawyer 1965). Hydrodynamic changes resulting in habitat changes were also considered responsible for the expansion of *Ulva* on the mudflats in Langstone Harbour, south coast of England (Lowthion et al. 1985). Finally, it is perhaps pertinent here to report of larger enclosed bodies of water, including the Mediterranean Sea (Bellan and Peres 1972), the northern Adriatic Sea (Revelante and Gilmartin 1976; Justic 1987; Degobbis 1989; Marchetti et al. 1989) and the Baltic (Dybern 1972; Fonselius 1972; Cederwall and Elmgren 1980; Larsson et al. 1985; Bonsdorff et al. 1991) becoming eutrophic.

1.3   Secondary Effects of Nutrient – Carrying Waters

Any consideration of the ecological impact of nutrients derived from these anthropogenic sources, particularly those, for example, discharged with sewage effluent, must take into account the effect of other potential pollutants contained within the carrying waters (O'Sullivan 1971). Both sewage effluent and agricultural runoff will contain fresh water, suspended solids and biological waste material. In addition, sewage contains various surfactants/dispersants and occasionally traces of heavy metals, whilst agricultural runoff can include quantities of herbicides/pesticides as well as hydrocarbons originating from roads (Bell et al. 1989). Many of these pollutants are known to exert a marked effect on marine benthic communities, particularly marine benthic algae. For example, suspended particulate matter including organic material will increase the turbidity of the water, with a concomitant reduction in light penetration (Bellamy and Whittick 1964; Bellamy et al. 1972; Bellan 1967b; O'Sullivan 1971; North et al. 1972). The photosynthetic activity of benthic algae would thus be reduced (O'Sullivan 1971) under conditions of high sediment loading and this could prove limiting for some benthic algae, particularly those in the sublittoral zone (Grenager 1957; Rueness 1973). For example, Burrows (1971) and Burrows and Pybus (1970, 1971), revealed a critical light requirement for gametophyte fertilization in *Laminaria saccharina*. Reported effects of reduced light penetration due to pollutants include a decline in *Fucus* biomass in Kiel Bight, Western Baltic (Vogt and Schramm 1991), the absence of *Fucus serratus* in effluent receiving fjord waters in Norway (Bokn 1979), a decline in *Macrocystis* beds in the vicinity of sewage outfalls in California (Devinny and Volse 1978) and a restricted distribution of *Laminaria saccharina* on the Northeast coast of England (Burrows and Pybus 1970). Competitive advantage would also be given to low-light-adapted algae, as suggested, for example, for *Phyllophora truncata*,

Phycodrys rubens and *Polysiphonia nigrescens* which have apparently widely replaced *Fucus spp.* communities in Kiel Bight below 2 m depths (Vogt and Schramm 1991). Advantage would also be afforded to algae which can make efficient use of variable irradiances as shown for *Ulva rotundata* (Henley and Ramus 1989).

Also reported are changes in algal zonation patterns and depth distributions as a result of increased sediment loading (Gilet 1960; North et al. 1972; Devinny and Volse 1978; Meistrell and Montagne 1983; Kautsky et al. 1986; Breuer and Schramm 1988; Rosenberg et al. 1990; Rueness and Fredriksen 1991; Vogt and Schramm 1991; Munda 1993). For example, a reduction in the kelp depth range has been reported in the vicinity of sewage outfalls with plants generally restricted to shallow water (Clendenning and North 1959; Bellamy et al. 1972) and very few new kelps present under the old ones (Clendenning and North 1959). Decreased water transparency has also been implicated in changes in the depth penetration of eelgrass, from 10–12 to 5–6 m from 1900 to 1989, in the South Kattegat region of the Baltic (Nielson and Knudsen 1990), in decreased depth penetration of *Fucus vesiculosus* in the Baltic, e.g. from a maximum depth of 11 to 2 m between 1944 and 1984 (Vogt and Schramm 1991; Kautsky et al. 1986, 1992) and in vertical distribution changes of macroalgae on the Swedish west coast (Rosenberg et al. 1990).

Silt deposition will also be increased in the vicinity of sewage outfalls and this can exert a number of detrimental influences on marine benthic algal communities. The sediments can cover all available substrata interfering with the processes of spore (and larval) attachment/recruitment (Devinny and Volse 1978). They can smother young germlings (Devinny and Volse 1978) and inhibit their growth and development. Combined with water movement, sediments can also abrasively scour surfaces of settled spores. As deposits on algal thalli they will also reduce photosynthetic activities, at the same giving competitive advantages to more adapted species/life-forms of algae. All the above-described detrimental influences of increased sedimentation on benthic communities have been well-recognized (Clendenning and North 1964; Rueness 1973) and have been offered as explanations for the decline of kelp beds off the coast of southern California (Clendenning and North 1959, 1964; North 1963; Devinny and Volse 1978; Neushal et al. 1976; Meistrell and Montagne 1983; James et al. 1990). They also find support in the experimental studies of Devinny and Volse (1978) who revealed that sediments interfere with *Macrocystsis* gametophyte development. A similar detrimental effect of silt on zoospore development was shown for *Laminaria saccharina* by Burrows (1971). Increased sedimentation has also been implicated in the decline of *Posidonia* beds (Ruiz et al. 1990), in the decline of the red crustose coralline alga

Lithothamnion (Ffrench-Constant 1981), in restricting the distribution of *Laminaria* spp. (Burrows and Pybus 1970; Ebling et al. 1960) and in the general absence/ impoverishment of algae, leaving only a few selected species/ life-forms, in the vicinity of sewage outfalls (North 1963; Clendenning and North 1964; Borowitzka 1972; Kindig and Littler 1980).

The high organic loading of many effluent-derived sediments, particularly those originating from sewage treatment works, also exerts a number of secondary influences on marine benthic communities. Probably the most notable of these is the role played by organic matter as an additional food source to benthic animal communities, especially filter feeders (Bellamy et al. 1972). For example, a number of authors have noted an increase in the predominance and biomass of mussels in polluted waters, especially in the vicinity of sewage outfalls. Bellan and Bellan-Santini (1972), Bellan-Santini (1969) and Bellan (1967a; 1970) reported an increase in *Mytilus gallo-provincialis* in the vicinity of a sewage outfall near Marseilles, whilst there have been numerous similar reports for the species *Mytilus edulis*, especially in the Baltic (Bleguad 1932; Ehrhardt 1968; Johnston 1971/72; Riggio et al. 1990; Vogt and Schramm 1991; Kautsky et al. 1992). In situations of pollution abatement, for example, as a result of redirecting sewage effluent into deeper water or by improving treatment practices, a concomitant reduction in organic loading has been matched by a decline in *Mytilus edulis* populations (Bokn et al. 1992). This competitive advantage given to mussels in the vicinity of sewage outfalls has significant ecological connotations, especially for marine benthic algae. For example, reports suggest that mussels are unsuitable substrata for many algae (Klavestad 1967); they compete for available space (Johnston 1971/72), have been reported to cover surfaces of algae (Kautsky et al. 1992) and act as very efficient filters of potentially settling spores. Not surprisingly, therefore, several authors have linked increased mussel populations with declines in algal settlement and establishment, leading to their impoverishment and elimination (Johnston 1971/72; Riggio et al. 1990; May 1985; Vogt and Schramm 1991; Kautsky et al. 1992). Note, however, Letts and Richards (1911) reported that mussels provided firm anchorage to *Ulva* plants in Belfast Lough, Ireland with their attaching byssal threads holding onto the fronds.

There have also been many reports of increased grazing activity in the vicinity of many sewage outfalls, with sea urchins, in particular, deriving nourishment from the organic component of the sewage (North 1963, 1979, 1983; Devinny and Volse 1978; Dean et al. 1984; James et al. 1990). The increased number of grazers has been reported to cause a decline in *Macrocystis* kelp beds in California (Clendenning and North 1959; North 1963; Devinny and Volse 1978) by subjecting old kelp plants to high grazing pressure, attacking holdfasts and chewing the stipe leading to the loss of the upper blade portions etc. (Devinny and Volse 1978; North 1983; Dean et al. 1984; James et al. 1990). Young plants are also extremely vulnerable and this is borne out by reports of very few young kelps being present near the outfalls (Clendenning and North 1959). Other changes as a result of increased pollution-related grazer activity include reports of increased numbers of detritivorous snails e.g. *Hydrobia* (Kautsky et al. 1992) and enhanced grazing pressure by the isopod *Idothea baltica* on *Fucus ves-iculosus* (Kangas et al. 1982; Hällfors et al. 1984; Kangas and Niemi 1985; Wallentinus 1991), both reports being for the Baltic. The latter was attributed to the increased abundance of the *Idothea*, possibly due to nutrient-related changes in the phenolic content of the *Fucus* plant making the latter more vulnerable to grazing activity (see Wallentinus 1991 for discussion).

Apart from the above-described secondary effects of organic enrich-ment, in cases of severe pollution, with high organic loading, a number of authors have reported a much reduced and specialized flora. Notably, there appears to be a predominance of blue-green algae, especially species of the genera *Lyngbya* and *Phormidium* (Nasr and Aleem 1948; Grenager 1957; Golubic 1970; Munda 1974; Littler and Murray 1974; Murray and Littler 1974, 1976, 1983). This predominance of blue-green algae is probably the result of the anaerobic conditions and low pH values present (Otte 1979). It is also noteworthy here that sewage-derived sediments might well have toxic ingredients (Clendenning and North 1964; Devinny and Volse 1978). This was revealed for older, deeper sediments in experimental studies using gametophytes of the kelp *Macrocystis* (James et al. 1990); re-exposure of these contaminated sediments by disturbance would, therefore, have unde-sirable ecological consequences. Similar inhibitory properties of silt on algae have been recorded by Pybus (1971). Finally, with the possible presence of conservative materials such as heavy metals, various organic chemicals, pesticides, surfactants, detergents, chlorine etc., in sewage effluent (Bell et al. 1989; O'Sullivan 1971) it is not surprising that the latter has proved inhibitory to marine organisms. For example, Ogawa (1983) revealed municipal waste water to be inhibitory to three *Sargassum* spp, in culture. Borowitzka (1972) also considered that observed reductions in the numbers of brown and red algal species near outfalls might be due to the toxic effects of detergents or surfactants present, whilst Kindig and Littler (1980) and Edwards (1972), using experimental culture studies, conclude that toxic chemicals are associated with sewage outfalls. These conclusions concur with the findings of Burrows (1971) that detergents affected the growth of *Laminaria saccharina* in laboratory culture; detergents have also been implicated in reducing settlement density in this species (Pybus 1973). Certainly it is likely that the levels of anionic detergents present in sewage

(5-10 ml) will have inhibitory effects (Clendenning and North 1964). Pertinent to this discussion are reports of the presence of tumours on algae near sewage outfalls (Katayama and Fujiyama 1957; North et al. 1972; North 1979).

1.4   Eutrophication – Influence on Marine Benthic Algae

Van Bennikom et al. (1975) defined eutrophication as "an acceleration of chemical inputs that favour photosynthesis and influence algal popula-tions". Probably the most important chemicals required by algae are sour-ces of nitrogen and phosphorus. As nitrogen is very often considered to be the major limiting nutrient for plant growth in coastal waters, any additi-onal inputs, for example, from sewage effluent or agricultural runoff, would, therefore, be expected to exert a major influence on local plant communities.

Probably the most widely noted effect of coastal eutrophication is an increase in the productivity of a region. Provided other pollutants present are not limiting and away from the immediate and usually turbid, inhibi-tory environment of a sewage discharge point (Munda 1974), a number of authors have reported an increase in primary productivity in the vicinity of the nutrient source, both with respect to pelagic phytoplankton communi-ties (Revelante and Gilmartin 1976) and macroalgal benthic communities (Causey et al. 1945; McNulty 1959; Clendenning and North 1964; Subbaramaiah and Parekh 1966; O'Sullivan 1971; Johannes 1972; Tewari 1972; Anonymous 1973; Saunders and Lindsay 1979; Bell et al. 1989). In general, the increased phytoplankton activity is considered to be detrimental to the growth of benthic algae. For example, it was reported to have increased the number of mussels which compete with the macroalgae for space, whilst the associated increase in turbidity was reported to have reduced the lower depth distribution of *Fucus* in the Baltic (Kautsky et al. 1992). The increased productivity of benthic macroalgae, however, has been manifest in reports of larger plants (Nasr and Aleem 1948; Johannes 1972), an increase in *Laminaria hyperborea* stipe production (Bellamy and Whittick 1964), more primary biomass (Jeffrey et al. 1992) and generally more abundant life (McNulty 1959). This quantitative increase in benthic algae is, however, tempered by a number of qualitative changes, with many authors reporting a general impoverishment of the flora and an overall reduction in species numbers and diversity (Gamulin-Brida et al. 1967; Smyth 1968; Johnston 1971/72; Borowitzka 1972; Littler and Murray 1974; Munda 1974, 1993; Enright 1978; Hirose 1978; Ffrench-Constant 1981). It is these qualitative changes associated with eutrophicated waters which have caught the attention of authors and caused concern to ecologists and environmentalists.

Probably one of the most noticeable floristic changes associated with eutrophication is a reported decline in the number of perennial macroalgae contributing to the so-called climax communities of both the intertidal and subtidal regions throughout the world. *Fucus* communities appear to be particularly vulnerable with several *Fucus* spp., and in particular *Fucus vesiculosus*, being reported to be replaced or reduced in numbers in the North Atlantic and Baltic as a result of waters being contaminated with sewage effuent (Grenager 1957; Johnston 1971/72; Rueness 1973; Russell 1974; Bokn 1979; Hallfors et al. 1987; Vogt and Schramm 1991; Park 1992; Kautsky et al. 1992). There are also some reports of a decline in North Atlantic populations of the knotted wrack *Ascophyllum nodosum* as a result of increased eutrophication (Rueness 1973; Bokn and Lein 1978; Bokn 1979). Work by Bokn and Lein (1978) and Rueness (1973) has indicated that dense populations of green algae provide unfavourable competition to these fucoids in Oslofjord, Norway, probably by hindering the establishment of germlings. A similar decline in fucoids has been reported in the Mediterranean, notably species of the genus *Cystoseira* (Bellan-Santini 1969; Golubic 1970; Munda 1974, 1993; Sfriso 1987), e.g., *Cystoseira fimbriata* and *C. barbata* in the Venice Lagoon (Sfriso 1987), *Cystoseira stricta* at Marseilles (Bellan and Bellan-Santini 1972; Bellan 1970) and *Fucus virsoides* in the northern Adriatic (Munda 1974, 1993). There are also reports of a decline in the number of *Sargassum* plants in Japan (Hirose 1978) and the northern Adriatic (Munda 1993). Some reports, however, refer to a pollution-tolerant fucoid, *Fucus distichus* ssp. *edentatus* whose occurrence and distribution in the North Atlantic appears to be associated with sewage contaminated water (Powell 1963; Bokn and Lein 1978). *Fucus evanescens* and *Fucus spiralis* were similarly reported by Bokn (1979) and Bokn and Lein (1978) to be more common than other fucoids in the sewage-polluted inner part of Oslofjord, Norway. In this respect, it is noteworthy that with improved sewage treatment and discharge practices, Bokn et al. (1992) reported an improvement in the fucoid vegetation in the inner Oslofjord.

There are also several reports from around the world of a decline in populations of seagrasses as a result of eutrophication (McNulty 1959; North 1963; Bellan 1970; Munda 1974; Guist and Humm 1976; Harlin and Thorne-Miller 1981; Murray and Littler 1983; Sfriso 1987; Ruiz et al. 1990). This has been particularly well documented for the Venice Lagoon, northern Adriatic which has seen a gradual decline in populations of the sea grasses *Cymodocea nodosa* and *Zostera noltii* over recent years (Sfriso 1987). Other phanerograms reported to be reduced in abundance near sewage outfalls include *Phyllospadix torreyi* in California (Littler and Murray 1975), *Posidonia oceanica* in Egypt (Nasr and Aleem 1948), *P. oceanica* in southeast Spain (Ruiz et al. 1990) and *Zostera muelleri* in Tasmania (Buttermore 1977). The loss of these perennial macroalgae and phanero gams is particularly detrimental to the local ecology of the waters as they provide a stable environment and invariably support a wide variety of associated marine benthic plant and animal communities (collectively called *Cystoseira* and seagrass associations respectively). The loss of these associations can be particularly damaging to local fisheries as they frequently act as nursery grounds for small fish; indeed, it is pertinent here to point out that increased interest is being shown in the artificial establishment of seaweed beds to enhance local fisheries in Japan (Largo and Ohno 1993).

Although the loss of these communities of large macroalgae and sea-grasses in eutrophicated waters can frequently be attributed to secondary pollutants, e.g. suspended particulate material, toxic ingredients etc. (see above), a number of reports have suggested that they are ousted by algae more readily adaptable to the high nutrient loadings. Some examples of sewage related floristic changes include those of Sfriso (1987), who reported a decline/disappearance of a number of algae belonging to the *Cystoseira* association in the Venice Lagoon, including the brown algae *Dictyopteris membranacea*, *Cladostephus verticillatus* and *Taonia atomaria* and the red genera *Laurencia*, *Dasya* and *Polysiphonia*, and their replacement by the algae *Dictyota dichotoma*, *Punctaria latifolia*, *Codium fragile*, *Ulva fasciata*, *Petalonia zosterifolia* and *Corallina officinalis*. Munda (1974) reported the complete deterioration of the *Fucus virsoides* and *Cystoseira* spp. associations (e.g. *Cystoseira barbata*) in the polluted waters of the northern Adriatic around Rovinj and their replacement by a prolific growth of *Halopteris scoparia* along with *Ulva lactuca* and *Codium tomentosum*. Littler and Murray (1974, 1975) reported an increase in the mean algal cover of *Gelidium pusillum* in an intertidal community near a sewage outfall on San Clemente Island, California, whilst McNulty (1959) reported *Gracilaria blodgetti* and *Agardhiella tenera* to be common inhabitants of sewage outfall areas in Biscayne Bay, Florida. An increase in the cover of *Gelidium pusillum* near a sewage outfall was also reported by Murray and Littler (1974) on San Clemente Island, along with blue-green algae, *Ulva californica*, *Pterocladia capillacea* and *Petalonia fascia*; this community replaced the more macroscopic and less pollution tolerant communities of *Egregia laevigata* subsp. *borealis*, *Halidrys dioica*, *Sargassum agardhianum*, the sea grass *Phyllospadix torreyi* and a number of smaller associated algae such as *Gigartina canaliculata*, *Gelidium robistum* and the coralline alga *Lithophyllum decipiens*. Gamulin-Brida et al. (1967), working in the Bay of Naples, also reported a reduction in the number of species with increased pollution leading to an increase in a number of tolerant species such as *Hypnea musciformis*. Grenager (1957), in a study of the inner polluted harbour of Oslofjord, Norway, observed no macroscopic Phaeophyceae with the vegetation only represented by a few Rhodophyceae (*Bangia fuscopurpurea*, *Ceramium strictum*), Chlorophyceae (*Enteromorpha intestinalis*, *E. crinita* and *Urospora penicilliformis*) and a cover of Cyanophyceae.

In general, the great majority of reports refer to an increase in the number of green algae associated with eutrophicated waters (Wilkinson 1963; Subbaramaiah and Parekh 1966; Ehrhardt 1968; Smyth 1968; Borowitzka 1972; Buttermore 1977; Bellan and Bellan-Santini 1972; Munda 1974; Littler and Murray 1975; Sfriso 1987; Sfriso et al. 1987; Wennberg 1987; Raffaelli et al. 1989; Wilson et al. 1990; Jeffrey et al. 1992; Munda 1993), this usually being at the expense of red algae (particularly fleshy reds) (North 1963; Munda 1974; Sfriso 1987) and brown algae (Munda 1974). Particularly common green algae include species of the genera *Enteromorpha*, *Ulva*, *Chaetomorpha* and *Cladophora*. Commonly recorded brown algae, however, include species of the genera *Ectocarpus* (Wilson et al. 1990; Jeffrey 1993) and *Pilayella* (Wilce et al. 1982), whilst frequently recorded red algae include species of the genera *Gracilaria*, *Gracilariopsis*, *Corallina*, *Ceramium*, *Gelidium* and *Bangia*. More recently, Sfriso (1987) recorded 95 species in the polluted Venice Lagoon compared to 141 and 104 recorded by Schiffner and Vatova (1938) and Pignatti (1962) in the 1930s and 1960s respectively. This represented a 2% increase in the Chlorophyta and a 65–51% decrease in the Rhodophyta. Occasionally, however, there are reports of large red algae (e.g. *Furcellaria lumbricata* and *Phycodrys rubens* increasing in abundance as a result of eutrophication (Baden et al. 1990). The Cyanophyceae, or blue-green algae, are also well represented, particularly in regions of high pollution and during the warmer summer months (Nasr and Aleem 1948; Grenager 1957; Littler and Murray 1974; Munda 1974, 1993; Murray and Littler 1974, 1983). Table 1.1 lists some of the genera and species of marine macroalgae described for sewage-polluted waters.

Many of these algae, especially the green algae, are often described by authors as annual, ephemeral, fast growing and opportunistic (Littler and Murray 1974; Murray and Littler 1974; Bokn et al. 1977; Smith et al. 1981; Kangas et al. 1982; Hällfors et al. 1987; Wennberg 1987, 1992; Elmgren 1989; Isaksson and Pihl 1990; Rosenberg et al. 1990; Vogt and Schramm 1991; Kautsky et al. 1992) and are well known as pioneering organisms on recently exposed or denuded surfaces. Essentially, the 'polluted' environment maintains an immature community characteristic of an early seral stage (referred to as a 'disclimax' community by Littler and Murray 1974). Such a simple, low turf forming, non-complex, unstratified community is

Table 1.1. Marine macroalgae, as named by the authors, characteristic of eutrophicated waters

Alga	Reference
CHLOROPHYTA	
Blidingia minima	Munda (1974); Bokn et al. (1977); Bokn and Lein (1978)
Bryopsis plumosa	Hirose (1978)
Caulerpa prolifera	Nasr and Aleen (1948)
Chaetomorpha linum	Perez-Ruzafa et al. (1991); Hällfors and Heikkonen (1993)
Chaetomorpha sp.	Johnston (1971/72); Reise (1983)
Cladophora spp.	Ehrhardt (1968); Borowitzka (1972); Johnston (1971/72); Bokn and Lein (1978); Hirose (1978); Reise (1983); Rosenberg (1985); Sfriso (1987); Wennberg (1992); Munda (1993)
Codium spp.	Golubic (1970)
Enteromorpha spp.	Grenager (1957); Wilkinson (1963); Johnston (1971/72); Borowitzka (1972); Perkins and Abbott (1972); Munda (1974); Guist and Humm (1976); Bokn and Lein (1978); Hirose (1978); Soulsby et al. (1978); Thorne-Miller et al. (1983); Raffaelli et al. (1989); Jeffrey (1993)
Ulva spp.	Nasr and Aleem (1948); Wilkinson (1963); Subbaramaiah and Parekh (1966); Borowitzka (1972); Perkins and Abbott (1972); Anonymous (1973); Littler and Murray (1974, 1975); Bokn and Lein (1978); Soulsby et al. (1978); Kindig and Littler (1980); Reise (1983); Pugnetti et al. (1990); Hällfors and Heikkonen (1993); Jeffrey (1993); Munda (1993)
Urospora penicilliformis	Grenager (1957)
Valonia aegrophila	Hällfors and Heikkonen (1993)
PHAEOPHYTA	
Chorda fillum	Munda (1974)
Colpomenia	Anonymous (1973)
Cutleria	Anonymous (1973)
Dictyota	Munda (1993)
Ectocarpus spp.	Grenager (1957); Munda (1974); Murray and Littler (1974); Wilson et al. (1990); Jeffrey (1993)
Fucus distichus sub. *edent.*	Powell (1963)
Fucus evanescens	Bokn (1979); Bokn and Lein (1978)
Fucus spiralis	Bokn (1979); Bokn and Lein (1978)
Halopteris scoparia	Munda (1974, 1993)
Petalonia fascia	Borowitzka (1972)
Pilayella littoralis	Munda (1974)
Pseudolithoderma nigra	Murray and Littler (1976, 1983)
Punctaria latifolia	Sfriso (1987)
RHODOPHYTA	
Agardhiella tenera	Littler and Murray (1975)
Bangia fuscopurpurea	Grenager (1957); Johnston (1971/72); Borowitzka (1972)

Table 1.1. (*Contd.*)

Alga	Reference
Bossiella	Dawson (1959)
Corallina spp.	Dawson (1959, 1965); North et al. (1972); Murray and Littler (1983); Bellan (1967b, 1970); Bellan and Bellan-Santini (1972)
Ceramium spp.	Grenager (1957); Johnston (1971/72); Fletcher (1974)
Chondrus crispus	Johnston (1971/72)
Dumontia incrassata	Johnston (1971/72)
Gelidium pusillum	Littler and Murray (1974, 1975); Kindig and Littler (1980); Murray and Littler (1974, 1976)
Gigartina sp.	Johnston (1971/72); Murray and Littler (1983)
Gracilaria spp.	Nasr and Aleem (1948); McNulty (1959); Sfriso (1987); Sfriso et al. (1991); Hällfors and Heikkonen (1993)
Hypnea musciformis	Gamulin-Brida et al. (1967)
Jeannerettia	Anonymous (1973)
Polysiphonia spp.	Johnston (1971/72)
Porphyra spp.	Johnston (1971/72); Reise (1983)
Pterocladia	Borowitzka (1972); Murray and Littler (1974)
Rhodoglossum	Anonymous (1973)
CYANOPHYTA	
Lyngbya spp.	Nasr and Aleem (1948)
Phormidium spp.	Nasr and Aleem (1948); Munda (1974)
Unidentified	Nasr and Aleem (1948); Grenager (1957); Munda (1974)

considered to be more adaptable to the environmental stress and instability caused by periodic sewage influx.

Another notable group of algae which are reported to be particularly resistant to sewage pollution are calcareous red algae (Dawson 1959, 1965; North et al. 1972; Kindig and Littler 1980). Dawson (1965), for example, reported *Bossiella* and *Corallina* to have a particularly high biomass near a domestic sewage outfall in California. North et al. (1972) referred to articulated *Corallina* spp. as a possible indicator of pollution, whilst Murray and Littler (1983) observed that cleared surfaces near a sewage outfall in California were dominated by a small number of algae which included *Corallina officinalis* var. *chilensis* and *Corallina vancouveriensis*. These results found support in the experimental studies of Kindig and Littler (1980), who revealed that *Amphiroa zonata*, *Bossiella orbigniana* and *Corallina officinalis* var. *chilensis* all demonstrated enhanced growth in the presence of primary sewage. *Corallins* cf. *officinalis* was also reported to be the most dominant alga near a sewage outfall near Marseilles (Bellan 1970; Bellan and Bellan-Santini 1972; Bellan-Santini 1969). Also often described as abundant in sewage polluted waters are finely branched red algae, in particular *Ceramium* spp., *Polysiphonia* spp. and *Bangia* sp. (Grenager 1957; Murray and Littler 1983; Kautsky et al. 1992; Bellan 1970).

However, it is only a relatively small number of algal genera which are universally and commonly reported in eutrophicated waters. With the exception of the brown alga *Ectocarpus* and to a lesser extent *Pilayella*, they are almost exclusively representatives of the Chlorophyta, and notably species of the genera *Chaetomorpha*, *Cladophora*, *Enteromorpha* and *Ulva*. As the only algae present at the polluted sites, they frequently occur in large quantities. For example, Rigoni et al. (1989) and Sfriso (1987) reported 10–13 kg/m wet weight of *Ulva rigida* in the Venice Lagoon, Briand and Morand (1987) reported *Ulva* in quantities reaching 240 kg/m, forming 1-m-thick tides washed up on Brittany shores, whilst *Enteromorpha prolifera* has been reported on mud flats in Scotland in quantities of approximately 1 kg/m^2 (Coleman and Stewart 1979). When these excessive, mat-like growths of algae are washed up periodically on the tide line they are descriptively termed "green tides". Table 1.2 tabulates many of the world-wide reports of these "green tides".

Table 1.2. World-wide reports of "green tides"

Alga	Locality	Reference
CHLOROPHYTA		
Chaetomorpha linum	Peel Inlet, Australia	Lavery and McComb (1991)
Chaetomorpha linum	Mar Menor Lag., Spain	Perez-Ruzafa et al. (1991)
Chaetomorpha linum	Orielton Lag., Tasmania	Buttermore (1977)
Chaetomorpha linum	Corsica	Grimes and Hubbard (1971)
Cladophora albida	Peel Inlet, Australia	Gordon et al. (1981)
Cladophora glom.	Laholm Bay, Sweden	Fleischer et al. (1985); Rosenberg (1985); Wennberg (1992)
Cladophora mont.	Peel Inlet, Australia	Gordon et al. (1985)
Cladophora sp.	Is. of Silt Germany,	Grenager (1957); Bokn and Lein (1978);
Cladophora sp.	Oslofjord, Norway	Klavestad (1978)
Enteromorpha int.	Medway Est., England	Wharfe (1977)
Enteromorpha int.	Orielton Lag., Tasmania	Buttermore (1977)
Enteromorpha plumosa	Rhode Is., USA	Thorne-Miller et al. (1983)
Enteromorpha prolif.	Forfar Loch, Scotland	Coleman and Stewart (1979)
Enteromorpha sp.	Clyde Est., Scotland	Perkins and Abbott (1972)
Enteromorpha sp.	Ythan Est., Scotland	Raffaelli et al. (1989)
Enteromorpha sp.	Forfar Loch, Scotland	Ho (1979)
Enteromorpha sp.	Oslofjord, Norway	Rueness (1973); Grenager (1957); Bokn and Lein (1978); Klavestad (1978)
Enteromorpha sp.	Dublin Bay, Ireland	Wilson et al. (1990); Jeffrey (1993)
Enteromorpha sp.	Rogerstown Est., Ireland	Fahy et al. (1975)
Enteromorpha sp.	Langstone Harb., England	Nicholls et al. (1981);

Table 1.2. (*Contd.*)

Alga	Locality	Reference
		Dunn (1972); Anonymous (1976a); Soulsby et al. (1985); Montgomery and Soulsby (1980)
Enteromorpha sp.	Portsmouth Harb., England	Soulsby et al. (1985)
Enteromorpha sp.	The Wash, England	Anonymous (1976b)
Enteromorpha sp.	Coos Bay Est., USA	Marshall Pregnall and Rudy (1985)
Ulva lactuca	Roskilde Fjord, Denmark	Geertz-Hansen and Sand-Jensen (1990)
Ulva lactuca	Odense Fjord, Norway	Frederiksen (1987)
Ulva lactuca	Tauranga Har., New Zealand	Park (1992)
Ulva lactuca	Lagune du Provost, France	Casabianca-Chassany (1983, 1989)
Ulva lactuca	Alexandra Harb., Egypt	Nasr and Aleem (1948)
Ulva lactuca	Hong Hong Island	Ho (1986a,b)
Ulva lactuca	Thanet Bays, England	Fletcher, (1974)
Ulva lactuca	Naples Bay, Italy	Golubic (1970)
Ulva lactuca	Chesapeake Bay, USA	Hanks (1966), White and Highlands (1968)
Ulva lactuca	Avon-Heathcote Est., NZ	Steffensen (1976)
Ulva fasciata	Vewraval, India	Subbaramaiah and Parekh (1966)
Ulva latissima	Belfast Lough, Ireland	Cotton (1910, 1911); Letts and Richards (1911)
Ulva pertusa	Tomioka Bay, Japan	Tsutsumi (1990)
Ulva rigida	Venice Lagoon, Italy	Sfriso (1987); Sfriso et al. (1987, 1991)
Ulva sp.	Dublin Bay, Ireland	Wilson et al. (1990); Jeffrey (1993)
Ulva sp.	Alfaques Bay, Spain	Martinez-Arroyo and Romero (1990)
Ulva sp.	Boston Harb., USA	Sawyer (1965)
Ulva sp.	Portsmouth Harb., England	Soulsby et al. (1985)
Ulva sp.	Langstone Harb., England	Soulsby et al. (1985)
Ulva sp.	Holes Bay, Poole, England	Southgate (1972)
Ulva sp.	Tampa Bay, Florida	Guist and Humm (1976)
Ulva sp.	Veerse Meer Lag., Netherlands	Nienhuis (1992)
Ulva sp.	Saint-Brieuc Bay, Brittany	Briand and Morand (1987); Le Bozec (1993)
PHAEOPHYCEAE		
Ectocarpus sp.	Dublin Bat, Ireland	Jeffrey et al. (1992); Jeffrey (1993)
Ectocarpus sp.	Oslofjord, Norway	Grenager (1957)
Pilayella litt.	Natiant Bay, USA	Wilce et al. (1982)
RHODOPHYTA		
Furcellaria lumb.	Laholm Bay, Sweden	Wennberg (1987)
Phycodrys rubens	Laholm Bay, Sweden	Wennberg (1987)

1.5   Characteristics of Green Tide Forming Algae

Not surprisingly, particular interest has been directed towards obtaining a better understanding of the biology of the relatively small number of genera (and species) which contribute to the formation of these "green tides". In general, they can be termed opportunistic and are frequently reported as early colonizers or pioneering organisms on newly immersed, denuded or disturbed substrata (Murray and Littler 1976; Wennberg 1992). For this reason, they often occur as important marine fouling organisms and as early colonizers of substrata denuded by natural disturbances e.g. sand scouring, storm damage, or by man's activities, e.g. oil pollution treatment (Nelson-Smith 1972; Fletcher 1988). This pioneering role appears to be aided by a wide geographical distribution, a common local abundance and wide seasonal occurrence, a copious production of reproductive spore bodies (Littler and Murray 1974), a short, simple life history (Littler and Murray 1974; Murray and Littler 1974) and an inherent rapid rate of colonization and growth (Vogt and Schramm 1991; Park 1992). All these biological features ensure a rapid and successful colonization of a wide range of habitats and environments and provide a competitive edge over many other algae.

The floristic components of these "green tides" can also be described as stress-tolerant and physiologically hardy (Kindig and Littler 1980). Genera such as *Ulva* and *Enteromorpha*, in particular, exhibit a wide tolerance to adverse environmental conditions of temperature, light intensity and salinity (Gordon et al. 1980; Chan et al. 1982; Vermaat and Sand-Jensen 1987). *Ulva*, for example, has been reported to show high adaptability to different irradiances, being especially tolerant of low light conditions (Ramus 1983; Vermaat and Sand-Jensen 1987; Sand-Jensen 1988; Henley and Ramus 1989; Martinez-Arroyo and Romero 1990); it is also markedly tolerant of anoxic conditions, presence of sulphide etc. (Vermaat and Sand-Jensen 1987; Sand-Jensen 1988) and to extended periods of desication (Beer and Eshel 1983). The ability of *Ulva* to adapt to low irradiance levels, with its low compensation light intensity, will optimize production rates in tidal waters (Henley and Ramus 1989), including those with high sediment loading, and should prove particularly beneficial for plants forming dense growths similar to those contributing to the "green tides" (Lavery and McComb 1991). *Ulva rotundata* is also known to photoacclimate (Henley and Ramus 1989), whilst the red alga *Gracilaria confervoides* is known to be markedly resistant to hydrogen sulphide (Stokke 1956). Both the ability to photoacclimatize and tolerate H_2S would be particularly beneficial to plants existing in large, dense mats. Additional adaptive features of these "green tide"-forming algae include vegetative fragmentation/propagation and the ability to form free-floating or loose-lying populations capable of almost unlimited vegetative growth (Gordon et al. 1985). Certainly vegetative fragmentation is characteristic of plants living in calm-water conditions (Norton and Mathieson 1983), whilst Langton et al. (1977) observed that fragmentation in *Hypnea musciformis* was related to high ammonium-N concentrations. Some reports also suggest that sporulation is relatively uncommon in unattached plants (Cotton 1911; Gordon et al. 1985; Park 1992) which might be related to ambient nutrient levels (Park 1992).

Probably one of the most important attributes of these algae, however, is their competitive response to the high nutrient content of the surrounding waters. Apart from an ability to tolerate high nutrient levels, many authors, have reported these algae to be capable of high and rapid inorganic nitrogen uptake and storage, i.e. they have a high nitrogen saturation level (Letts and Richards 1911; Wilkinson 1963; Subbaramaiah and Parekh 1966; Baalsrud 1967; North et al. 1972; Chapman and Craigie 1977; Wallentinus 1981; Gerard 1982; Kautsky 1982; Rosenberg et al. 1984; Fujita 1985; Lapointe 1985; Soulsby et al. 1985; Ramus and Venable 1987; Fujita et al. 1988, 1989; Smith and Horne 1988; Pugnetti et al. 1990; Vogt and Schramm 1991). Indeed, Rosenberg and Ramus (1982) concluded that fast N uptake rates are a necessary strategy for fast growing, opportunistic macroalgae and that this must be followed by rapid growth before losses occur from grazing and abrasion. Very small increases in nutrients have also been reported to cause massive growth increases in *Ulva* (Harlin and Thorne-Miller 1981). *Ulva* is also opportunistic in that it is highly adapted for rapid uptake and storage of nutrients received intermittently (Rosenberg and Ramus 1984; Fujita 1985; Ramus and Venable 1987; Fujita et al. 1988, 1989; Duke et al. 1989; Lundberg et al. 1989). It is essential, however, that the nutrients are supplied continuously to such algae: long-term storage is not possible compared to the degree which occurs in the slower-growing, longer-living, climax components of algal communities, e.g. the high NO_3 concentration factor mentioned above for *Laminaria longicruris* which sustains growth throughout the nitrogen-deficient summer periods (Chapman and Craigie 1977; Chapman et al. 1978).

Although it is generally recognized that NO_3 is the main available source of N for macroalgae (Anderson 1942; Fries 1963; Prince 1974; Wallentinus 1991), some genera such as *Ulva*, *Enteromorpha* and *Gracilaria* can also rapidly take up and utilize ammonia (Wilkinson 1963; Waite and Gregory 1969; Neish and Fox 1971; Waite and Mitchell 1972a,b; Guist and Humm 1976; Harlin et al. 1978; Rosenberg and Ramus 1984; Duke et al. 1989). In some experimental studies, for example, many green algae have been shown to exert a preference for it over NO$_3$, with the algae exhibiting faster growth rates, rendering them particularly adaptable to sewage effluent (Letts and Richards 1911; Foster 1914; Wilkinson 1963; Waite and Mitchell 1972a; Waite et al. 1972; Harlin and Thorne-Miller 1981; Vandermeulen and Gordin 1990). Certainly, it has been shown that *Ulva* can store molecular nitrogen intracellularly, whilst Fujita et al. (1988) revealed it is converted into macromolecular nitrogen. This stimulatory effect of ammonia might well be related to the latter's ease of assimilation (Bongers 1956; Syrett 1962). It would also be advantageous to be capable of using ammonia in late summer when nitrate is depleted (Prince and Kingsbury 1973; Topinka and Robbins 1976).

Also pertinent to this discussion are reports showing that a number of genera such as *Ulva*, *Corallina* and *Lithothrix* have the ability to use organic sources of nitrogen, either as the amide or amino group, including urea, amino acids etc, and organic sources of carbon including acetate, glucose etc. as a nutrient source. (Foster 1914; Fries 1963; North 1963; Iwasaki 1967; Fries and Pettersson 1968; Nasr et al. 1968; Neish and Fox 1971; Mohsen et al. 1974; DeBoer et al. 1978; DeBoer 1981; Markager and Sand-Jensen 1990; Park 1992). With the large quantities of soluble organic compounds discharged into some inshore waters, North et al. (1972) suggest that this organic waste may explain the excessive growths of some algae reported. The rapid uptake systems in *Ulva*, *Enteromorpha*, *Lithothrix* and *Corallina* were also considered by North et al. (1972) to be partially responsible for the success of these algae in polluted environments. They further point out that algae with little or no uptake ability (fleshy brown and red algae) are conspicuously absent from contaminated waters. For example, algae such as *Ulva fasciata* were shown to exhibit higher growth rates when supplied with urea compared with either NH$_4$ or NO$_3$ (Mohsen et al. 1974; DeBoer 1981). Many of the organic nitrogen sources are commonly discharged in sewage effluent and it is interesting to add that *Ulva lactuca* has been shown to take up acetate which is a by-product of bacterial decomposition of organic matter (Gemmill and Galloway 1974). Also pertinent is the report of Markager and Sand-Jensen (1990) who revealed *Ulva lactuca* to grow significantly faster in cultures supplied with glucose and acetate in the dark and/or dim light, explaining its ability to adapt to prolonged periods of low light conditions and maintain an intact photosynthetic apparatus. It has also been suggested that this ability to take up organic substrates may explain *Ulva*'s ability to tolerate extended periods of burial (Vermaat and Sand-Jensen 1987). North et al. (1972) also demonstrated that a very rapid uptake of organic compounds occurred in *Ulva* and *Enteromorpha*, this being important because of limited tidal cycle exposure. This ability to rapidly absorb nitrogen from seawater offers some potential for the use of these "green tide" components in waste water treatment. Particularly noteworthy in this respect are the studies of Langton et al. (1977), Chan et al. (1982), Lehnberg and Schramm (1984), Krom et al. (1989), Vandermeulen and Gordin (1990) and Neori et al. (1991).

A number of workers have drawn attention to the morphology of many of the "green tide" components and the role this might play in nutrient absorption. They suggest that the frequently recorded filamentous, sheet-like, tubular and ball-like nature of their thalli produces a high ratio of surface to volume which ensures maximum nutrient uptake (Espinoza and Chapman 1983; Rosenberg and Ramus 1984; Wallentinus 1984, 1991). Indeed, Parker (1981) suggests that algal shape, water motion and nutrient content all interact in controlling nutrient uptake and growth patterns. Certainly it is known that the water velocity can determine nutrient uptake and supply to algal thalli (Harlin and Thorne-Miller 1981; Norton and Mathieson 1983) and it is perhaps not surprising that the morphological features of macroalgae can play a major influencial role in this process by creating turbulence at the thallus surface, thereby increasing nutrient availability (as shown for *Macrocystis integrifolia* and *Nereocystis leutkeana* by Hurd et al. (1993). By hanging as vertical sheets (as reported for the Venice Lagoon), it was further suggested that *Ulva* can increase its potential exchange surface with passing water (Welsh 1980). Not unrelated to this discussion was the observation that filamentous algae tend to have very high specific carbon fixation rates (Leskinen et al. 1992). Finally, it is possible that nutrient ecotypes have developed for "green tide" algae in response to the high ambient nutrient levels in some eutrophicated coastal waters similar to that reported for *Laminaria longicruris* in Nova Scotia (Espinoza and Chapman 1983).

The effect of nutrients on benthic community structure was particularly well demonstrated by Harlin and Thorne-Miller (1981) who treated seagrass beds with nutrients. Additions of NH_4 and NO_3 resulted in the occurrence of dense beds of free-floating *Enteromorpha plumosa* and *Ulva lactuca* plants. In general, they also noted that low nutrient concentrations favoured vascular plants and that NO_3 may even be toxic to *Zostera*. Raymont (1947) also observed that the addition of nitrate and phosphate fertilizer to a Scottish sea loch encouraged an extremely heavy growth of algae.

1.6   Impact of Green Tides

Whilst at first acting beneficially as nutrient strippers – e.g. Welsh (1980)
described *Ulva* as having a great capacity to trap nutrients during short-
term intensity releases – in many situations, the excessive growths of algae
eventually have adverse ecological effects. For example, the sheer volume of
algae present can restrict water movement and velocity (Waite and Mitchell
1972a; White and Highlands 1968; Rasmussen 1990; Park 1992). This can
lead to increased sedimentation rates (Rasmussen 1990), it can interfere
with oxygen transport (Wharfe 1977; Lindahl and Hernroth 1983; Reise
1983; Rosenberg 1985; Frederiksen 1987; White and Highlands 1968; Sfriso
1987; Rasmussen 1990; Park 1992), and it can also restrict plankton
transport, with obvious implications in the supply of food for suspension
feeders (Park 1992). It will also restrict larval movement, preventing the
settlement and recruitment of polychaete and bivalve larvae (Price and
Hylleberg 1982; Reise 1983; Thrush 1986; Olafson 1988). The algal mats
would also interfere with surface feeding deposit feeders (Park 1992) and
influence epifaunal associations. Raffaelli et al. (1991) and Raffaelli et al.
(1989), for example, reported the disappearance of *Corophium volutator*
under high weed biomass in the Ythan estuary in Scotland as a result of the
algal mats interfering with feeding behaviour. Other authors have noted
unusually high numbers of epibenthic fauna such as *Peloscolex benedeni*
and *Hydrobia ulvae* associated with the algal mats (Wharfe 1977;
Montgomery and Soulsby 1980; Nicholls et al. 1981; Aneer 1987; Hull 1987)
with obvious ecological repercussions on food chains (Raffaelli et al. 1991).

The large mats of algae would also shade and reduce the photosynthetic
activity of underlying benthic seagrass and algal communities (Rasmussen
1990) as well as physically press down and abrade them (Rasmussen 1990;
Guist and Humm 1976). There are also reports of thick mats of the alga
Dictyosphaeria cavernosa being formed in sewage contaminated waters and
smothering all underlying reef organisms in Kaneohe Bay, Hawaii (Banner
1974; Smith et al. 1981). Reports also suggest that toxic exudates are
released from these loose-lying algal mats (Johnston and Welsh 1985;
Magre 1974) and that these might be responsible, for example, for the
high mortality of Baltic herring (*Clupea harengus*) eggs laid/deposited on
the surface of the algae (Aneer 1987; Rajasilta et al. 1989). In this respect, it
is interesting to note Magre's (1974) suggestion that *Ulva lactuca* might
produce substances which are toxic to *Balanus*. Finally, the physical
presence of these loose-lying algal mats on mud flats has been reported to
deter feeding wild fowl (Nicholls et al. 1981) and to interfere with
fishing methods with resultant loss of catches (Taylor and Bernatowicz 1969;

Norton and Mathieson 1983; Rosenberg 1985; Perez-Ruzafa et al. 1991; Park 1992).

Additional problems are reported to occur when these algal mats are removed by tide/current movements and drift to new localities. For example, they can be deposited in deeper waters (Rasmussen 1990), this representing a short-term nutrient loss to the local ecosystem; however, as shown for the Baltic, nutrients accumulated in deep waters are eventually mixed up to the surface layers through thermohaline convection (Fonselius 1969). The mats can also be deposited on the shoreline, including bathing beaches and spoiling them with resultant aesthetic, ecological and economic problems (Sawyer 1965; Hanks 1966; Horne and Nonomura 1976; Casabianca-Chassany 1984; Rosenberg 1985; Jeffrey et al. 1992; Park 1992; Jeffrey 1993), notably the release of pungent smells during their decay (Wilkinson 1963; Sawyer 1965; Hanks 1966; Southgate 1972; Fletcher 1974; Guist and Humm 1976; Horne and Nonomura 1976; Buttermore 1977; Sfriso et al. 1987; Park 1992) and expensive clearance operations associated with their removal (Fletcher 1974). There have also been reports of large floating drift weed deposits blocking power-station intake ducts (Southgate 1972) and intake pipes on ships (Park 1972).

A number of authors do, however, note some beneficial aspects of the loose-lying, algal mats (Cowper 1978). They do, for example, represent a very important sink for nutrients and, therefore, play a vital role in nutrient recycling processes (Birch et al. 1981; Sfriso et al. 1988b; Pugnetti et al. 1990). They add considerably to the detritus system and contribute to the dissolved organic carbon pool (Cowper 1978). They can also physically provide food, habitat and shelter to many small invertebrates and fish and act as important spawning and nursery grounds for many small fish (Kimura et al. 1958; Cowper 1978; Norton and Mathieson 1983).

Particular problems arise, however, during the subsequent and sometimes catastrophic decomposition of the algal mats. This usually occurs during periods of high temperatures and calm conditions when oxygen demand dramatically increases e.g. during night-time respiration (Perkins and Abbot 1972; Casabianca-Chassany 1989; Rasmussen 1990). Widespread axonic conditions ensue which have considerable ecological consequences (Frederiksen 1987; Sfriso et al. 1987, 1988b; Rasmussen 1990). For example, this reduction in oxygen and the associated excessive development of sulphide on the underlying mud surface, is reported to have a detrimental effect on the underlying mud infauna. Perkins and Abbot (1972) noted that the bivalve molluscs *Cardium (Cerastoderma) edule* and *Macoma balthica* moved up to the surface to avoid deeper anaerobic conditions; *Nephtys* also emerged. Many of these animals died, or being unprotected, were predated by wading birds. Similar events were described for *Mya arenaria* and *Cardium edule* under *Cladophora* deposits in Sweden (Fahy et al. 1975; Rosenberg and Edler 1981; Rosenberg 1985; Rosenberg and Loo 1988; Baden et al. 1990; Rosenberg et al. 1990) and for the soft shell clam *Mya arenaria* and *Mercenaria mercenaria* under *Ulva lactuca* by White and Highlands (1968) in the USA. In general, a number of authors have reported a reduction in the numbers and diversity of mud infauna (Ryther 1954; Baalsrud 1967; Perkins and Abbott 1972; Wharfe 1977; Dauer and Connor 1980; Nicholls et al. 1981; Soulsby et al. 1982; Price and Hylleberg 1982; Reise 1983; Rosenberg 1985; Thrush 1986; Olafson 1988; Park 1992), with severe implications to the local fishing industries and bird populations (Perkins and Abbott 1972; Fahy et al. 1975; Tubbs 1977; Montgomery and Soulsby 1980; Tubbs and Tubbs 1980; Nicholls et al. 1981).

Usually there is a replacement of the fauna by a small number of pollution-tolerant species. For example, Wharfe (1977), Montgomery and Soulsby (1980) and Nicholls et al. (1981) reported a reduced biomass of infauna under *Enteromorpha* mats on mud flats on the south coast of England with high numbers of more pollution-tolerant species such as the polychaetes *Capitella capitata* and *Scolelepis fuliginosa* and the oligochaete *Peloscolex benedenii*. The worm *Peloscolex* was also shown to be encouraged by algal mats on Bull Island, Ireland, to the exclusion of other forms of interstitial life (see Fahy et al. 1975). Perkins and Abbott (1972) noted that the lugworm *Arenicola marina* was able to tolerate anaerobic conditions and the presence of sulphide under *Ulva* and *Enteromorpha* on mud flats in the Firth of Clyde, Scotland. The decomposing weed has also been implicated in the release of toxic products (Magre 1974; Johnson and Welsh 1985).

Additional effects of this anaerobic decomposition process include the reported high sulphide content of surface waters giving them an opaque and creamy coloured appearance, referred to as "milky water" (Hanks 1966), the release of noxious smells of hydrogen sulphide (Letts and Richards 1911; Sawyer 1965; Hanks 1966), the blackening of lead pigment paints on local buildings (Wilkinson 1963; Sawyer 1965; Hanks 1966) and the rapid proliferation of chironomid flies, whose larval stages can survive the anoxic conditions because they have an anaerobic metabolism. The latter was a particular problem in the Venice Lagoon during 1988 when the mosquito *Chironomus salinarius* Kieff invaded the city and islands in swarms, hindering the railway and airport facilities; a similar problem, with swarms of the flying midge *Dicrotendipes conjunctus* has been reported for the Orielton Lagoon, Tasmania (Buttermore 1977).

An important secondary effect of the decomposition process is the release of nutrients back into the ecosystem. Drift algal mats do act as a temporary reservoir of nutrients which are quickly released to the overlying waters at the sediment surface replenishing the superficial sediment nutrient load (Nicholls and Keeney 1973; Birch et al. 1981; Gabrielson et al. 1983; Zimmermann and Montgomery 1984; Sfriso et al. 1987, 1988b; Pavoni et al. 1990; Pugnetti et al. 1990; Lavery and McComb 1991; Hanisak 1993; Sfriso and Marcomini 1994). Essentially this acts as an organic input into the sediments (Zimmermann and Montgomery 1984); it also releases nutrients back into the water column (Gabrielson et al. 1983) which can be manifested in the occurrence of phytoplankton blooms (Sfriso et al. 1987, 1989). Pertinent to this discussion was the report of Gabrielson et al. (1983) that *Cladophora* tissue decomposed into the usable forms ammonium and orthophosphate. Also interesting was Smith and Horne's (1988) report that *Ulva* had a greater affinity for nitrogen than do some plankton, thereby depressing phytoplankton growth.

1.7   Treatment of Green Tides

Not surprisingly the world-wide occurrence and ecological impact of "green tides" has aroused considerable concern and much consideration has been given towards the development and implementation of various control methods. Although there is some evidence that grazing by invertebrates can exert some control on biomass development (Geertz-Hansen and Sand-Jensen 1990), in the majority of "green tide" situations and during the peak periods of biomass production, such natural biological control methods are totally inadequate and treatment is required.

Probably the most widely used treatment involves the physical removal of the troublesome algal biomass. This particularly applies to situations where large quantities of weed are present e.g. 85 000 m³ of harvestable *Ulva* deposits on beaches in Brittany (Dion 1994) and an estimated 1–1.2 million tons of *Ulva* growing in the Venice Lagoon (Orlandini 1988; Sfriso et al. 1989). Usually, the algae are mechanically removed following its accumulation on the tide line, and transported to land-fill sites (Merrill and Fletcher 1991). In some situations, the algal biomass is removed in situ. For example, special 'reaping' machines were used to remove *Ulva* from Odense Fjord in Denmark (Frederiksen 1987), whilst harvesting boats were used to remove floating masses of *Ulva* from the Venice Lagoon (Orlandini 1988; Merrill and Fletcher 1991). On one occasion, an unsuccessful attempt was made to treat *Ulva* mats by chemical means using quicklime and copper sulphate in Winthrop Harbour, USA (Sawyer 1965). Certainly, harvesting provides an efficient remedial control method and offers additional benefits in the potential utilization of the algal biomass obtained. For example, consideration is being given to anaerobic digestion processes to produce biogas (Wise et al. 1979; Habig et al. 1984; Croatto 1985; Missoni and Mazzagardi 1985; Nicolini and Viglia 1985; DeBusk et al. 1986; Briand and Morand 1987; Frederiksen 1987; Rigoni et al. 1989), aerobic digestion processes to produce soil compost (Gladstone and Lopez-Real 1993; Maze et al. 1993), use as animal/human feed (Bonalberti and Croatto 1985) and the industrial extraction of fine chemicals, lectins, polysaccharides, B-carotene etc.

Other control methods which have been considered include:

(1) improving water circulation and current flow and thereby, preventing the build up of nutrients (Buttermore 1977);
(2) lowering the salinity of the contaminated waters by controlled freshwater influx in order to kill the algae (Buttermore 1977);
(3) removal of nutrients from effluents at the tertiary sewage treatment stage or at least regulating the nutrient composition to produce an effluent less favourable to algal growth (Frederiksen 1987; Bokn et al. 1992). Nutrient extraction and regulation is, however, an expensive process and would not be appropriate in regions where the nutrient loading was derived principally from agricultural runoff (e.g. 80% of nitrogen entering Odense Fjord, Denmark; Frederiksen 1987) or if the bulk of nutrients is from diffuse sources via river run-off (Wilson et al. 1990); the latter would require long-term reductions in the use of fertilizers.

One interesting prospect is the removal of nutrients by mariculture techniques. For example, Merrill (1994) and Merrill and Fletcher (1991) suggest that eutrophicated regions could provide suitable conditions for the commercial production of economically important species.

References

Alfimov NN (1959) Methods of hydrobiological analysis of the sanitary quality of littoral seawater. Tr Vses gidrobiol Obua 9: 360–366

Anderson M (1942) Einige ernährungsphysiologische Versuche mit *Ulva* und *Entero-morpha*. K Fysiogr Sallsk Lund Forh 12: 42–52

Aneer G (1987) High natural mortality of Baltic herring (*Clupea harengus*) eggs caused by algal exudates? Mar Biol 94: 163–169

Anonymous (1973) Environmental study of Port Phillip Bay. Report on Phase One 1968–1971. Melbourne, 372pp

Anonymous (1976a) Langstore Harbour Study – the effect of sewage effluent on the ecology of the harbour. Rep to Southern Water Authority. Portsmouth Polytechnic, Portsmouth, 356pp

Anonymous (1976b) The wash water storage scheme feasibility study, a report on the ecological studies. NERC Publ Ser C 15. Natural Environment Research Council, London, 36pp

Baalsrud K (1967) Influence of nutrient concentrations on primary production. In: Olsen TA, Burgess FJ (eds) Pollution and marine ecology. Wiley, New York, pp 159–169

Baden SP, Loo L-O, Pihl L, Rosenberg R (1990) Effects of eutrophication on benthic communities including fish: Swedish West Coast. Ambio 19: 113–122

Banner AH (1974) Kaneohe Bay: urban pollution and a coral reef ecosystem. Proc 2nd Int Symp Coral Reefs, Brisbane 2: 685–702

Beer S, Eshel A (1983) Photosynthesis of Ulva spp. I. Effects of desiccation where exposed to air. J Exp Mar Biol Ecol 70: 91–97

Bell PRF, Greenfield PF, Hawker D, Connell D (1989) The impact of waste discharges on coral reef regions. Water Sci Technol 21: 121–130

Bellamy DJ, Whittick A (1964) Problems in the assessment of the effects of pollution on inshore marine ecosystems dominated by attached macrophytes. Field Stud 2: 49–54

Bellamy DJ, John DM, Jones DJ, Starkie A, Whittick A (1972) The place of ecological monitoring in the study of pollution of the marine environment. In: Ruivo M. (ed) Marine pollution and sea life. FAO Publication. Fishing News (Books), London, pp 421–425

Bellan G (1967a) Pollution et peuplements benthiques sur substrat meuble dans la region de Marseille. Première partie. Le secteur de Cortiou. Rev Int Oceanogr Med 6–7: 53–87

Bellan G (1967b) Pollution et peuplements benthiques sur substrat meuble dans la région de Marseille. Deuxième partie. L'ensemble portuaire Marseillais. Rev Int Oceanogr Med 8: 51–95

Bellan G (1970) Pollution by sewage in Marseilles. Mar Pollut Bull Ns 1: 59–60

Bellan G, Bellan-Santini D (1972) Influence de la pollution sur les peuplements marins de la region de Marseille In: Ruivo M (ed) Marine pollution and sea life. FAO Publication. Fishing News (Books), London, pp 396–401

Bellan G, Peres J-M (1972) La pollution dans le Bassin Mediterraneen (quelques aspects en Mediterranee Nord–occidentale et an Haute adriatique: leurs enseignements, pp 32–35

Bellan-Santini D (1969) Contribution à l' etude des peuplements infralittoraux sur substrat rocheux (Etude qualitative et–quantitative de la frange supérieure). Recl Trav Stn Mar Endoume (63–47): 5–294

Birch PB, Gordon DM, McComb AJ (1981) Nitrogen and phosphorus nutrition of Cladophora in the Peel-Harvey system. Bot Mar 24: 281–387

Bleguad H (1932) Investigations of the bottom fauna at outfalls of drains in the Saund. Rep Dan Biol Stat 37: 1–20

Bokn T (1979) Brukav tang som overvakings parameter i en naeringsrik fjord. In: Over vaking ar uattenomraden. 15. Nordiska symposiet om Vatlenforskning. NORDFORSK, Miljovards sekr publ 1979 2: 181–200

Bokn T, Lein TE (1978) Long-term changes in fucoid association of the inner Oslofjord, Norway. Norw J Bot 25: 9–14

Bokn T, Kirkerud L, Krogh T, Nilsen G, Magnusson J (1977) Undersokelse au hudrogragisje og biologiske forholdi Indre Oslofjord. Overvakingsprogram. Arstrapport 1975–76. Norska institutt for vannforskning, 0–71160, 119 pp

Bokn TL, Murray SN, Moy FE, Magnusson JB (1992) Changes in fucoid distributions and abundances in the inner Oslofjord, Norway: 1974–80 versus 1988–90. Acta Phytogeogr Suec 78: 117–124

Bonalberti E, Croatto V (1985) Use of algal systems as a source of fuel and chemicals. In: Palz W, Coombs J, Hall DO (eds) Energy from biomass. Proc Int Conf on Biomass. Elsevier, London, pp 158–163

Bongers LH (1956) Aspects of nitrogen assimilation by cultures of green algae. Meded Landbouwhogesch Wageningen 56: 1–52

Bonsdorff E, Aarnio K, Sandberg E (1991) Temporal and spatial variability of zoobenthic communities in Archipelago waters of the northern Baltic Sea – consequences of eutrophication? Int Rev ges Hydrobiol 76: 433–449

Borowitzka MA (1972) Intertidal algal species diversity and the effect of pollution. J Mar Freshwater Res 23: 73–84

Braarud T (1945) A phytoplankton survey of the polluted waters of inner Oslofjord. Hvalradets Skr ifter. Norske Videnskaps Akad, Oslo no 28

Breuer G, Schramm W (1988) Changes in macroalgal vegetation of Kiel Bight (western Baltic Sea) during the past 20 years. Kiel Meeresforsch Sonderh 6: 241–255

Briand X, Morand P (1987) Ulva, stranded algae: a way of depollution through methaniz-ation. In: Biomass for energy and industry. Elsevier, London pp 834–839

Burrows EM (1971) Assessment of pollution effects by the use of algae. Proc R Soc Lond B 177: 295–300

Burrows EM, Pybus C (1970) Culture of Laminaria in relation to problems of marine pollution. Proc Challenger Soc IV 2: 80–81

Burrows EM, Pybus C (1971) Laminaria saccharina and marine pollution in north-east England. Mar Pollut Bull 2: 53–56

Buttermore RE (1977) Eutrophication of an impounded estuarine lagoon. Mar Pollut Bull 8: 13–15

Casabianca-Chassany M-L (1983) Relations entre la production algale macrophytique et le degré deutroehication du milieu dans une lagune Méditerranéenne. Rapp Comm int Mer Médit 28: 359–363

Casabianca-Chassany M-L (1984) Analyse de problèmes écologiques liés à la récolte de biomasse algale en milieu lagunaire. Rapport de contract n° 81.G.0983. Action Concertée: écologie et aménagement rural. M R T Paris, 80 pp

Casabianca-Chassany M-L (1989) Dégradation des ulves (Ulva rotundata, lagune du Prevost, France). C R Acad Sci Paris 308: 155–160

Causey NB, Pryterck JP, McCaskill, J Humm HJ, Wolf FA (1945) Influence of environmental factors on the growth of Gracilaria confervoides. Bull Duke Univ Mar Stn 3: 19–24

Cederwall H, Elmgren R (1980) Biological effects of eutrophication in the Baltic Sea, particularly the coastal zone. Ambio 19: 109–112

Chan K, Wong PK, Ng SL (1982) Growth of Enteromorpha linza in sewage effluent and sewage effluent-seawater mixtures. Hydrobiologia 97: 9–13

Chapman ARO, Craigie JJ (1977) Seasonal growth in Laminaria longicruris: relations with dissolved inorganic nutrients and internal reserves of nitrogen. Mar Biol 40: 197–205

Chapman ARO, Markham JW, Luning K (1978) Effects of nitrate concentration on the growth and physiology of Laminaria saccharina (Phaeophyta) in culture. J Phycol 14: 195–198

Clendenning KA, North WJ (1959) Effects of wastes on the giant kelp, Macrocystis pyrifera. Proc 1st Int Conf Berkeley, pp 82–91

Clendenning KA, North WJ (1964) An investigation of the effects of discharged wastes on kelp. The resources Agency of California. State Water Quality Control Board, Publication o 26, 121 pp

Cole HA (1972) North sea pollution. In: Ruivo M (ed) Marine pollution and sea life. FAO Publication. Fishing News (Books), London, pp 3–10

Coleman NV, Stewart WDP (1979) Enteromorpha prolifera in a polyeutrophic loch in Scotland. Br Phycol J 14: 121

Cotton AD (1910) On the growth of Ulva latissima in water polluted by sewage. Bull Misc Inform R Bot Gard Kew, pp 15–19

Cotton AD (1911) On the growth of *Ulva latissima* in excessive quantity, with special reference to the *Ulva* nuisance in Belfast Lough. Royal Commission on sewage disposal, 7th Rep II (Appendix IV), HMSO, London, pp 121–142

Cowper SW (1978) The drift algae community of seagrass beds in Redfish Bay, Texas. Contrib Mar Sci 21: 125–132

Croatto U (1985) Improved technologies in biogas production from algae of the Venice Lagoon and waste treatment. Mar Pollut Bull 22: 577–579

Dauer DM, Connor WG (1980) Effects of moderate sewage input on benthic polychaete populations. Estuar Coast Mar Sci 10: 335–346

Dawson EY (1959) A preliminary report on the benthic marine flora of southern California. In: Oceanographic survey of the Continental Shelf area of southern California. Publ Calif State Water pollut Control Board 20: 169–264

Dawson EY (1965) Intertidal algae. In: An oceanographic and biological survey of the southern California Mainland Shelf. Publ Calif State Water pollut Quality Control Board 27: 220–231, 351–438

Dean TA, Schroeter SC, Dixon JD (1984) Effects of grazing by two species of sea urchins (*Strongylocentrotus franciscanus* and *Lytechinus anamesus*) on recruitment and survival of two species of kelp (*Macrocystis pyrifera* and *Pterygophora californica*) Mar Biol 78: 301–313

DeBoer JA (1981) Nutrients. In: Lobban CS, Wynne MJ (eds) The biology of seaweeds. Botanical monographs, vol 17. Blackwell, Oxford, pp 356–392

DeBoer JA, Guigli HJ, Israel TL, D'Elia CF (1978) Nutritional studies of two red algae. I. Growth rate as a function of nitrogen source and concentration. J Phycol 14: 261–266

De Busk TA, Blakeslee M, Ryther JH (1986) Studies on the outdoor cultivation of *Ulva lactuca*. Bot Mar 24: 381–386

Degobbis D (1989) Increased eutrophication of the northern Adriatic Sea. Mar Pollut Bull 20: 452–457

Devinny JS, Volse LA (1978) Effects of sediments on the development of *Macrocystis pyrifera* gametophytes. Mar Biol 48: 343–348

Dion P (1994) Present status of the *Ulva* mass blooms on the Britanny coasts. In: Delepine R, Morand P (eds) Production and exploitation of entire seaweeds, proc 3rd COST 48, Subgroup 3 Worksh. EEC, Brussels, pp 34–36

Duke CS, Litaker W, Ramus J (1989) Effect of temperature on nitrogen-limited growth rate and chemical composition of *Ulva curvata* (Ulvales: Chlorophyta). Mar Biol 100: 143–150

Dunn J (1972) A general survey of Lanstore Harbour with particular reference to the effects of sewage. Rep to the Hampshire River Authority and Hampshire County Council. Portsmouth Polytechnic, Portsmouth, 79 pp

Dybern BI (1972) Pollution in the Baltic. In: Ruivo M (ed) Marine pollution and sea life. FAO Publication. Fishing News (Books), London pp 15–23

Ebling FJ, Sleigh MA, Sloane JF, Kitching JA (1960) The ecology of Lough Ine VII Distribution of some common plants. J Ecol 48: 29–53

Edwards P (1972) Cultured red alga to measure pollution. Mar Pollut Bull 3: 184–188

Ehrhardt J (1968) Note pour l' identification biologique des eaux polluées marines et saumâtres. Rev corps santé Armees 9: 89–103

Elmgren R (1989) Man's impact on the ecosystem of the Baltic Sea: energy flows today and at the turn of the century. Ambio 18: 326–332

Enright CT (1978) Competitive interaction between *Chondrus crispus* (Rhodophyceae) and *Ulva lactuca* (Chlorophyceae) in *Chondrus* aquaculture. In: Jensen A, Stein JR (eds) Proc 9th Int Seaweed Symp. Science Press, Princeton pp 209–218

Espinoza J, Chapman ARO (1983) Ecotypic differentiation of *Laminaria longicruris* in relation to seawater nitrate concentration. Mar Biol 74: 213–218

Fahy E, Goodwillie R, Rochford J, Kelly D (1975) Eutrophication of a partially enclosed estuarine mudflat. Mar Pollut Bull 6: 29–31

Ffrench-Constant RH (1981) An assessment of the effects of a sewage outlet on the inter-tidal macro-organisms of a rocky shore. Unpublished manuscr University of Exeter, 15 pp

Fitzgerald WJ (1978) Environmental parameters influencing the growth of *Enteromorpha clathrata* (Roth) J Ag. in the intertidal zone on Guam. Bot Mar 21: 207–220

Fleischer S, Rydberg L, Stibe L, Sundberg J (1985) Temporal variations in nutrient transport to the Laholm Bay. Vatten 41: 29–35

Fletcher RL (1974) *Ulva* problem in Kent. Mar Pollut Bull 5: 21

Fletcher RL (1988) Brief review of the role of marine algae in biodeterioration. Int Biodeterior Bull 24: 141–152

Fonselius SH (1969) Hydrography of the Baltic deep Basin III. Fishery Board of Sweden Ser Hydrography No 23

Fonselius SH (1972) On eutrophication and pollution in the Baltic sea. In: Ruivo M (ed) Marine pollution and sea life. FAO Publication. Fishing News (Books) London, pp 23–28

Fonselius St. H (1978) On nutrients and their role as production limiting factors in the Baltic. Acta Hydrochim Hydrobiol 6: 329–339

Foster GL (1914) Indications regarding the sources of combined nitrogen for *Ulva lactuca*. Ann Miss Bot Gard 1: 229–235

Frederiksen OT (1987) The fight against eutrophication in the inlet of "Odense fjord" by reaping of sea lettuce (*Ulva lactuca*) Water Sci Technol 19: 81–87

Fries L (1963) On the cultivation of axenic red algae. Physiol Plant 16: 695–708

Fries L, Pettersson H (1968) On the physiology of the red alga *Asterocystis ramosa* in axenic culture. Br Phycol Bull 3: 417–422

Friligos N (1987) Nutrient conditions in the Patraikos Gulf. Mar Pollut Bull 18: 558–561

Fujita RM (1985) The role of nitrogen status in regulating transient ammonium uptake and nitrogen storage by macroalgae. J Exp Mar Biol Ecol 92: 283–301

Fujita RM, Wheeler PA, Edwards RL (1988) Metabolic regulation of ammonium uptake by *Ulva rigida* (Chlorophyta): a compartmental analysis of the rate-limiting step for uptake. J Phycol 24: 560–566

Fujita RM, Wheeler PA, Edwards RL (1989) Assessment of macroalgal nitrogen limitation in a seasonal upwelling region. Mar Ecol Prog Ser 53: 293–303

Gabrielson JO, Birch PB, Dolin KS, Hamel DK (1983) Decomposition of *Cladophora*. II In-vitro studies of nitrogen and phosphorus regeneration. Bot Mar 26: 173–179

Gamulin-Brida H, Giaccone G, Golubic S (1967) Contribution aux études des biocenoses subtidales. Helgol Wiss Meeresunters 15: 429–444

Geertz-Hansen O, Sand-Jensen K (1990) Nitrogen limitation of growth, and grazing control of abundance of the marine macroalga *Ulva lactuca* L. in a eutrophic Danish estuary. Presentation made at the 25th Eur Mar Biol Symp Sept 10–15, 1990, Ferrara, Italy

Gemmill ER, Galloway RA (1974) Photoassimilation of ^{14}C-acetate by *Ulva lactuca*. J Phycol 10: 359–366

Gerard VA (1982) Growth and utilization of internal nitrogen reserves by the giant kelp *Macrocystis pyrifera* in a low-nitrogen environment. Mar Biol 66: 27–35

Gilet R (1960) Water pollution in Marseilles and its relations with flora and fauna. Proc Int Conf Waste Disp Mar Envir 1: 39–56

Gladstone J, Lopez-Real JM (1993) Recycling and composting of seaweed wastes. Phycologist 34: 27

Golubic S (1970) Effect of organic pollution on benthic communities. Mar Pollut Bull NS 1: 56–57

Gordon DM, Birch PB, McComb AJ (1980) The effect of light, temperature and salinity on photosynthetic rates of an estuarine *Cladophora*. Bot Mar 23: 749–755

Gordon DM, Birch PB, McComb AJ (1981) Effects of inorganic phosphorus and nitrogen on the growth of an estuarine *Cladophora* in culture. Bot Mar 24: 93–106

Gordon DM, van den Hoek C, McComb AJ (1985) An aegagropiloid form of the green alga *Cladophora montagneana* Kutz. (Chlorophyta, Cladophorales) from southwestern Australia. Bot Mar 27: 57–65

Grenager B (1957) Algological observations from the polluted area of the Oslofjord. Nytt Mag Bot 5: 41–60

Grimes BH, Hubbard JCE (1971) A comparison of film type and the importance of season for interpretation of coastal marshland vegetation. Photogrammetric Rec 7: 213–222

Guist GG, Humm HJ (1976) Effects of sewage effluent on growth of *Ulva lactuca*. Florida Sci 39: 267–271

Habig C, DeBusk TA, Ryther JH (1984) The effect of nitrogen content on methane production by the marine algae *Gracilaria tikvahiae* and *Ulva* spp. Biomass 4: 239–251

Hällfors G, Heikkonen K (1993) *Chorda tomentosa* Lyngbye in Finnish coastal waters. Acta Phytogeogr Suec 78: 79–84

Hällfors G, Kangas P, Niemi A (1984) Recent changes in the phytal at the south coast of Finland. Ophelia Suppl 3: 51–59

Hällfors G, Viitasalo I, Niemi A (1987) Macrophyte vegetation and trophic status of the gulf of Finland – a review of Finnish investigations. Meri 13: 111–158

Hanisak MD (1993) Nitrogen release from decomposing seaweeds: species and temperature effects. J Appl Phycol 5: 175–181

Hanks RW (1966) Observations on 'milky' water in Chesapeake Bay. Chesapeake Sci 7: 175–176

Harlin MM, Thorne-Miller B (1981) Nutrient enrichment of seagrass beds in a Rhode Island coastal lagoon. Mar Biol 65: 221–229

Harlin MM, Thorne-Miller B, Thursby GB (1978) Ammonium uptake by *Gracilaria* sp. (Rhodophyceae) and *Ulva lactuca* (Chlorophyceae) in closed system fish culture. In: Jensen A, Stein JR (eds) Proc 9th Int Seaweed Symp. Science Press, Princeton pp 285–292

Henley WJ, Ramus J (1989) Time course of physiological response of *Ulva rotundata* to growth irradiance transitions. Mar Ecol Prog Ser 54: 171–177

Hirose H (1978) Composition of benthic marine algae in relation to pollution in the Seto Inland sea, Japan. In: Jensen A, Stein JR (eds) Proc 9th Int Seaweed Symp. Science Press, Princeton pp 173–179

Ho YB (1979) Inorganic mineral nutrient level studies on *Potamogeton pectinatus* L. and *Enteromorpha prolifera* in Forfar Loch, Scotland. Hydrobiologia 62: 7–15

Ho YB (1986a) Changes in the intertidal algal species in Victoria Harbour, Hong Kong, over the past 50 years. Mem Hong Kong Nat Hist Soc 17: 99–102

Ho YB (1986b) Common intertidal algae of the southern part of Hong Kong Island. Mem Hong Kong Nat Hist Soc 17: 103–106

Hoogweg PHA, Wulffrat KJ, Van de Wetering BGM (1991) North Sea strategies. Mar Pollut Bull 23: 57–61

Horne AJ, Nonomura A (1976) Drifting macroalgae in estuarine waters: interactions with salt marsh and human communities. Univ Calif (Berkeley) Sanitary Eng Environ Health Res Lab Regst 76–3, 76 pp

Hull S (1987) Macroalgal mats and species abundance: a field experiment. Estuar Coast Shelf Sci 25: 519–532

Hurd CL, Harrison PJ, Druehl LD (1993) The effect of current speed on nutrient uptake by morphologically distinct forms of kelp obtained from sheltered and exposed sites. Phycologist 34: 29

Isaksson I, Pihl L (1990) Eutrophication-induced structural changes in benthic macro-vegetation and associated epibenthic faunal communities. Paper presented at Symp Plant-animal interactions in the marine benthos Sept 18-21 1990, Liverpool

Iwasaki H (1967) Nutritional studies of the edible seaweed *Porphyra tenera*. II. Nutrition of *Conchocelis*. J Phycol 3: 30–34

James DE, Stull JK, North WJ (1990) Toxicity of sewage-contaminated sediment cores to *Macrocystis pyrifera* (Laminariales, Phaeophyta) gametophytes determined by digital image analysis. Hydrobiologia 204/205. In: Lindstrom SC, Gabrielson PW (eds) Proc 13th Int Seaweed Symp. Kluwer Dordrecht, pp 483–489

Jeffrey DW (1993) Sources of nitrogen for nuisance macroalgal growths in Dublin Bay, Republic of Ireland. Phycologist 34: 30

Jeffrey DW, Madden B, Rafferty B, Dwyer R, Wilson J, Allott N (1992) Dublin Bay. Water quality management plan. Technical Report No 7. Algal growths and foreshore quality. Environmental Research Unit, Dublin, 168 pp

Johannes RE (1972) Coral Reefs and Pollution. In: Ruivo M (ed) Marine pollution and sea life. FAO Publication Fishing News (Books), London, pp 364–374

Johnson DA, Welsh BL (1985) Detrimental effects of *Ulva lactuca* (L.) exudates and low oxygen on estuarine crab larvae. J Exp Mar Biol Ecol 86: 73–83

Johnston CS (1971/72) Macroalgae and their environment. Proc R Soc Edinb 71: 195–207

Justic D (1987) Long-term eutrophication of the northern Adriatic Sea. Mar Pollut Bull 18: 281–284

Kangas P, Niemi A (1985) Observations on recolonization by the bladder wrack *Fucus vesiculosus* on the southern coast of Finland. Aqua Fenn 15: 133–141

Kangas P, Autio H, Hallfors G, Luther H, Niemi A, Salemaa H (1982) A general model of the decline of *Fucus vesiculosus* at Trarminne, south coast of Finland in 1977–81. Acta Bot Fenn 118: 1–27

Katayama T, Fujiyama T (1957) Studies on the nucleic acid of algae with special reference to the desoxyribonucleic acid contents of crown-gall tissues developed on *Porphyra tenera* Kjellm. Bull Jpn Soc Sci Fish 23: 249–254

Kautsky L (1982) Primary production and uptake kinetics of ammonium and phosphate by *Entermorpha compressa* in an ammonium sulphate industry outlet area. Aquat Bot 12: 23

Kautsky N, Kautsky H, Kautsky U, Waern M (1986) Decreased depth penetration of *Fucus vesiculosus* L. since the 1940s indicates eutrophication of the Baltic Sea. Mar Ecol Prog Ser 28: 1–8

Kautsky H, Kautsky L, Kautsky N, Kautsky V, Lindblad C (1992) Studies on the *Fucus vesiculosus* community in the Baltic Sea. Acta Phytogeogr Suec 78: 33–48

Kimor B (1991) Changes and stress signs in plankton communities as a result of man-induced perturbations in enclosed coastal seas (Mediterranean, Baltic). Mar Pollut Bull 23: 171–174

Kimura K, Hotta H, Fukushima S, Odate S, Fulkuhara A, Naito M (1958) Study of the Pacific Saury spawning on the drifting seaweeds in the sea of Japan. Bull Tohoku Reg Fish Res Lab 12: 34–39

Kindig AC, Littler MM (1980) Growth and primary productivity of marine macrophytes exposed to domestic sewage effluents. Mar Environ Res 3: 81–100

Kinsey DW, Davies PJ (1979) Effects of elevated nitrogen and phosphorus levels on coral reef growth. Limnol Oceanogr 24: 935–940

Klavestad N (1967) Undersokelser over benthos-algevegelasjonen i indre Oslofjord i 1962–1965. In: Oslofjorden og dens forurensningsproblemer. 1. Undersokelsen 1962–1965. Rep Norw Inst Water Res 9: 1–119

Klavestad N (1978) The marine algae of the polluted inner part of the Oslo-Fjord. A survey carried out 1962–1966. Bot Mar 21: 71–97

Krom MD, Neori A, Van Rijn J (1989) Importance of water flow rate in controlling water quality processes in marine and fresh water fish ponds. Bamidgeh 41: 23–33

Langton RW, Haines KC, Lyon RE (1977) Ammonia nitrogen production by the bivalve mollusc *Tapes japonica* and its recovery by the red seaweed *Hypnea musciformis* in a tropical mariculture system. Helgol Wiss Meeresunters 30: 217–229

Lapointe BE (1985) Strategies for pulsed nutrient supply to *Gracilaria* cultures in the Florida keys: interactions between concentration and frequency of pulses. J Exp Mar Biol Ecol 93: 211–222

Largo DB, Ohno M (1993) Constructing an artificial seaweed bed. In: Ohno M, Crichley AT (eds) Seaweed cultivation and marine ranching. JICA, pp 113–130

Larsson V, Elmgren R, Wulff F (1985) Eutrophication and the Baltic Sea: causes and consequences. Ambio 14: 9–14

Lavery PS, McComb AJ (1991) Macroalgal-sediment nutrient interactions and their importance to macroalgal nutrition in a eutrophic estuary. Estuar Coast Shelf Sci 32: 281–295

Le Bozec S (1993) Green Tides on the Brittany Store: the case of the Bay of Saint-Brieuc. Phycologist 34: 23

Lehnberg W, Schramm W (1984) Mass culture of brackish-water adapted seaweeds in sewage-enriched seawater. I. productivity and nutrient accumulation. Hydrobiologia 116/117: 276–281

Leskinen E, Makinen A, Fortelius W, Lindstrom M, Salemaa H (1992) Primary production of macroalgae in relation to the spectral range and sublittoral light conditions in the Tvarminne archipelago, northern Baltic Sea. Acta Phytogeogr Suec 78: 85–93

Letts EA, Richards EH (1911) Report on green seaweeds (and especially *Ulva latissima*) in relation to the pollution of the waters in which they occur. Royal Commission on sewage disposal 7th Rep vol 11, Appendix III. HMSO, London

Lindahl O, Hernroth L (1983) Phyto-zooplankton community in coastal waters of western Sweden – an ecosystem off balance. Mar Ecol Prog Ser 10: 119–126

Littler MM, Murray SN (1974) Section 5. Primary productivity of macrophytes. In: Murray SN, Littler MM (eds) Biological features of intertidal communities near the US Navy seawage outfall, Wilson Cove, San Clemente Island, California. US Navy NUC T P 396 pp 67–85

Littler MM, Murray SN (1975) Impact of sewage on the distribution, abundance and community structure of rocky intertidal macro-organisms. Mar Biol 30: 277–291

Lowthion D, Soulsby PG, Houston MCM (1985) Investigation of a eutrophic tidal basin. 1. Factors affecting the distribution and biomass of macroalgae. Mar Environ Res 15: 263–284

Lumb CM, Fowler SL (1989) Assessing the benthic impact of fish farming. In: McManus J, Elliott M (eds) Developments in estuarine and coastal study techniques. EBSA 17 symp Olsen and Olsen, Fredensborg, Denmark, pp 75–78

Lundberg P, Weich RG, Jensen P, Vogel HJ (1989) Phosphorus-31 and nitrogen-14 NMR studies of the uptake of phosphorus and nitrogen compounds in the marine macroalgae *Ulva lactuca*. Plant Physiol 89: 1380–1387

Magre EJ (1974) *Ulva lactuca* L. negatively affects *Balanus balanoides* (L.) (cirrinedia thoracica) in tidepools. Crustaceana 27: 231–234

Marchetti R, Provini A, Crosa G (1989) Nutrient-load carried by the river Po into the Adriatic Sea, 1968–1987. Mar Pollut Bull 20: 168–172

Markager S, Sand-Jensen K (1990) Heterotrophic growth of *Ulva lactuca* (Chlorophyceae) J Phycol 26: 670–673

Marshall Pregnall A, Rudy PP (1985) Contribution of green macroalgal mats (*Enteromorpha* spp.) to seasonal production in an estuary. Mar Ecol Prog Ser 24: 167–176

Martinez-Arroyo A, Romero J (1990) Some features of *Ulva* spp.: Photosynthesis and nutrient uptake in a semi estuarine embankment. Presentation made at the 25th Eur Mar Biol Symp, Sept 10–15, 1990, Ferrara, Italy

May FE (1985) Utbredelse av *Fucus serratus* L. i indre Oaslofjord relatert til forekomsten av *Mytilus edulis* L., samfunnsanalyse og felteksperimenter. Thesis, Univ Oslo 135 pp

Maze J, Morand P, Potoky P (1993) Stabilisation of 'green tides' *Ulva* by a method of composting with a view to pollution limitation. J Appl Phycol 5: 183–190

McNulty JK (1959) Ecological effects of sewage pollution in Biscayne Bay, Florida: sediments and the distribution of benthic and fouling macro-organisms. In: Tarzwell CM (ed) Trans 2nd Seminar Biol problems in water pollution VS Public Health Service

Meistrell JC, Montagne DE (1983) Waste disposal in southern California and its effects on the rocky subtidal habitat. In: Bascom W (ed) The effects of waste disposal on kelp communities. So Calif Coastal Water Res Proj Long Beach, pp 84–102

Merrill JW (1994) Mariculture techniques for use in reducing the effects of eutrophication. In: Delepine R, Morand P (eds) Production and exploitation of entire seaweeds, Proc 3rd COST 48, Subgroup 3 Workshop. EEC, Brussels, pp 52–56

Merrill J, Fletcher R (1991) Green tides cause major economic burden in Venice Lagoon, Italy. Appl Phycol Forum 8: 1–3

Missoni G, Mazzagardi M (1985) Production of algal biomass in Venice lagoon. Environmental and energetic aspects. In: Pals W, Coombs J, Hall DO (eds) Energy from biomass. 3rd EC Conference. Elsevier, Amsterdam, pp 384–386

Mohsen AF, Khaleafa AF, Hashem M, Metwalli A (1974) Effect of different nitrogen sources on growth, reproduction, amino acid, fat and sugar contents in *Ulva fasciata* Delile. Bot Mar 17: 218–222

Montgomery HA, Soulsby PG (1980) Effects of eutrophication on the intertidal ecology of Langstone Harbour, UK, and proposed control measures. Prog Water Technol 13: 287–294

Munda I (1974) Changes and succession in the benthic algal associations of slightly polluted habitats. Rev Int Oceanogr Med 34: 37–52

Munda IM (1993) Changes and degradation of seaweed stands in the northern Adriatic. In: Chapman ARO, Brown MT, Lahaye M (eds) Proc 14th Int Seaweed Symp. Hydrobiologia 260/261: Kluwer, Dordrecht pp 239–253

Murray SN, Littler MM (1974) Section 3. Analyses of standing stock and community structure of macro-organisms. In: Murray SN, Littler MM (eds) Biological features of intertidal communities near the US Navy seawage outfall, Wilson Cove, San Clemente Island, California US Navy NUC T P 396 pp 23–85

Murray SN, Littler MM (1976) An experimental analysis of sewage-impact on a macrophyte-dominated rocky intertidal community. J Phycol (Suppl 12): 15–16

Murray SN, Littler MM (1983) Seaweed standing stocks and successional events in a rocky intertidal habitat near the Whites Point sewage outfall, California, USA. Presentation made at the XIth Int Seaweed Symp, June 19–25, 1983, Qingdao, China

Nasr AA, Bekheet IA, Ibrahim RK (1968) The effects of different nitrogen and carbon sources on amino acid synthesis in *Ulva*, *Dictyota*, and *Pterocladia*. Hydrobiologia 31: 7–16

Nasr AH, Aleem AA (1948) Ecological studies of some marine algae from Alexandria. Hybrobiologia 1: 251–281

Neish AC, Fox CH (1971) Greenhouse experiments on the vegetative propagation of *Chondrus crispus* (Irish moss). National Research Council, Canada, Atlantic Regional Laboratory Technical Report Ser No 14, Halifax, 25 pp

Nelson-Smith A (1972) Oil pollution and marine ecology. Elek Science, London 260 pp

Neori A, Cohen I, Gordin H (1991) *Ulva lactuca* biofilters for marine fishpond effluents II. Growth rate, yield and C:N ratio. Bot Mar 34: 483–489

Neushal M, Foster MS, Coon DA, Woessner JW, Harger BW W (1976) An in situ study of recruitment growth and survival of subtidal marine algae: techniques and preliminary results. J Phycol 12: 397–408

Nicholls DS, Keeney DR (1973) Nitrogen and phosphorus release from decaying water milfoil. Hydrobiologia 42: 509–525

Nicholls DJ, Tubbs CR, Haynes FN (1981) The effect of green algal mats on intertidal macrobenthic communities and their predators. Kiel Meeresforsch Sonderh 5: 511–520

Nicolini S, Viglia A (1985) Anaerobic digestion of macroalgae in the lagoon of Venice: experiences with a 5 m³ capacity pilot reactor. In: Palz W, Coombs J, Hall DO (eds) Energy from biomass. Proc Int Conf on Biomass. Elsevier, London, pp 614–616

Nielson K, Knudsen L (1990) Eutrophic effects on the marine vegetation, southern Kattegat. Presentation made at the 25th Eur Mar Biol Sym, Sept 10–15, 1990, Ferrara, Italy

Nienhuis PH (1992) Ecology of coastal lagoons in the Netherlands (Veerse Meer and Grevelingen). Vie Milieu 42: 59–72

North WJ (1963) Ecology of the rocky nearshore environment in southern California and possible influences of discharged wastes. Int J Air Water Pollut 7: 721–736

North WJ (1979) Adverse factors affecting giant kelp and associated seaweeds. Experientia 35: 445–447

North WJ (1983) The sea urchin problem. In: Bascom W (ed) The effects of waste disposal on kelp communities. So Calif Coastal Water Res Prof, Long Beach, pp 147–162

North WJ, Stephens GC, North BB (1972) Marine algae and their relations to pollution problems. In: Ruivo M (ed) Marine pollution and sea life. FAO Fishing News (Books), London, pp 330–340

Norton TA, Mathieson AC (1983) The biology of unattached seaweeds. Prog Phycol Res 2: 333–386

Ogawa H (1983) Effects of treated municipal wastewater on the earlier development of Sargassaceous plants. Presentation made at the 11th Int Seaweed Symp, June 19–25, 1983 Qingdao, China

Olafson EB (1988) Inhibition of larval settlement to a soft bottom community by drifting algal mats: an experimental test. Mar Biol 97: 571–574

Orlandini M (1988) Harvesting of algae in polluted lagoons of Venice and Orbetello and their effective and potential utilization. In: de Wart J, Nienhuis PH (eds) Aquatic primary biomass (marine macroalgae); biomass conversion, removal and use of nutrients. Proc 2nd COST 48 Workshop. EEC, Brussels, pp 20–23

O'Sullivan AJ (1971) Ecological effects of sewage discharge in the marine environment. Proc R Soc Lond B 177: 331–351

Otte G (1979) Investigations on the effects of domestic sewage on the benthic ecosystem of marine intertidal flats. Helgol Wiss Meeresunters 32: 73–148

Park S (1992) Ulva lactuca monitoring programme. Bay of Plenty Regional Council, Techn Rep Number 32, 56 pp

Parker HS (1981) Influence of relative water motion on the growth, ammonium uptake and carbon and nitrogen composition of Ulva lactuca (Chlorophyta). Mar Biol 63: 309–318

Pavoni B, Calvo C, Sriso A, Orio AA (1990). Time trend of PCB concentrations in surface sediments from a hypertrophic, macroalgae populated area of the lagoon of Venice. Sci Tot Envir 91: 13–21

Perez-Ruzafa A, Marcos-Diego C, Ros JD (1991) Environmental and biological changes related to recent human activities in the Mar Menor (SE of Spain). Mar Pollut Bull 23: 747–751

Perkins EJ, Abbott OJ (1972) Nutrient enrichment and sand-flat fauna. Mar Pollut Bull 3: 70–72

Pignatti S (1962) Associazioni di alghe marine sulla costa veneziana. Mem Ist Veneto Lattere Sci Arti 32: 1–134

Piriou JY, Menesguen A (1990) Environmental factors controlling the Ulva sp. blooms in Brittany (France)

Powell HT (1963) New records of Fucus distichus subspecies for the Shetland and Orkney Islands. Br Phycol Bull 2: 247–254

Price LH, Hylleberg J (1982) Algal faunal interactions in a mat of *Ulva fenestrata* in False Bay, Washington, Ophelia 21: 75–88

Prince JG (1974) Nutrient assimilation and growth of some seaweeds in mixtures of seawater and secondary sewage treatment effluents. Aquaculture 4: 69–79

Prince JS, Kingsbury JM (1973) The ecology of *Chondrus crispus* at Plymouth, Massachusetts, II. Field studies. Am J Bot 60: 964–975

Pugnetti A, Viaroli P, Ferrari I (1990) Processes leading to dystrophy in a Po river delta lagoon (Sacca di Goro): phytoplankton-macroalgae interactions. Presentation made at the 25th Eur Mar Biol Symp, Sept 10–15, 1990, Ferrara, Italy

Pybus C (1971) The use of *Laminaria saccharina* (L.) Lamour. as an indicator organism for marine pollution. PhD Thesis, University of Liverpool

Pybus C (1973) Effects of anionic detergent on the growth of *Laminaria*. Mar Pollut Bull 4: 73–76

Raffaelli D, Hull S, Milne H (1989) Long-term changes in nutrients, weed mats and shorebirds in an estuarine system. Cah Biol Mar 30: 259–270

Raffaelli D, Limia J, Hull S, Pont S (1991) Interactions between the amphipod *Corophium volutator* and macroalgal mats on estuarine mud flats. J Mar Biol Ass UK 71: 899–908

Rajasilta M, Eklund J, Kääriä, Ranta-aho K (1989) The deposition and mortality of the eggs of the Baltic herring, *Clupea harengus membras* L., on different substrates in the south-west archipelago of Finland. J Fish Biol 34: 417–427

Ramus J (1983) A physiological test of the theory of complementary chromatic adaption. II. Brown, green and red seaweeds. J Phycol 19: 173–178

Ramus J, Venable M (1987) Temporal ammonium patchiness and growth rate in *Codium* and *Ulva* (Ulvophyceae). J Phycol 23: 518–523

Rasmussen Bo M (1990) Environmental consequences of filamentous algae mats in the archipelago of south Funen, Denmark. Presentation made at the 25th Eur Mar Biol Symp, Sept 10–15, 1990, Ferrara, Italy

Raymont JEG (1947) A fish farming experiment in Scottish sea lochs. J Mar Res 67: 219–227

Reise K (1983) Sewage, green algal mats anchored by lugworms and the effects of *Turbellaria* and small polychaeta. Helg Meersunters 36: 151–162

Revelante N, Gilmartin M (1976) The effect of Po River discharge on phytoplankton dynamics in the northern Adriatic Sea. Mar Biol 34: 259–271

Riggio S, D'anna G, Sparla MP (1990) Coastal eutrophication and settlement of mussel beds and other filter feeders in N. W. Sicily. Presentation made at the 25th Eur Mar Biol Symp, Sept 10–15, 1990, Ferrara, Italy

Rigoni S, Rismondo R, Szpyrkowicz L, Zilio-Grandi F, Vigato PA (1989) Anaerobic digestion of nitrophillic algal biomass from the Venice lagoon. Biomass 23: 179–199

Rosenberg G, Ramus J (1982) Ecological growth strategies in the seaweeds *Gracilaria foliifera* (Rhodophyceae) and *Ulva* sp. (Chlorophyceae): soluble nitrogen and reserve carbohydrates. Mar Biol 66: 251–259

Rosenberg G, Ramus J (1984) Uptake of inorganic nitrogen and seaweed surface area: volume ratios. Aquat Bot 19: 65–72

Rosenberg G, Probyn TA, Mann KH (1984) Nutrient uptake and growth kinetics in brown seaweeds: response to continuous and single additions of ammonium. J Exp Mar Biol Ecol 80: 125–146

Rosenberg R (1985) Eutrophication. The future marine coastal nuisance? Mar Pollut Bull 6: 227–231

Rosenberg R (1990) Eutrophication effects on the Swedish west coast. Presentation made at the 25th Eur Mar Biol Symp, Sept 10–15, 1990, Ferrara, Italy

Rosenberg R, Edler L (1981) Laholmsbuklen-ovis framtid for stord miljo. Forsk Framsteg 3: 36–39

Rosenberg R, Loo LO (1988) Marine eutrophication induced oxygen deficiency: effects on soft bottom fauna, western Sweden. Ophelia 29: 213–225

Rosenberg R, Elmgren R, Fleischer S, Jonsson P, Persson G, Dahlin H (1990) Marine eutrophication case studies in Sweden. Ambio 19: 102–110

Rueness J (1973) Pollution effects on littoral algal communities in the inner Oslofjord, with special reference to *Ascophyllum nodosum*. Helgol Wiss Meeresunters 24: 446–454

Rueness J, Fredriksen S (1991) An assessment of possible pollution effects on the benthic algae of the outer Oslofjord, Norway. Ophelia 17: 223–235

Ruiz JM, Marin A, Ramirez L (1990) Eutrophication processes associated to the presence of coastal constructions on Aguilas Bay (Murcia, SE of Spain). Presentation made at the 25th Eur Mar Biol Symp, Sept 10–15, 1990 Ferrara, Italy

Russell G (1974) *Fucus distichus* communities in Shetland. J Appl Ecol 11: 679–684

Ruud JT (1968) Introduction to the studies of pollution in the Oslofjord. Helgol Wiss Meeresunters 17: 455–461

Ryther JH (1954) The ecology of phytoplankton blooms in Moriches Bay and Great South Bay, Long Island, New York. Biol Bull Mar Biol Lab Woods Hole 106: 198–209

Ryther JH, Dunston WM (1971) Nitrogen, phosphorus and eutrophication in the coastal marine environment. Sci NY 171: 1008–1013

Sand-Jensen K (1988) Photosynthetic responses of *Ulva lactuca* at very low light. Mar Ecol Prog Ser 50: 195–201

Saunders RG, Lindsay JG (1979) Growth and enhancement of the agarophyte *Gracilaria* (Florideophyceae) In: Jensen A, Stein JR (eds). Proc 9th Int Seaweed Symp. Science Press, Princeton, pp 249–255

Sawyer CN (1965) The sea lettuce problem in Boston harbour. J Water Pollut Control Fed 37: 1122–1133

Schiffner V, Vatova A (1938) Le alghe della Laguna: Chlorophyceae, Phaeophyceae, Rhodophyceae, Myxophyceae. In: Mino M (ed) La Laguna di Venezia, 3. Venezia 250 pp

Sfriso A (1987) Flora and vertical distribution of macroalgae in the lagoon of Venice: a comparison with previous studies. G Bot Ital 121: 69–85

Sfriso A, Marcomini A (1994) Gross primary production and nutrient behaviour in a shallow coastal environment. Biores Tech 47: 59–66

Sfriso A, Marcomini A, Pavoni B (1987) Relationships between macroalgal biomass and nutrient concentrations in a hypertrophic area of the Venice Lagoon. Mar Environ Res 22: 297–312

Sfriso A, Pavoni B, Marcomini A (1989) Macroalgae and phytoplankton standing crops in the central Venice Lagoon: primary production and nutrient balance. Sci Tot Env 13: 139–159

Sfriso A, Pavoni B, Marcomini A, Orio AA (1988a) Annual variations of nutrients in the Lagoon of Venice. Mar Pollut Bull 19: 54–60

Sfriso A, Pavoni B, Marcomini A, Orio AA (1988b) Macroalgal biomass production and nutrient recycling in the lagoon of Venice. Ing Samitaria 5: 255–266

Sfriso A, Raccanelli S, Pavoni B, Marcomini A (1991) Sampling strategies for measuring macroalgal biomass in the shallow waters of the Venice Lagoon. Environ Technol Lett 12: 263–269

Smith DW, Horne AJ (1988) Experimental measurement of resource competition between planktonic microalgae and macroalgae (seaweeds) in mesocosms simulating the San Fransisco Bay-Estuary, California. Hydrobiologia 159: 259–268

Smith SV, Kimmerer WJ, Laws EA, Brock RE, Walsh TW (1981) Kaneohe Bay Sewage Diversion Experiment: Perspectives on Ecosystem responses to nutrional perturbation. Pac Sci 35: 279–380

Smyth JC (1968) The fauna of a polluted shore in the Firth of Forth. Helgol Wiss Meeresunters 17: 216–223

Soulsby PG, Lowthion D, Houston MCM (1978) Observations on the effects of sewage discharged into a tidal harbour. Mar Pollut Bull 9: 242–245

Soulsby PG, Lowthion D, Houston M (1982) Effects of macroalgal mats on the ecology of intertidal mudflats. Mar Pollut Bull 13: 162–166

Soulsby PG, Lowthion D, Houston M, Montogomery HAC (1985) The role of sewage effluent in the accumulation of macroalgal mats on intertidal mud flats in two basins in Southern England. Neth J Sea Res 16: 257–263

Southgate BA (1972) Langstone Harbour study. Rep to Hampshire River Authority and the Hampshire County Council. The Counties Public Health Laboratories, London, 20pp

Steffensen DA (1976) Morphological variation in *Ulva* in the Avon-Heathcote Estuary, Christchurch. NZJ Mar Freshwater Res 10: 329–341

Stokke K (1956) The resistance of *Gracilaria confervoides* to hydrogen sulphide. In: Braarud T, Sørensen NA (eds) Proc 2nd Int Seaweed Symp Trondheim Pergamon Press, London, pp 210–214

Subbaramaiah K, Parekh RG (1966) Observations on a crop of *Ulva fasciata* growing in polluted sea water. Sci Cult 32: 370

Syrett PJ (1962) Nitrogen assimilation. In: Lewin RA (ed) Physiology and biochemistry of algae. Academic Press, New York, pp 171–188

Taylor WR, Bernatowicz AJ (1969) Distribution of marine algae about Bermuda. Spec Publ No 1. Bermuda Biological Station for Research, St George's West, Bermuda, 42 pp

Tewari A (1972) The effect of sewage pollution on *Enteromorpha prolifera* growing under natural habitat. Bot Mar 15: 167

Thorne-Miller B, Harlin MM, Thursby GB, Brady-Campbell MM, Dworetzky BA (1983) Variations in the distribution and biomass of submerged macrophytes in five coastal lagoons in Rhode Island, USA. Bot Mar 26: 231–242

Thrush SF (1986) The sublittoral macrobenthic community structure of an Irish Sea lough: effect of decomposing accumulations of seaweed. J Exp Mar Biol Ecol 96: 199–212

Topinka JA, Robbins (1976) Effects on nitrate and ammonium enrichment on growth and nitrogen physiology in *Fucus spiralis*. Limnol Oceanogr 21: 659–664

Tsutsumi H (1990) Population persistance of *Capitella* spp. (Polychaeta, Capitellidae) on a mud flat subject to environmental disturbance by organic enrichment. Mar Ecol Prog Ser 63: 147–156

Tsutsumi H, Kikuchi T, Tanaka M, Higashi T, Imasaka K, Miyazaki M (1991) Benthic faunal succession in a cove organically polluted by fish farming. Mar Pollut Bull 23: 233–283

Tubbs CR (1977) Widfowl and waders in Langstone Harbour. Br Birds 70: 177–199

Tubbs CR, Tubbs JM (1980) Waders and shelduck feeding distribution in Langstone Harbour, Hampshire. Bird Study 27: 239–248

Van Bennekom AJ, Gieskes WWC, Tijssen SB (1975) Eutrophication of Dutch coastal waters. Proc R Soc Land 189: 359–374

Vandermeulen H, Gordin H (1990) Ammonium uptake using *Ulva* (Chlorophyta) in intensive fishpond systems: mass culture and treatment of effluent. J Appl Phycol 2: 363–374

Vermaat JE, Sand-Jensen K (1987) Survival, metabolism and growth of *Ulva lactuca* under winter conditions: a laboratory study of bottlenecks in the life cycle. Mar Biol 95: 55–61

Vogt H, Schramm W (1991) Conspicuous decline of *Fucus* in Kiel Bay (Western Baltic): what are the causes? Mar Ecol Prog Ser 69: 189–194

Vukadin I (1991) Impact of nutrient enrichment and their relation to the algal bloom in the Adriatic Sea. Mar Pollut Bull 23: 145–148

Waite T, Gregory C (1969) Notes on the growth of *Ulva* as a function of ammonia and nitrogen. Phycologia 18: 65–69

Waite T, Mitchell R (1972a) The effect of nutrient fertilization on the benthic alga, *Ulva lactuca*. Bot Mar 15: 151–156

Waite TD, Mitchell R (1972b) Role of benthic plants in a fertilized estuary. J Sanit Eng Div Proc Am Soc Civ Eng 98: SA5 763–770

Waite TD, Spielman LA, Mitchell R (1972) Growth rate determinations of the macrophyte *Ulva* in continuous culture. Environ Sci Technol 6: 1096–1100

Wallentinus I (1981) Chemical constituents of some Baltic macroalgae in relation to environmental conditions. In: Levring T (ed) Proc 10 Int Seaweed Symp. de Gruyter, New York, pp 363–370

Wallentinus I (1984) Comparisons of uptake rates for Baltic macroalgae with different thallus morphologies. Mar Biol 80: 215–225

Wallentinus I (1991) The Baltic Sea gradient. In: Mathieson AC, Nienhuis PH (eds) Ecosystems of the World-24 Intertidal and littoral ecosystems. Elsevier, Amsterdam, pp 83–108

Welsh BL (1980) Comparative nutrient dynamics of a marsh mud flat system. Estuar Coast Mar Sci 10: 143–165

Wennberg T (1987) Long term changes in the composition and distribution of the macroalgal vegetation in the southern part of Laholm Bay, south west Sweden, during the last thirty years. Swedish Environmental Protection Agency Rep 3290, pp 1–47

Wennberg T (1992) Colonization and succession of macroalgae on a breakwater in Laholm Bay, a eutrophicated brackish water area (SW Sweden). Acta Phytogeogr Suec 78: 65–77

Wharfe JR (1977) The intertidal sediment habitats of the lower Medway Estuary, Kent. Environ Pollut 13: 79–91

Wharfe JR, Hutchings BJ, Jowett PE (1981) Some effects of the discharge of chlorinated sewage from a short sea-outfall, on intertidal macro-organisms. Environ Pollut 25: 9–17

White JT, Highlands NJ (1968) The destruction of clams by sea lettuce. Underwater Nat 5: 27

Wilce RT, Schneider CW, Quinlan AV, van den Bosch K (1982) The life history and morphology of free living *Pilayella littoralis* (L.) Kjellm (Ectocarpaceae, Ectocarpales) in Natiant Bay, Massachusetts. Phycologia 21: 336–354

Wilkinson L (1963) Nitrogen transformations in a polluted estuary. Int J Air Water Pollut 7: 737–752

Wilson JG, Jeffrey DW, Madden B, Rafferty (1990) Algal mats and eutrophication in Dublin Bay, Ireland. Presentation made at the 25th Eur Mar Biol Symp, Sept 10–15, 1990, Ferrara, Italy

Wise DL, Augenstein DC, Ryther JH (1979) Methane fermentation of aquatic biomass. Resour Recovery Conserv 4: 217–237

Zimmermann CF, Montgomery JR (1984) Effects of a decomposing drift algal mat on sediment pore water nutrient concentration in a Florida seagrass bed. Mar Ecol 19: 299–302

Zou J, Dong L (1983) Preliminary studies on eutrophication and the red tide problems in Bohai Bay. Presentation made at the 11th Int Seaweed Symp, 19–25 June, 1983, Qingdao, China

2 The Management of Eutrophicated Waters

R.H. Charlier[1] and Th. Lonhienne[2]

2.1 Introduction

Eutrophication, blooms and green tides have over the few recent decades become scourges, indicating problems of substantial economic magnitude. Various types of pollution have enriched waters with nutrients. These in turn have favored blooms, particularly of algae, seemingly a mechanism of nature to "clean up". However, the cleansing agent has become itself a nuisance and a source of pollution, in turn fouling water, beaches and the atmosphere.

Fauna has been threatened by algal proliferation, posing a problem for fisheries, and strandings have had adverse effects on tourism. The areas affected by the development of an exploding algal population are expanding. The phenomenon has been observed worldwide and has triggered major concern in most EC countries (Fig. 2.1). Can this scourge be turned into a resource? Algae, no longer a commercially viable source of iodine, have nevertheless a wide range of uses. Bioconversion is often proposed. Yet, except for the some very specific applications, particularly in the food industry, major undertakings appear economically non-competitive.

However, wait-and-see attitudes should no longer prevail, and plans for the management of eutrophication and its consequences should be designed and implemented.

If algae appear currently to be a scourge, their economic significance is, it ought to be stressed, not always negative. Their use seems to be often a fashion of the times witness the recent cartoon in the American press where a spouse having prepared a meal involving algae tells her

[1]Free University of Brussels (Earth Sciences, VUB), 2 Avenue du Congo (Box 23), 1050 Brussels, Belgium
[2]State University of Liège, Liège, Belgium

Ecological Studies, Vol. 123
Schramm/Nienhuis (eds) Marine Benthic Vegetation
© Springer-Verlag Berlin Heidelberg 1996

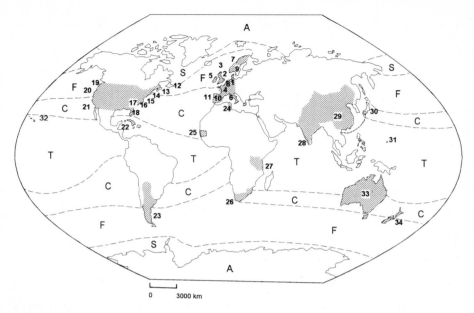

Fig. 2.1. Worldwide geographic distribution of algal proliferations. (After Briand 1987)
Climates: *A* Arctic; *S* subarctic; *F* microthermal; *C* mesothermal; *T* intratropical. Legend:
1 Germany; 2 England and Wales; 3 Scotland; 4 France; 5 Ireland; 6 Italy; 7 Norway; 8
The Netherlands; 9 Sweden; 10 Spain; 11 Portugal. 12–21 USA: 12 Massachusetts; 13
Rhode Island; 14 Connecticut; 15 New York; 16 North Carolina; 17 South Carolina; 18
Florida; 19 Washington; 20 Oregon; 21 California. 22 Cuba; 23 Argentina; 24 Tunisia; 25
Senegal; 26 Republic of South Africa; 27 Tanzania; 28 India; 29 China; 30 Japan; 31 Guam;
32 Hawaii; 33 Australia; 34 New Zealand

dismayed husband: "Don't think of it as sea-weed, but as sea-lettuce".
Indeed if records relate that dulse was eaten fourteen centuries ago,
currently the use of algae in food is expanding, particularly in the health-
food industry; the European-grown *Undaria pinnatifida* is now commonly used as human food. The ancient Greeks fed seaweeds to cattle in
times of scarcity. And the press hails at regular intervals sea plants as the
saviors of humanity the day lack of food will starve people.

One may wonder, however, considering that sea plants are hard to
digest and have a high mineral content whether consumption as food
and/or feed will make a large dent in the algae "oversupply". Agricultural use, pharmaceutical preparations, packing materials are common
but limited markets. Agricultural use has even been in decline because in
industrial countries chemical fertilizers are cheaper.

A case has been made about the use of algae in cosmetology (XIIth
International Seaweed Symposium [1984]; Y. De Roeck-Holtzhauer in Seaweed Resources in Europe [1991]; L. Evin in Bull. Soc. Dermopharm.

[1959]) but it is not clearly separated from applications in thalassotherapy wherein the range is wide.

While production remains geared towards providing commercial compounds there are a few more uses which will be touched upon later. Here it is however necessary to point out that all uses taken together, their importance will not suffice to counter the problems and worries about algal proliferation.

2.2   Eutrophication

The importance of phytoplankton and phytobenthos is impressive: their yearly global production is estimated at 160×10^9 t dry matter, and is known to exceed 53×10^9 t dry matter. Proliferation of biomass in some geographical areas has been a subject of concern.

Two most conspicuous features of eutrophication are an increased incidence of phytoplankton blooms followed by an increase in the macrobenthic biomass. Briand (1987, 1988) lists as common characteristics of blooms which have reoccurred: (1) morphological simplicity of the species; (2) vegetative reproductive systems, allowing large productivity; (3) fragility, enhancing proliferation (viz blooms); and (4) nutrient absorption capacity, favoring a high growth rate. The tolerance of algae to wide ranges of salinity, temperature, light and pollution helps their spreading and survival.

Webster defines eutrophication – derived from the classical Greek "entropos" ($\varepsilon \gamma \theta \rho o \pi o \varsigma$)" – generally as the fact of being well nourished, and particularly for a body of water rich in dissolved nutrients, such as phosphates, but often shallow and seasonally deficient in oxygen.

Eutrophication is an increase in the availability of nutrients essential to the growth of algae and other flora, viz. nitrogen (biologically the most active by way of N_2, NH_4^+, NO_3^-), phosphorus (PO_4^{3-}), silica (SiO_2), vitamins, traces of metals, and some others.

At first sight, eutrophication may seem a positive occurrence as plants need nutrients and the entire food chain benefits so that fishing productivity has increased in some sites substantially (e.g. the Baltic Sea). Instead, it is the main cause of excessive microalgal blooms which are, in the long run, detrimental to marine biological resources. Macroalgal proliferation ensues equally. The negative effects include: (1) ailments and/or mortality of flora, benthos and pelagic fishes, in pisciculture and mariculture basins; (2) discharge into the atmosphere of sulfur compounds, by bacteria which decompose organic matter of algal origin.

These compounds are, in part responsible for acid rain; (3) heavy beach pollution resulting from algal strandings.

With increased eutrophication, "specialists" in resource competition will be replaced by phytoplankton "generalists", then by poorly edible primary producers. The positive aspect of nutrient increase leading to more algae, more zooplankton, and hence more fishes certainly does not hold true in highly eutrophicated areas (Riegman 1991).

There is no absolute evidence whether changes recorded in recent years in the shelf system are the consequences of excess nutrient input, the results of climatic cycles, or simply an effect of the improvement in observation coverage. What is generally agreed upon are the considerable detrimental effects of eutrophication. They outweigh by far the beneficial effects such as an increased productivity or the removal of atmospheric carbon dioxide by way of burial of organic matter. Increased fish yields have been considered a positive effect, but realistic limits should notwithstanding be set to allowable chemical pollution (Tatara 1981; Clark 1989).

Examining the causes of eutrophication, we note that nitrogen rather than phosphorus is probably the limiting plant nutrient in the sea. Agriculture, the development of river beds and, in catchment basins, soil erosion are major sources of eutrophication materials. Among other factors involved are acceleration of river flow and drainage of wetlands, with accumulation of sediments in estuaries as a result. Remedies needed so that navigation is not impeded include dredging, and the deeper water allows leaching of nutrients. Car emissions and energy conversion cause "inadvertent" fixing of nitrogen, while pollution by toxins results in a reduction of freshwater plants, and eventually the harmful, substances are picked up by the marine phytoplankton. Additionally, touristic activities and the development of sandy areas generate quantities of untreated wastes and lead, in the long term, to severe consequences.

2.2.1 Eutrophication, Dystrophication and Mediators

Two concepts should be introduced here. Dystrophication is the modification of the bacterial flora with the concurrent development of an anoxic environment and hydrogen sulfide production, the final stages in the eutrophication process. Toxic "red tides" result which may kill marine biota and even produce problems for humans. Plant- or animal-produced chemical "mediators" dispersed in the marine milieu influence the behavior and physiology of these, or other species. Some are called telemediators because they act from a considerable distance.

While it is widely held that the excessive contribution of nutrients to sea waters are at the basis of marine eutrophication, some researchers,

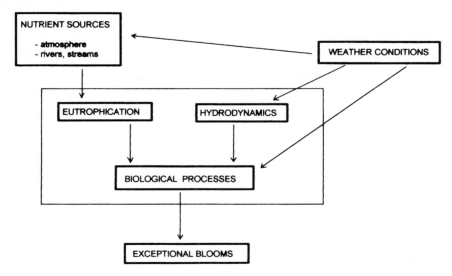

Fig. 2.2. Major factors leading to algal blooms

like M. Aubert (1990) have held for some time that such an explanation is too "simplistic" in view of the complexity of the natural phenomena.

In his "theory on eutrophication", the processes are enhanced by mediators secreted by specific biota that condition relations between marine organisms. Ecological disorders ensue when pollutants that alter the activities of mediators disrupt the normal phytoplanktonic cycle. Eutrophication and dystrophication then occur.

"Exceptional blooms" designate algal proliferation that have harmful effects for man. Blooms may be punctual (single occurrence) or recurrent. They are linked to biological processes influenced by three parameters (Fig. 2.2): (1) eutrophication; (2) hydrodynamics; and (3) weather conditions, which also have an impact on the two other parameters. Hydrodynamic parameters include currents, water temperature and salinity, tides and the position of the thermal and haline picnolines. Weather conditions influence nutrients' presence, hydrodynamics and biological processes through air temperature, isolation precipitation, speed and direction of winds.

Exceptional blooms depend upon complex biological processes, at the level of the first trophic system in the ecosystem. They include photosynthesis and growth of the plankton, interspecies competition, predatory behavior of zooplankton, molluscs and crustaceans, algal sedimentation, and bacterial decomposition. The so-called Lake of Tunis, for example, is the site of extreme eutrophication during the summer months, where, in calm weather, the whole water column may become anaerobic. Nitrite concentrations may reach 50 times those of enclosed bays in Greek

waters. Incidentally occurring toxic blooms are a function of meteorological conditions.

Eutrophication, as stated before, has been reported worldwide and is
more intense in nearly closed basins with a narrow access to the open
sea, such as gulfs, bays and fjords, although it is not necessarily absent
along relatively open coasts. The coastlines and coastal lagoons of Europe are, of course, no exception. Proliferation of algae is a consequence of
this phenomenon (Belkhir 1981; Chassany de Casabianca 1984;
Montgomery et al. 1985; Virnstein and Carbonara 1988).

The eutrophication state of a lagoon can be ascertained by measurements of salinity, temperature, oxygen and nutrients; these measurements will also give information on the existence of anoxic conditions
and stratification.

Nearshore primary producers include phytoplankton, marine phanerogams, benthic micro- and macroalgae. Plant growth is an initial response to
excess nutrients; this may lead to development of the benthic suspension-
feeding bivalves. The initial direct effect of eutrophication is change in
species composition, and next altered distribution patterns such as those of
fucoid algae (Clark 1989; Gray 1991). However, not all algal blooms should
be ascribed to increased nutrients and entrophication. Habitat destruction
probably plays a role. Algal blooms are often sources of toxins or volatile
sulfur compounds; many species are unpalatable to herbivores.

2.3   Eutrophication in European Sea Bodies

Alarming signs of change in nutrient and productivity patterns of ecosystems have been reported for many European coastal waters. They include
increased phytoplankton production and mass development of algae, and
a fortiori green tides (Schramm 1991). The need to protect the systems
from eutrophication and pollution may attract anew attention to marine
macrophytes as candidates for tertiary wastewater treatment; seaweeds are
suited for purification of nitrogen-rich domestic and urban sewage, as well
as for some agricultural and industrial waste effluents (Schramm 1991).

Proliferation of green algae affects particularly *Ulva, Enteromorpha*
and *Cladophora*, and among red algae *Gracilaria* and *Porphyra*. Though
proliferation refers mostly to macroalgae and blooms of microalgae, the
rule is not absolute. The extent of proliferation has been estimated by
Briand (1987, and in his thesis at Lille in 1989) at 1.3×10^6 metric tonnes,
but is probably higher.

Algal blooms are currently common in Europe along coasts in the Mediterranean, Atlantic, The Channel, North Sea and Baltic. The Venice lagoon has a $15 \, kg \, m^{-2}$ biomass (dry matter) and mixed-bag annual algal production of 1 million metric tonnes dry matter. On the French Atlantic coasts densities may reach $240 \, kg \, m^{-2}$ and a thickness of $1 \, m$ (Figs. 2.3, 2.4). On some parts of the Mediterranean densities of $5 \, kg \, m^{-2}$ are not unusual; densities may reach $10 \, kg \, m^{-2}$ and algal decay may destroy an important part of the fauna. Biomass densities off the coast of south-east England are $2 \, kg \, m^{-2}$, and off Ireland are $45 \, kg \, m^{-2}$. In Germany, the Isle of Sylt is particularly affected, and the literature is replete with examples from Norway, Scotland, The Netherlands, Spain and Portugal (Reise 1983; Chassany de Casabianca 1984; Lowthion et al. 1985; Montgomery et al. 1985).

In Italy, severe problems have been reported for example in Orbetello, north of Rome, and in the Venice lagoon. In Orbetello, algal proliferation competes with fisheries and is detrimental to the tourist trade of the area. Algae, in particular *Chaetomorpha* or *Cladophora*, are a mere nuisance, there being no economic utilization of their biomass. Cut, picked up, transferred to trucks, they are dumped inland, making heaps infested by insects and left to rot. Near the city of the doges, *Gracilaria* is manually harvested and used for agar-agar extraction. Besides the types mentioned, blooms of *Ulva* and *Valonia* species are not uncommon.

Eutrophication in European waters is generally believed to be mainly man-caused by increase in wastes. Nutrients taken on by the seas have two major and one less important origin: (1) nutrients carried by streams and rivers and coming from domestic and industrial "used" water, as well as the leaching by rainwater of fertilized soil; (2) nutrients contributed by the atmosphere contained in rainwater, and by nitrogen fixation by blue-green algae or cyanobacteria; and (3) nutrients contributed by artificial fish ponds, namely, fish excreta and food surplus.

Surrounded by industrialized countries, with a restricted communication with the ocean, the Baltic Sea is rich in nutrients brought by rivers and by the atmosphere (NH_4^+, NO_3^-). Nitrogen fixation by blue-green algae is not negligible. It is estimated that since 1900, inorganic phosphorus has increased eightfold and inorganic nitrogen fourfold. Phosphate increases have been stemmed through a concerted effort to treat used waters. Fertilizers used on farms remain the culprits for a continued increase in nitrogen.

Blooms are recurrent during the entire vegatative period, i.e. spring and summer. They were already reported at the end of the nineteenth century, and were similar to those of the 1980s.

Fig. 2.3. The "green tide" strikes at Hillion Bay (France). In the background, mixed with sawdust. (Photo: Ph. Morand)

Fig. 2.4. Atlantic coast of Brittany, bulldozers are called in to remove seaweed that accumulates daily on beaches

For Riegman (1991), novel algal blooms may be the result of shifts in the N/P and NH_4^+/NO_3^- ratios rather than the effect of N/P enrichment. Measures should be region specific, otherwise the risk is run of new nuisances caused by algal blooms.

Depletion in oxygen at greater depths and on the bottom is due to increased sedimentation of organic matter which augments bacterial concentration and benthic macrofauna, leading to temporary or permanent anoxic conditions. The situation is worsened by the Baltic's permanent halocline inhibiting oxygen mixing in deep waters; it results from the sea's narrow inlets, from a considerable freshwater input and from a "positive balance", viz. the incoming flux is less than the loss of fresh water by evaporation.

Fucus and some red algae, deprived of light and starved of oxygen, are yielding to filament-type algae which feed on the excess nutrients and develop very fast. On the benefit side of the ledger is bacterial denitrification in deep waters; nitrates are transformed into nitrogen which dissipates into the atmosphere.

Severe eutrophication has been observed in intermediary zones between the North and Baltic seas, viz. the Skagerrak and Kattegat, the so-called Belt Sea and the Sund (Sound). Researchers have placed a "factor 3" on the increase in nutrients concentrations spanning the last three decades. Recently published data (Elmgren 1989) reflect an increase in primary pelagic production in the Baltic of 30 to 70%, while oxygen depletion in deeper waters resulted in biological losses over $20\,000\,km^2$ of bottom.

The development of exceptional blooms has been favoured by meteorological and hydrographic conditions. Indeed, these water zones are areas where North Sea salty water masses mix both horizontally and vertically with the brackish waters of the Baltic Sea.

During the northern hemisphere summer, winds blow with less force so that the mixing action is greatly reduced thereby allowing water stratification, and brackish water floats on top of the saltier water. This layering of water masses usually develops during May when winds are either gentle or absent. The situation is further enhanced both by solar radiation which warms up the sea surface and by the reduction of the salinity of the brackish Baltic waters due to a larger discharge of fresh water from rivers. The stratification of waters nurtures phytoplankton development as nutrients have a tendency to sink towards the pycnocline which results in greater water transparency, and thus in a better transport of solar energy to the phytoplankton (Fig. 2.5).

In these intermediary zones as well as in the Baltic Sea itself, water stratification leads to the formation of anoxic zones at greater depths beneath the pycnoline. This development resulting from exceptional algal

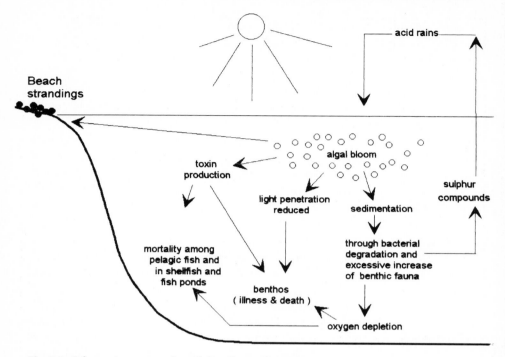

Fig. 2.5. Schematic presentation of the effects of algal blooms

blooms is particularly harmful. Oxygen concentration at the bottom of such zones has steadily decreased over the last 30 years.

Specifically in 1978, a first exceptional bloom occurred on the western coast of Sweden, in the northern sector of the Kattegat. The "population explosion" of the non-toxic alga *Ceratium* resulted in a 59% mortality among fishes being raised in Gullmar Fjord, and killed many benthic dwellers in shallow bays. In the autumn of 1980 another bloom, again with *Ceratium* as the dominant species, developed in the southern area of the Kattegat. The oxygen depletion that ensued was at the origin of heavy mortality among benthic organisms. A year later in fall 1981, at the Skagerrak level on the west coast of Sweden, an exceptional bloom of *Gyrodinium aureolum* caused numerous deaths in the fish raising and aquaculture basins, among some pelagic fish species and benthic organisms living below the 10-m isobath. Simultaneously, still on the Swedish coast but on the Kattegat, another *Ceratium*-dominated exceptional bloom killed many benthic organisms as a result of oxygen depletion. The worst ever algal bloom took place in May–June 1988: the sudden proliferation of *Chrysochromulina polylepis*, a microalga, affected extensive areas of the Belt Sea, the Sund, Kattegat and Skagerrak all the way

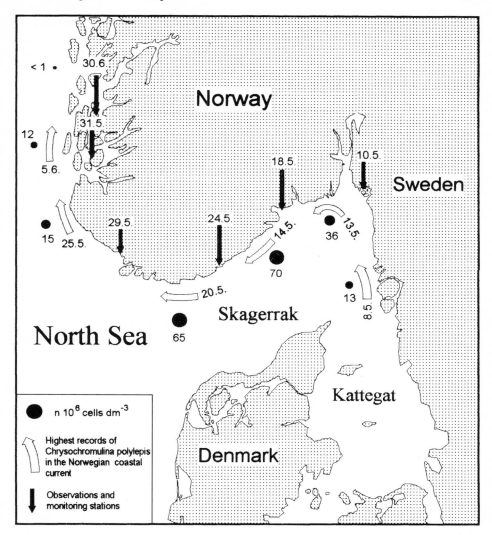

Fig. 2.6. Occurrence of the "red tide" algae *Chrysochromulina polylepis* in Kattegatt, Skagerrak, and adjoining waters in May – June 1988. (After Underdal et al. 1989)

up to the southern tip of Norway's west coast (Fig. 2.6). Ideal weather conditions fostered the bloom: extensive insolation, high temperatures (about 23°C) and gentle winds. The coastal current traveling from the southwest coast of Sweden towards the southwest coast of Norway and thus the North Sea, contributed to the spreading of the bloom. Never before had the consequences been so severe. The mortality covered a wide spectrum of marine organisms among which several species of rocky bottom macroalgae, especially red algae, numerous invertebrates, and at least ten fish species. In the fish-raising basins rainbow trout and salmon suffered huge losses, mussels became toxic, and the pelagic fish biomass dropped drastically during the periods of low oxygen concentrations.

Chronic eutrophication of several coastal areas in the North Sea led to both an increase in primary production and the doubling of the macro-benthic biomass in the Wadden Sea. The algal biomass increases and thereby increases the amount of organic matter that sinks to the bottom. The thus enhanced activity of the benthos results in an increased oxygen demand of the sediments (Lancelot et al. 1987; Smayda 1990).

Although eutrophication has occurred in the past, as shown by the historical record, its increase is significant during the last 20 years (Billen 1991a,b). It is perhaps premature to preliminarily conclude that the phenomenon is not man-induced. Based upon an idealized river-ecology model, phosphate concentrations increase as coastal areas develop, the ratio P:N climbs and nitrogen becomes the limiting factor for algal growth. Concentration increases with urbanization but may drop as a consequence of denitrification. Hence coastal anthropogenic eutrophica-tion may well develop in two stages: urbanization first, secondary waste water purification next.

During the last decade, nutrient presence has practically trebled in the North Sea. The major part came from large rivers crossing industrial regions, and is accumulated by the water masses moving from the South-ern sector of the channels to the German Bight (Fig. 2.6). Exceptional blooms have been due to *Phaeocytis*, an alga that develops each year in ever-increasing quantity, specifically along the coasts of Belgium, The Netherlands, and Germany (Fig. 2.7). *Phaeocytis pouchetti* is implicated in DMS production (Fig. 2.8). Increased eutrophication has been held responsible for repeated blooms in the Kiel Bight since May 1983. A new skeletonless form of *Dictyocha speculum*, resulting from adaptation to higher N/Si ratios, is responsible for large fish kills.

Large blooms, followed by thanatocoenosis, have been signaled in the Belt Sea and the German Bight. As remains of *Ceratium* accumulate on the sea floor anoxia results. In addition, blooms of *Ceratium tripos* have been detected in the New York area. While causes have not been eluci-dated, there is a strong suspicion that eutrophication is the culprit.

Weather has an important influence. 1988 was named "Red Seas" because blooms were exceptional, obviouly favored by a mild winter with abundant precipitation. Rain and higher temperatures favored *Phaeocytis* blooms. This alga changes from a unicell to form large flagellum-less colonies (5 to 10% of individuals) enrobed in a self-secreted polysacchar-

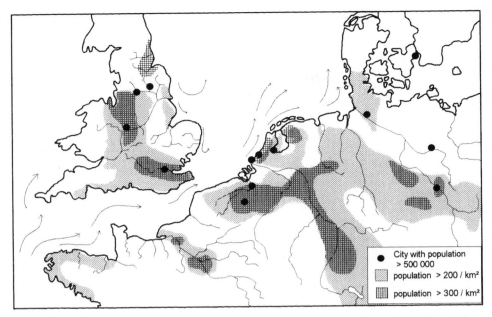

Fig. 2.7. Major rivers discharging into the Channel and the Southern Bight of the North Sea, population of their watersheds, and general residual circulation of the seawater masses. *Arrows* indicate direction of water movement. (After Lancelot et al. 1987)

ide mucus-like mass. Microzooplankton feed on individuals, and can hardly ingest the colonial form. As a result, the largest amount of organic matter produced by *Phaeocytis* is at best microbiologically decomposed or else deposited on sea bottoms or, worse, on beaches.

Phaeocytis blooms commonly spread from the Dover Straits to the German Bight. Extent and frequency have steadily increased over the last two decades. They come on the heels of diatom developments. As the algae extract silica from the waters, high nitrate concentrations become dominant, and blooms endure until the nutrient is exhausted. The process is common to coastal areas and the open ocean, however close to shore *Phaeocytis* accounts for 80% of the algal biomass due to enrichment of water by anthropogenic nitrogen (Fig. 2.8).

Eutrophication occurs mainly along the coasts between France and Denmark. The phenomenon is somewhat different in the open ocean from in waters closer to the coast. In the open ocean, measurements covering 50 years show periods of numerous blooms rather than a trend in their increase. Heavy rains and larger river discharges favor higher nutrient concentrations and the added quantity of fresh water permits water stratification which provides better light conditions for

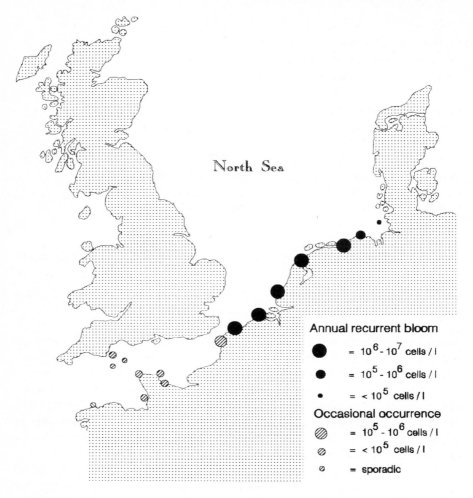

Fig. 2.8. Number of *Phaeocystis* sp. (cells per liter) recorded at a few stations in the Channel and the Southern Bight of the North Sea. (After Lancelot et al. 1987)

phytoplankton. Bloom intensity is maximal during periods of best insolation and calm winds, usually in summer.

Phaeocytis blooms were reported prior to eutrophication. It has been advanced that their biomass, not their presence is enhanced by eutrophication. Biomass has increased tenfold in The Netherlands and blooms last for longer periods (Riegman 1991). Single individuals are common on the Belgian coast, attributed to the denitrification in the Scheldt estuary; the Dutch coast is the locale of blooms of colonial forms (Billen et al. 1985).

On the Atlantic coast, serious eutrophication problems have been reported in Brittany with dire consequences for local tourism. *Ulva* proliferation is at the origin of the expression "green tides"; areas such as Lannion Bay and the bay of St. Brieuc are particularly well documented cases (Kopp 1977). The phenomenon has been recorded for over 20 years since 1968, and is expanding in Lannion Bay to such an extent that there alone, 10 years ago, some 25 000 m^3 of *Ulva lactuta* and *U. rigida* were gathered in a single season. These algae can reproduce throughout the year and the freeing of reproductive cells is apparently linked with the tidal cycle (Brault 1983). *Ulva lactuta* is, incidentally, spread over all oceans. It is thus not surprising that it has caused a series of problems in Denmark, where it proliferated in Ödense Fjord.

In France still, algal proliferation has increased along the coasts of the Languedoc-Roussillon and Corsica. The affected areas have extended to St. Malo and the southwest (La Rochelle, salt water lakes of the Landes) on the Atlantic and to Sète on the Mediterranean. In addition, the phenomenon has expanded in time as it now occurs in May through November.

No phytoplankton blooms have been reported for the Greek eastern Mediterranean, viz. the Aegean, Ionian and Levantine seas, though high nutrient content has been observed in several gulfs (Kavala, Patraikos, Pagassitikos). Algal blooms are frequent in Saronikos Gulf where eutrophication is common. Intensive agriculture contributes to the situation (Panagotidis and Pagou 1990). Heavy concentrations of nutrients, attributed to run-off and river discharge have been studied in the Gulf of Alexandroupolis (Friligos and Karydis 1988). Red tides are not infrequent in this lagoon (1951, 1989).

Urbanization, industrialization, agriculture, harbor and canal construction, tourism, drilling and marine mining have modified the Black Sea's phytoplankton over the last decade and a half (Table 2.1). A considerable biomass increase has been observed and blooms have become more frequent (Bodeanu 1984, 1992). Reduction of nutrient influx is an absolute necessity to curb the blooms, the massive algal development and the eutrophication (Bodeanu 1992).

An ecological decline is threatening the Romanian coast. During 1989, blooms attained their highest development ever (Bodeanu and Roban 1989; Skolka and Bolaga 1990). The strength of blooms and the intensity of eutrophication had already been reported in 1984 (Bologa 1985/1986). Benthic seaweeds showed a gradual, but continuous qualitative and quantitative decline since mid-century (Bologa 1989). A very high chlorophyll *a* content in the pre-Danubian Black Sea sector, observed for some 20 years, is due to eutrophication (Bologa et al. 1985). Second only

Table 2.1. Human, social and economic impact upon eutrophication

Human activity	Agents/consequences	Coastal zone degradation problems
Urbanization and transport	Land-use changes; road, rail and air congestion; dredging and disposal of harbour sediments; spills at sea; ports; water abstraction; waste water and waste disposal	Eutrophication; loss of habitats and biological diversity; visual intrusion; lowering of groundwater table; salt water intrusion; water pollution; human health risks
Agriculture	Land reclamation; fertilizer and pesticide use; livestock densities; water abstraction; river channelization	Eutrophication; loss of habitats and biological diversity; water pollution; reduction of freshwater inputs to coastal waters
Tourism, recreation and hunting	Land-use changes; road, rail and air congestion; ports and marinas; water abstraction; waste water and waste disposal	Eutrophication; visual intrusion; lowering of groundwater table; salt water intrusion in aquifers; water pollution; human health risks; loss of habitats and biological diversity
Fisheries and aquaculture	Port construction; fish processing facilities; fishing gear; fish farm effluents	Overfishing; impacts on non-target species; litter and oil on beaches; sediment disturbance; water pollution; eutrophication
Industry, including energy production	Land-use changes; power stations; extraction of natural resources; process effluents; cooling water; windmills; river impoundment; tidal barrages	Water pollution; decreased input of freshwater and sediment to coastal zone; thermal pollution; eutrophication; visual intrusion; loss of habitats and biological diversity; coastal erosion

Source: The ELOISE Programme, European Commission, EU, Brunels, 1994.

to coastal and hinterland development, eutrophication is the most important threat caused by pollution. The northern Adriatic Sea has been subjected to occurrences of mucilagenous materials and is the site of an almost permanent algal bloom, particularly along the Emilia-Romagna coast. The Sea of Marmara, the Saronikos Gulf, the bays of Izmir, Iskanderum and Alexandria (Mediterranean) suffer from chronic dystrophia. Proliferation but also degradation of algae are at the basis of the death of many animals (Morand et al. 1992). Beaches had to be closed, fishing interrupted, and governments had to intervene due to eutrophication along the coasts of France, Spain, and Italy. The situation has been particularly bad south of the Po River and in the Venice Lagoon where disappearance or decline of many algae has been reported. Ulvaceae have colonized large areas because of their adaptability to polluted conditions as reported by Sfriso and his collaborators (Sfriso 1987; Sfriso et al. 1987).

Cooperative studies into the mechanism and dynamics of the impact of marine eutrophication on benthic macrophytes in European coastal waters have been launched. EUMAC is a cooperative EU project of several universities and research centers. The effort is bringing a welcome relief to those attempting to solve the problem – and not very successfully – on a local short-term basis. As the phenomenon is on a global scale and is no longer an open-sea matter only, a multinational approach seemed required. Adequate management of the problems remains thus far impossible as, among other reasons, balanced models exist only on a local scale. Their development is hampered by a paucity of knowledge about the taxonomy and physiology of macroalgae. Local data commonly target instantaneous nutrient concentrations and details on the dynamic facets are rare, with causal relations between these and the eutrophication process rarely documented. EUMAC proposes identification of the dominant nuisance macroalgae and in vitro identification of their nutrient kinetic physiological characteristics. Not only will the green tides be fought and attempts been made at preventing further eutrophication by biomanagement, but the product, i.e., the nuisance algae, can conceivably be put to use for waste water treatment or nutrient recycling, or processed into food, feed or fertilizer, or used as a source of specific chemical compounds and for bioenergy purposes (Charlier and Justus 1993).

Already twenty years ago, Waite and Mitchell (1972) described green tides as occurring in numerous estuaries as well as along coasts and ascribed them to human causes. Discharged waters, derived from farming and industry, are responsible for eutrophication, since they are rich in nitrates, phosphates and ammonium, which can be directly assimilated by algae. The salts are concentrated in waters that are not constantly renewed, and in some locations dominant currents carry floating biomass onto the shore. To a point a "catch 22" situation may develop; unless stranded algae are removed, they will ferment releasing H_2S^- and remineralization processes may take place, providing renewed fertilization of the waters which, in turn, may accelerate eutrophication.

Coastal urbanization by engendering eutrophication has contributed to primary production and to increased chances of summer distrophies. Other causes are biological, meteorological, oceanographic and geomorphological. Though not the only culprits, sport fishing and black tides (oil spills) have brought about a reduction of the grazers' population. Weather has undergone changes reflected in modification of water temperature, insolation, amount and distribution of precipitation and direction of predominant winds. The path of marine currents and the shape of the coastline play a significant role in "green tide" occurrence.

Concentrations, in aquaculture basins, of feces and pseudofeces (undigested food) also contribute to the problems, as in the case of culture of mussels. In addition, the nitrogen-rich food provided for salmon also contributes to the eutrophication of fjords. Additional factors are intensification of fishing and the decimation of marine mammals.

Algae, foreign to European waters, have added to the nuisance caused by native species. The Japanese *Sargassum muticum* invaded Atlantic and Mediterranean littorals, apparently when Japanese oysters were imported (*Crassostrea gigas*) because the "Portuguese" oyster could no longer cope with rising pollution. In some instances small pleasure craft are hampered by the sometimes massive *Sargassum* accumulations.

The danger of an accidental introduction of a foreign species is not illusory. In the case of *Caulerpa taxifolia*, not less than two ministries (Ministries of the Sea and of the Environment), one governmental agency (Maritime Affairs), four scientific and research organizations (IFREMER, SRETIE, DRAE PACA and CEVA) and the University of Nice-Sophia Antipolis are involved in the search for a solution to the blooms of this alga. In 1984, the tropical *Caulerpa taxifolia* (Vahl) *C. Agardh*, normally displayed in the aquarium of the Musée Océanographique de Monaco, was accidentally introduced into the sea where it soon thrived in the waters at the foot of the museum. Unexpectedly it resisted the low temperatures prevailing there and colonized a wide variety of substrates (rock, sand, mud). By 1990, it had spread not only to Cap Martin which is only 3 km from Monaco, but also to Toulon at some 150 km. It is found as deep as 30 m. Sexual and efficient vegetative reproduction permit us to expect further spreading. Its toxin (caulerpenyn) is dangerous for grazers, epiphytes and competitors. Furthermore it is eaten only by *Sarpa salpa* (Mediterranean bream); consumption of the fish is then poisonous to humans (Cigautera poisoning symptoms). Aquaria, but also mariculture installations thus pose a threat of pollution, under specific circumstances, of the native waters (Meinesz and Hesse 1991). Because of oyster culture in some areas of the Mediterranean, other species have spread, particularly in the Languedoc ponds, and then drifted towards Spain on the open sea, e.g. *Sargassum muticum, Undaria japonica* and *Laminaria japonica*. Meinesz (pers. comm.) estimates that the speed of spreading reaches 5 cm/year. This exponential proliferation threatens to cover the entire Mediterranean and to modify all littoral ecosystems of the region. There seem to be no scientific grounds to expect an end to the spreading.

Diatom blooms occurred in 1989 in the Adriatic Sea, with eutrophication suspected as the cause, even though similar blooms were already observed in 1911. Another "slime" – producing diatom *Ascino duscus walesii* moved into the Channel and the North Sea areas. It originates from the Pacific Ocean (Smayda 1989).

Gyrodinium auredum, a newcomer to the North Sea and the Norwegian coasts, is harmful for agriculture. Extensive blooms have been reported on the French and British coasts and, it should be noted, at sites far away from possible human influence.

2.4   Relationship Eutrophication – Environment

Eutrophication is thus enhanced by man-made discharges as well as by natural input of nutrients, organic materials, and pollutants. It affects particularly estuaries and coastal gulfs and bays that are rather shallow, in narrow communication with main water bodies, and which are polluted. Water, sediment, flora, and fauna undergo its effects. Sediments are modified and toxicity is increased. The fauna is subjected to the presence of pollutants, additional organic matter and changes in oxygen supply because of sediment modifications, which in turn influence toxicity (toxins). The sediments are also affected by nutrients and carbon disulfide (CS_2). An increase in herbivorous dwellers is recorded, while algae including opportunistic species compete for substrate and consume oxygen. The atmosphere is fouled by released hydrogen sulfide (H_2S).

The ensuing situation may be worsened by both a trend to overfish herbivorous species, and by the introduction in the area of non-indigenous opportunistic algae. A process of successive steps sets in because of easy nutrient assimilation, successful adaptation to existing pollution and vegetative multiplication, proliferation occurs, leading to strandings with consequent decomposition, or harvesting followed by

Table 2.2. Strandings of macroalgae in Europe–geographical distribution

Country	Site	Species	Density g dry weightm^2	Annual production	Bloom period
Northern littoral					
France	Brittany	*Ulva* sp.	0.2–400 (2–100 approx)	35 000	June–September(1968)
	Bay St–Brieuc.	*Porphyra* sp.	(3–5 approx)	450	June
England	Langstone Harbour	*Enteromorpha* sp.	>2 (10–12 approx) 100 ha	6000	May–December
Scotland	Ythan estuary	*Enteronorpha* sp.	0.05–1.16		April–August
Ireland	Belfast Dublin	*Ulva lactuta* var. *latissina* *Enteromorpha compressa* *E. intestinalis*	45 (maxi) (20–90 approx)		Summer–fall
Germany	Island of Sylt	*Enteromorpha* sp. *Ulva* sp. *Cladophora* sp. *Chaetomorpha* sp. *Porphyra* sp.	(10–15 approx)		July–August
Low countries	Wadden Sea	Green algae			
Sweden		*Ulva* sp.			
Norway	Oslo–Fjord	*Enteromorpha* sp. *Ulva* sp.			July–August

Southern littoral

Italy	Venice lagoon	Ulva lactuta	15	1000 000
		Gracilaria verrucosa		
		Valonia agropila		
Spain		Ulva sp.		
		Enteromorpha sp.		
Portugal		Ulva sp.		
		Enteromorpha sp.		
France	Ponds and	Ulva lactuta		
	lagoons of	Ulva rigida	5–10	April–June
	Languedoc	Enteromorpha flexuosa		
	Roussilon	Enteromorpha prolifera	5	March–April

stockpiling which still leads to decomposition. Collection and dumping are a financial burden, strandings constitute a physical hindrance, and air fouling both cuts down tourism and is a nuisance for the local population. Median effects are that year after year algal blooms occur, relatively independently of the weather.

Amounts of stranded material increase yearly. In 1987 million metric tonnes were reported for Italy, 35 000 metric tonnes for France, and 6000 metric tonnes in England (Table 2.2). Outside the European Union examples abound: Senegal 150 000 metric tonnes, Tunisia 5000 metric tonnes, Australia 35 000 metric tonnes .

The coasts of California and Argentina are equally affected, as are those of India, New Zealand, South Africa, etc. Most blooms are from green (*Ulva*, *Enteromorpha*, *Cladophora*) and red algae (*Gracilaria*, *Porphyra*). The phenomenon is ascribed to environmental perturbation.

Phaeocytis not significantly entering the zooplankton food chain affects fish productivity and modifies the benthos as organic deposits on the sea floor cause oxygen depletion. Sulfurous volatile matter is generated and released into the atmosphere enhancing acid rain. In addition, sea foam is accumulated on beaches, occasionally reaching heights of 2 m.

Many bloom-forming flagellates are important sources of dimethylsulfides (DMS). When degradation of large quantities of algal matter leads to oxygen depletion, enhanced production of nitrous oxide (N_2O) is likely. Both *Prymnesiophite* and *Gerodinium* are major producers of DMS. Blooms caused by eutrophication play an important role in generation of these gases.

The geographical distribution of affected areas shows a parallelism with human occupation (viz. settling), as they are usually close to tilled land, with restricted communication between marine and lagoonal milieus.

Brown tides have been mentioned in northeastern United States. *Phaeocytis* engenders *Prymnesiophite* blooms resembling a foamy gelatinous mass (Cosper 1989). The press has had some news flashes of "killer algae": referred to kills of fish, even macroalgae by the toxic flagellate *Chrysochromulina polylepis*.

Red tides due to dinoflagellates have been spreading geographically. Their potent toxins have caused fish kills and human poisonings by shellfish consumption. Their occurrence is more frequent in eutrophicated bays.

2.5   Impact of Algal Pollution

There is a direct correlation between the intensity of eutrophication and algal proliferation. Purification capacity is credited to the algae which colonize the affected areas according to their own resistance to pollution: they use up and accumulate the nutrient surplus together with other pollutants. However this "beneficial role" ends when the algae are stranded, or are unable to cope with the degree of pollution, so that they die and degradation sets in. As the extent of the anaerobic surface grows, both water and air pollution ensue. The recycling of these nutrients favors and keeps up entrophication, and can maintain a high biomass productivity level in an ecosystem on its own.

Houvenaghel and Mathot (1982) in their study of a correlation between decaying biomass, general pollution and algal growth, showed that algae have economic value as a substrate or additive in methanization, as fertilizer, as antibiotics, and in food and feed preparation. They pointed to the anti-eutrophication and concentrator role of *Ulva*. In the Venice lagoon, collected algae release biogas by mesophilic fermentation, leading to methane production. Fuel production aids in pollution control (Croatto 1982).

The impact of algal "pollution" is economic as the very pulse of numerous affected regions is closely controlled by tourism. Long "strings" of stranded algae are not only an aesthetic deterrent to use of a beach, they make access difficult; in addition, hydrogen sulfide produces a foul odour, and battling algae in the water is not appreciated by bathers. To provide an idea of the cost of cleaning in- and onshore, the bill ran to US$ 200 000 for 1986 in the Armor region of France. Using the algal material as fertilizer may clear the beach, but results in stream and aquifer pollution, as liquefaction juices are loaded with volatile fatty acids.

More appropriate approaches are biogas extraction, mixing with domestic wastes to produce composts, manufacture of a pre-mix for aviculture, and making an organic-rich soil improver. Nitrogen, phosphorus and metals can be recycled.

Algal decomposition results in an inversion of the gradient concentration of nutrients in the sediment. Thus man-generated nitrogen and phosphorus add to the quantities of nutrients resulting from decomposition of algal deposits which keep up the eutrophication condition.

In the Bay of Lannion alone, an additional loss of 500 metric t/year of sand results as 2000 truck round-trips are necessary to remove the algae.

Although not well studied with regard to their precise effects, these activities contribute to a further disturbance of the ecological balance.

2.6 Managing the Algal Problem

Globally considered, algae represent a source of problems, a target of several nuisances, a decision and management tool, and a remedy for some recent developments (Charlier 1991).

In 1983, the world production of algae reached 4 833 046 metric tonnes of which 124 012 came from the EU, 7673 metric tonnes from the former USSR and 1 575 650 from the Chinese People's Republic. Close to 4 million metric tonnes were composed of brown algae. Alginates are the most common extracted product, followed by carageens, agar-agar and far behind, furcelareens (from FAO data). Some other products used in technology, foods, surgery and pharmacy are also extracted. Iodine extraction was halted roughly 40 years ago, while that of phycocolloids has been on the increase for some decades.

Algae can be used in foods and feeds, in industry, agriculture, pharmacy, medicine, thalassotherapy, cosmetology, energy generation, water purification and pollution abatement (Beavis and Charlier 1986; Charlier et al. 1986). The last use has been attempted for example near Kiel, Thessaloniki and Venice, and proposed for Brittany (Schramm 1991).

Seaweed biomass can also be used as a source of fuel or bioconversion products, coastal protection and fertilization or composts; hence such biomass can be used as an integrated system of production of food, feed, fish, fuel and raw material. Bioconversion can be achieved by thermal treatment, composting or methanization (Mancliére 1985; Morand et al. 1991). Methane production in a pollution abatement plant would be possible (Schramm and Lehnberg 1984; Morand et al. 1991).

The experiment in the Danish Ödense Fjord, where algae were harvested to remove nutrients from the water, and were then used for biogas production, proved feasible and efficient, but showed that at current energy prices, the undertaking was not economically profitable.

In Italy and France, researchers chose to methanize an *Ulva*-derived juice rather than use the entire plant, a tenfold production was obtained for an equal volume. The French opted for hydrolysis and pressing, the Italians for centrifugation.

The French undertaking, financed by public and private funds, would allow a 7-year investment return, for methanization, but pollution abatement would require 20 years. *Ulva* methanization has been carried out at the Centre d'Etude et de Valorisation des Algues at Pleubian, Brittany. Pollution abatement is, however, the primary aim. *Laminaria* has been methanized as part of an integrated economy concept on Sein Island (Brittanyi; Morand et al. 1990).

The higher productivity potential of seaweeds over land plants may encourage their use for, e.g., pollution abatement and energy generation. However, this, by itself, will not "solve" eutrophication problems as different kinds of seaweed have different economic advantages (caloric content, conversion efficiency, maximum yield). Undoubtedly, methanization is a reasonable choice under certain conditions, such as the right location, scale effect, and existence of an environmental problem. Any project should however, first define optimum choice for algae utilization, and reduce possibilities of setbacks.

The approach to gas production differs on each side of the Atlantic. Seaweed proliferation in Europe and pollution abatement, have prevailed over concerns with productivity and nutrient provision. Ecological and economic climates will determine whether research, with adequate financial backing, will be carried out or not.

Utilization of algal material in human food, though valued at more than % 700 million per year in Asia, is hampered by legal problems in Europe. In France, only *Laminaria digitata* is being utilized for a trial period. In Belgium, though somewhat less stringent, legislation is very specific (Charlier 1992).

In most countries throughout the world, the market for algae runs into technological, financial, commercial and even ecological constraints, though none of them are impossible to overcome.

2.7 Comprehensive Management

Solving the eutrophication problem is an undertaking currently handled on a local, or at best regional level. Now that eutrophication problems have been recurrent for several decades and are spreading steadily to more geographical areas, action on a larger scale appears indicated. It can hardly be delayed. In fact, an approach similar to that recommended for coastal zone management seems indicated; perhaps there is more than a parallelism, and eutrophication has become as much part of coastal zone management as coping with the natural hazards common to sea and shore. Furthermore, it is desirable to reject the mere observation of the problem, as is still occasionally expounded, and the time has probably come for a coordinated European policy (Ministère de l'Environnement France 1987).

It has been established that human activities and climatic modifications are leading causes of eutrophication and of the ensuing "green tides". Little, if anything, can be done about weather and climate. Notwithstanding sustained campaigns in favor of and pious promises to implement measures for environmental clean up and pollution abatement, it is highly improbable that discharges of domestic and industrial wastes will be so reduced that enrichment of the coastal waters in phosphates and nitrates will be eradicated. And yet it has been shown that river sand streams are the major nutrient contributors, particularly in the Baltic area where used water treatment drastically reduced phosphate discharge into the sea. Consequently, steps must be taken to approach the problem differently.

Toxic algal blooms have been reported in water masses with high nutrient and high irradiance levels, but the role of eutrophication is questioned (Smayda 1989). A reduction in the quantity of nutrients will not affect fish productivity, but rather that of filter feeders. Quantity and quality of nutrients delivered to the coastal zone depend strongly on the retention capacity of the river system, which is itself largely affected by land use and management of the watershed (Billen 1991a,b).

Bioconversion has been frequently suggested as a remedy for the accumulation of algae, as mentioned above. Indeed, it is important for the development of seaweed cultivation on an industrial scale, it can play a far from negligible role in the production of needed organic fertilizers, and it is a weapon against eutrophication. Yet, commercialization entails may aleas: if one were to consider economic exploitation of the strandings on the coasts of Brittany, one would be faced with an investment with a return time of 10 years. Leaving natural production, for cultivation it was once estimated that kelp farms along the Californian coast covering 400 000 ha (10^6 acres) could gross 0.2 quad of energy. However, the degree of uncertainty ranges from 0.1 to 10 times a given estimate (Jackson and North 1973). The net energy recovered is, according to some researchers, only a small fraction of the crop energy content (Charlier 1991).

Bioconversion appears well adapted for treating the proliferating algal biomass, for seaweeds (cultivated or harvested), for purifying the environment or for eliminating nutrients from waste waters.

No matter what the primary utilization of a coastal zone, attempts must be made to reduce and control the agents bringing about eutrophication. This entails a coastal management plan on a national scale, which does not overlook the international aspects, and which establishes a balance among uses of a given littoral fringe (Charlier 1992; Cendrero and Charlier 1989). The approach must be both multidisciplinary and interdisciplinary and economic use must not be the only determining factor. The causes of pollution must be identified, their geographical sources recognized, and efficacious measures must be taken to suppress them, or at least substantially reduce them. The problem will not be fully solved with these measures first because not all pollution can be stopped, any more than can human expansion and concentration on or near coasts (Charlier 1989), and second other factors, whether short or long term, cannot be modified.

An attempt can be made to turn a nuisance into a source of useful raw material. A current approach to algal proliferation and stranding has been to transform shore pollution into land pollution (Orbetello, St. Brieuc), or to make a timid attempt to use the algae as raw fertilizer. The latter approach, however, has resulted in contamination of aquifers and renewed transfer of nutrients to coastal waters, particularly in bays, coastal lagoons, lakes and the like.

The purification ability of algae has been established, but attempts at placing them into service are not numerous and results have been mostly due to natural processes. The EU international cooperation program COST-48, later absorbed in a new program (BRIDGE), provided the opportunity for a joint German – Greek project in the Bay of Thessaloniki.

The primary benefit from a lagooning plant, i.e. a water body acting in the same manner as a natural lagoon, installed alongside a biological purification plant, would be the return to the natural environment of water virtually free of any excess nutrients and, eliminating one of the causes of eutrophication.

In addition to controlled utilization of waste waters, another project targets the uncontrolled growth and proliferation of seaweed biomass in eutrophicated areas. The natural coastal ecosystems study (France and Italy) covers the ecophysiological aspects of macroalgal blooms, their causes, and the impact of harvesting, together with processing for biotechnical purposes.

In France, the Centre d'Entr'Aide par le Travail de Quatre-Vaulx (St. Cast-le-Guildo) has mixed collected algae with animal droppings and forest residuals and markets quite successfully a compost of high quality at competitive prices. However, this ensures the disposal of only quite limited quantities of algae. The Venice algae are used for agar-agar extraction, originally done in Padua, but now carried out nearer to Venice, with a greater economic return. The Quatre-Vaulx production includes a seaweed–wood waste compost, one for hobby use (gardening) and one for professional use. In addition, a Belgian company (SOPEX) has tested a digester in Morocco which extracts methane from processed algal residues. Other experiments were conducted on Sein Island.

Methanization remains one of the best modes of utilization of algae. To assign a commercial value to green tide algae has been discussed for the specific case of Brittany in the frame of a biotechnical seminar in Morocco (Brault and Briand 1984).

2.8   Conclusion

Obviously, processing plants for algae vie for space along coasts with other industries whose economic return is far better. However, because they act as a water cleanser, and because their removal and subsequent use enhances the touristic and recreational worth of a given coastline, a sound management plan should include such facilities.

Research is needed to assess what degree of benefit is gained from biomass harvesting and what is due to the whole ecosystem in nutrient removal. No treatment plant will see the day before it is established clearly whether seaweeds are an efficient tool for removal of nutrients.

Stabilization of *Ulva* green tides has been proposed and tested for Hillion Bay (France). Composting with a minimum of ligno–cellulosic substrates was successful and a pilot plant for the treatment of algal biomass is being considered for the area. (Morand et al. 1992).

As indirect effects of blooms on the ecosystem are far from well understood (e.g. energy transfer and production of calcifying organisms), elucidation of factors that determine the occurrence and species composition of phytoplankton blooms (flagellates proliferation) and their impact on the microbial food web should be pursued. Therefore we should search for the causes of the geographical spread of bloom-forming species. Systems ought to be developed to detect triggering conditions and to warn of their impact.

Removal of "nutrients production" at the source will reduce the load carried fluvially to the sea and thus treatment of used industrial waters, and action on domestic wastes should be an integral part of a management plan for "eutrophication control". In the Inland Sea of Japan, a reduction of organic input drastically reduced red tides. Obviously such prophylaxis is indicated.

The Commission of the European Communities (now European Union) sponsored under its BRIDGE (COST 48) program a Symposium-

Workshop on Production and Exploitation of Entire Seaweeds (St. Malo, France 1991). One third of its activities centered on eutrophication. Besides some concrete recommendations for research on the effects of uncontrolled eutrophication [···] on benthic primary biomass and the possibility of eutrophication [···] abatement by controlled harvesting or selective advancement (Schramm 1994a) six papers discussed the phenomenon. A brief overview (Fletcher 1994a) preceded an assessment of the situation on the Britanny coasts viewing nitrogen as an inhibitor (Dion 1994), an element also discussed for the Gulf of Thessaloniki (Kotropoulos et al. 1994), in the Western Baltic where since 1970 it has entrained a change in macrophytic vegetation (Schramm 1994b), and in the [Dutch] Veerse Meer (Nienhuis 1994). A practical suggestion for relief is presented by Merrill by mariculture which could possibly reduce eutrophication effects (1994).

This author once wrote, simultaneously cynical and discouraged, that coastal erosion is here to stay for any foreseeable future. The same can be said for eutrophication and, *par conséquence*, algal proliferation. Hence, there is a need for a management approach that covers extensive geographical surfaces and is integrated in the broader overall coastal zone management plan.

References

Aubert M (1990) Mediators of microbiological origin and eutrophication phenomena. Mar Poll Bull 21 (1): 24–29

Babenrd B (1990) Eutrophication-induced oxygen deficiency in the Belt sea area. In: Lancelot C, Billeng, Barth H (ed) Eutrophication and algal blooms in North Sea coastal zones, the Baltic and adjacent areas: prediction and assessment of preventive actions. Proc Worksh, EC, Brussels, 26–28 Oct 1989. Water pollution research report 12. Luxembourg, European Commission pp 215–224

Baron P (1988) La mer du Nord à bout de souffle. Sci Avenir 497 (19 July): 18–23

Batje M, Michealis H (1986) *Phaeocytis pouchetto* bloom in East Frisian coast with German Bight, North Sea. Mar Biol 93: 21–27

Beavis A, Charlier RH (1986) An economic appraisal for the onshore cultivation of *Laminaria* spp. Hydrobiologia: 151 – 152: 387 – 398

Belkhir M, Madj Ali Salem M (1981) Contribution à l'étude des mécanismes d' eutrophisation dans le lac de Tunis. Bull Inst Nat Sci Techn Oceanogr Pêche Salammbo 8: 81–98

Billen J (1991a) Historical retrospective of eutrophication. Abstract of communications at NATO-ARW: Global change and the coastal zone. Château de Bonas, France (handout)

Billen J (1991b) Retrospective analysis of coastal eutrophication: NATO ARW: Impact of global changes of coastal oceans, Abstr Bonas, France (handout)

Billen J et al. (1985) A nitrogen budget of the Scheldt hydrological basin Neth J of Sea Res 19, 314: 223–230

Bodeanu N (1984) Modifications sous l' influence anthropique dans le dévelopement quantitatif et dans la structure du phytoplankton du secteur roumain de la mer Noire. Trav Mus Hist Nat Grigore Antipa XXVI: pp 69–83

Bodeanu N (1992) Algal blooms and development of the main phytoplanktonic species at the Romanian Black Sea littoral in conditions of intensification of the eutrophication process. Science of the total environment. Supple. Elsevier, Amsterdam, pp 891–906

Bodeanu N, Roban A (1989) Les développéments massifs du phytoplancton des eaux du littoral roumain de la mer Noire au cours de l'année 1989: Cercetări marine (IRCM) 22: 127–146

Bokn T, Berge JA, Green N, Rygg B (1990) Invasion of the planktonic algae *Chrysomulina polylepis* along south Norway in June 1988, acute effects on biological communities along the coast. In: Lancelot C, Billen G, Barth H (eds) Eutrophication and algal blooms in North Sea coastal stones, the Baltic and adjacent areas: prediction and assessment of preventive actions. Proc Worksh, EC, Brussels, 26–28 Oct 1989. Water pollution research report 12, European Commission, Luxembourg, pp 183–194

Bologa AS (1985/1986) Planktonic primary productivity of the Black Sea: a review. Thalassia Jugosl 1/2: 1–22

Bologa AS (1989) Present state of seaweed production along the Romanian Black Sea shore. Vie Milieu 39, 2: 105–109

Bologa AS, Burlakova ZP, Tchmyr VD, Kholodov VI (1985) Distribution of chlorophyll *a*, phaeophytin *a* and primary production in the western Black Sea (May 1982). Cercetări Marine (IRCM) 18: 97–115

Brault M (1988) Epuration de milieu: exemple des Ulves: Biomasse Actualités no spéc 3(Suppl 12): 48–49

Brault D (1983) Les ulves. Epuration du milieu. Biomasse Actualités 3: 14–16

Brault D (1988) Epuration du milieu: exemple des ulves. Biomasse Actualités spéc no 3 (Suppl 12): 48–49

Brault D, Briand X (1984) Valorisation des algues marines, cas particulier lié avec marées vertes. Sém Biotechn, 28–29 Mai, Meknes, pp 15–18

Brault D, Briand X, Golven P (1985) Les marées vertes. Premier bilan concernant les essais de valorisation. In: Bases biologiques de l'aquaculture. Montpélier, 16–17 Dec 1983, Inst Franç Rech Exploit de la Mer. Act Coll I: 33–42

Briand X (1987) Les marées vertes. 4th Eur Conf. Biomass for energy and industry, Orleans, pp 98–95

Briand X (1988) L'exploitation des algues en Europe. EC Conf, Marine primary biomass, L' Houmeau Feb 1987; pp 57–71

Briand X, Morand p (1987) *Ulva* stranded algae; a way of depollution through methanization, In: Grazi G, Delmon B, Molle JF, Zibetta H (eds) 4th Eur Conf on Biomass for energy and industry, Orleans

Cendrero A, Charlier RH (1989) Resources, land-use and management of the coastal zone. Geolis III: 2, 40–60

Charlier RH, Beavis A, De Meyer CP (1986) *Laminaria* sp as energy source. Occans'86 Proc pp 621–626

Charlier RH (1989) Coastal zone occupance, management, and economic competitiveness. Ocean Shoreline Management 12: 383–402

Charlier RH (1991) Algae-resource of scourge? Part II: economics and environment. Int J Environ Stud 8: 237–250

Charlier RH (1992) Multiple use conflicts and management in the coastal zone. Pap Adv Res Works NATO Abst Auch, Château de Bordas

Charlier RH (1992) Use and harvesting of, and legislation pertaining to algae in food and feed in Belgium. In: Maleau S and Cavaloc E (eds) COST 48 Aquatic Primary Biomass. Marine Macroalgae. Proc 1st Workshop on food and feed from seaweeds, March 1992, Sorrento, pp 37–39

Charlier RH, Justus JR (1993) Ocean energies-environmental, economic and technological aspects of alternative power sources. Elsevier Oceanographic Series 56. Elsevier, Amsterdam, pp 16–17, 407–432

Chassany de Casabianca ML (1984) Analyse de problémes écologiques liés à la récolte de biomasse algale en milieu lagunaire. Missions et Recherches Technologiques Paris, pp 80

Chynoweth DP et al. (1987) Biological gasification of marine algae. In: Bird KT, Benson PH (eds) Seaweed cultivation for renewable resources. Elsevier, Amsterdam, pp 285–303

Clark RB (1989) Marine pollution. Oxford Scientific publications, Oxford

Cosper EM (1989) Primary production and growth dynamics of the brown tide in Long Islands embayments. In: Cosper EM, Bricelj VM, Charpenter EI (eds) Novel phytoplankton blooms. Coastal Estuarine studies, 35. Springer, Berlin Heidelberg New York, pp 139–158

Croatto U (1982) Energy from macroalgae of the Venice lagoon. In: Strub A, Chartier P, Schlesser G (eds) Proc 2nd Int conf, Energy from Biomass, Berlin, 20–23 sept 1982. Elsevier, London, pp 329–333

Dion P (1994) Present status of the Ulva mass bloom on the Britanny coast. In: Delépine R, Morand P and coll (eds) Production and exploitation of entire seaweeds. Commission of the European Communities, Brussels, pp 34–36

Elmgren R (1989) Man's impact on the ecosystems in the Baltic Sea: energy flows today and at the turn of the century. Ambio 18: 326–332

Fletcher RL (1994) The "green tide" problem. In: Delépine R, Morand P and coll. (eds) Production and exploitation of entire seaweeds. Commission of the European Communities, Brussels, pp 29–31

Friligos N (1989) Nutrients status in a eutrophic Mediterranean lagoon. Vie Milieu 39: 63–69

Friligos N, Karydis M (1988) Nutrients and phytoplankton distributions during spring in the Aegean Sea. Vie Milieu 38, 2: 133–143

Gray JS (1991) Marine eutrophication. 25 Eur Mar Biol Symp, Ferrara, Italy, 1991

Hauman L (1989) Algal blooms. In: Barth H, Nielsen A (eds) Water pollution research. Rep 10, Comm of the Europ Commun, Luxembourg, pp 9–19

Houvenaghel JT, Mathot JF (1982) The production of marine green algae in coastal waters and their culture in ponds enriched with waste waters. In: Strub A, Chartier P, Schlesser G (eds) Proc Int 2nd Conf, Energy from biomass, Berlin, 20–23 sept 1982. Elsevier, London, pp 308–312

Jackson GA, North WJ (1973) Concerning the selection of seaweeds suitable for mass cultivation in a number of large open ocean solar-energy facilities ("Marine Farms") in order to provide a source of organic matter for conversion of food, synthetic fuels and electric energy. US Naval Weapons Center, China Lake CA, USA

Kopp J (1977) Etude du phénoméne de marée verte affectant les baies de St Brieuc et de Lannion: étude de fécondité des ulves. ISTPM, Nantes, 26 pp

Kotropoulos D, Nikolaidis G, Haritonidis S (1994) Biomass response to the macrophyte Ulva species to the nitrogen enriched seawater. In: Delépine R, Morand P and coll. (eds) Production and exploitation of entire seaweeds. Commission of the European Communities, Brussels, pp 47–51

Lancelot C (1990) *Phaeocystis* blooms in the continental coastal areas of the channel and the North Sea. In: Lancelot C, Billen G, Barth H (eds) Eutrophication and algal blooms in the North Sea coastal zones, the Baltic, and adjacent areas: prediction and assessment of preventive actions. Water pollution research report no 12. EC, Luxembourg, pp 27–56

Lancelot C et al. (1987) *Phaeocystis* blooms and enrichment in the continental coastal zones of the North sea. Ambio 16: 37–46

Lancelot C, Billen G, Barth H (1990) Eutrophication and algal blooms in the North Sea coastal zones, the Baltic, and adjacent areas; prediction and assessment of preventive actions. Water Pollution Research Project no 12. Luxembourg, EC, 291 pp

Larsson U (1986) The Baltic Sea. In: Rosenberg R (ed) Eutrophication of marine waters surrounding Sweden. Natl Swed Env Prot BD Rep 1054: 16–70

Larsson U et al. (1985) Eutrophication and the Baltic Sea: causes and consequences. Ambio 14: 9–14

Levine HG (1984) The use of seaweeds for monitoring coastal water. In: Shubert LE (ed) Algae as ecological indicators. Academic Press, London, pp 183–212

Lowthion D et al. (1985) Investigation of a eutrophic tidal basin. Mar Environ Res 15: 263–284

Lundälv I (1990) Effects of eutrophication and plankton blooms in waters bordering the Swedish west coast: An overview. In: Lancelot C, Billen G, Barth H (eds) Eutrophication and algal blooms in the North Sea coastal zones, the Baltic, and adjacent areas: prediction and assessment of preventive actions. Water pollution research report 12. EC, Luxembourg, pp 195–214

Manclière P (1985) Méthanisation des algues: de l' énergie à revendre? Equinoxe 3: 7–12

Mazé J, Morand P, Potoky (1993) Stabilization of "green tides" *Ulva* by a method of composting with a view to pollution limitation. J Appl Phycol 5: 183–190

Meinesz A, Hesse B (1991) Introduction et invasion de l'algue tropicale *Caulerpa taxifolia* en Méditerranée nord-occidentale. Oceanol Acta 14, 4: 415–426

Merrill JE (1994) Mariculture techniques for use in reducing the effects of eutrophication. In: Delépine R, Morand P and coll. (eds) Production and exploitation of entire seaweeds. Commission of the European Communities, Brussels, pp 52–56

Ministére de l'Environnement, France (1987) Colloque International "Mer et Littoral: couple à rissque". La Documentation Française, Biarritz, pp 508

Montgomery HAC et al. (1985) Investigation of a eutrophic tidal basin; Part 2: nutrients and environmental aspects. Mar Environ Res 15: 285–302

Morand P (1985) Biomasse produite grâce au recylage d'éléments polluants. Biomasse Actualités 8 Nov: 7–10

Morand P, Charlier RH, Mazé J (1990) European bioconversion and realisations for microalgal biomass: Saint-Cast-le-Guildo, France. experiment Hydrobiologia 204/205, 301–308

Morand P, Carpentie B, Charlier RH et al. (1991) Bioconversion In: Guiry MD, Blunden J (eds) Seaweed resources in Europe: Uses and potential. John Wiley, Chichester, New York, pp 95–148

Morand P, Briand X, Orlandini M (1992) Macrolgae: pollution or help to the pollution abatement? Depollution planning of the Mediterranean Sea. Società Chimica Italiana, Roma, pp 57–72

Nielsen A, Richardson K (1990) *Chrysochromulina polylepsis* bloom in Danish, Swedish and Norwegian waters, May–June 1988; an analysis of extent, effects and causes. In: Lancelot C et al. (eds) Water pollution report 12 EC Luxembourg, pp 57–74

Nienhuis PH (1994) Production and biomass of *Ulva* species in an eutrophicated lagoon in The Netherlands. In: Delépine R, Morand P and coll. (eds) Production and exploitation of entire seaweeds. Commission of the European Communities, Brussels, pp 45–46

Panagotidés P, Pagou K (1990) Algal blooms and nutrient conditions in the Greek seas. In: Lancelot C et al. (eds) Water pollution report 12. EC, Luxembourg, pp 225–234

Reid PC (1990) The dynamics of algal blooms in the North Sea. In: Lancelot C et al. (eds) Water pollution report 12. EC, Luxembourg, pp 57–74

Reise K (1983) Sewage green mats anchored by lugworms and the effects on *Turbellaria* and small *Polychaeta*. Helgol Meeresunters 36: 151–162

Richardson K (1989) Algal blooms in the North Sea: The good, the bad and the ugly. Dana 8: 33–43

Riegman R (1991) Mechanisms behind eutrophication-induced novel algal blooms. Den Burg, Texel, Nederlands onderzoek (N10Z Rep 9)

Rönner R (1985) Nitrogen transformations in the Baltic proper: denitrification counteracts eutrophication. Ambio 14, 3: 134–138

Rosenberg R (1985) Eutrophication: The future marine coastal nuisance. Mar Poll Bull 6: 227–231

Rosenberg R, Elmgren E, Fleisher S, Jonsson P, Persson G, Dahlin H (1990) Marine eutrophication case studies in Sweden. Ambio 19, 3: 102–108

Rydberg L, Edler L, Floderus S, Granéli W (1990) Interaction between supply and nutrient, primary production, sedimentation and oxygen consumption in SE Kattegat. Ambio 19(3): 134–141

Schaumann K, Hesse KJ (1990) Frontal accumulation, and autochtonous eutrophication effect of a red tide in the German Bight/North Sea. In: Lancelot C, Billen G, Barth H (eds) Water pollution research report 12. EC, Luxembourg, pp 235–244

Schraeder H (1990) Eutrophication symptoms of northern European seas and actions to alleviate them. In: Lancelot C et al. (eds) Water pollution research report 12. EC, Luxembourg, pp 235–244

Schramm W (1991) Seaweeds for waste water treatment and recycling of nutrients. In: Guiry D, Blunden G (eds) Seaweed resources in Europe: uses and potential. John Wiley, Chichester, pp 149–168

Schramm W (1994a) Removal and use of nutrients, 1986–1989. Activities of Working Group [COST 48, Subgroup 3]. In: Delépine R, Morand P and coll. (eds) Production and exploitation of entire seaweeds. Commission of the European Communities, Brussels, pp 15–19

Schramm W (1994b) Eutrophication and recent changes in macrophytic vegetation in the Western Baltic (Kiel Bay). In: Delépine R, Morand P and coll. (eds) Production and exploitation of entire seaweeds. Commission of the European Communities, Brussels, pp 34–36

Schramm W, Lehnberg W (1984) Mass-cultivation of brackish water adapted seaweeds in sewage-enriched seawater. II. Fermentation for biogas production. In: Bird CJ, Ragan MA (eds) Proc Int Seaweed Sym Junk, Dordrecht, pp 282–287

Secrétariat d'Etat à la Mer (Fr) (1985) Une stratégie pour l'environnement. Groupe de Travail no 2: les végétaux marins et l'environnement. Les algues, source de nuisances; Les végétaux marins, outil de la restauration du milieu. Biomasse Actualités-Pyc, Paris pp. 13–23, 30–31

Sfriso A (1987) Ulvaceae colonization in Venice lagoon. G Bot Ital 121: 1–2, 69–85

Sfrizo A, Marcomini A, Pavoni B (1987) Relationship between macroalgal biomass and nutrient concentration in a hypertrophic area of the Venice lagoon. Mar Environ Res 3: 297–312

Skolka HM, Bologa AS (1990) Phytoplankton production and eutrophication off the Romanian Black Sea shore in 1989. Abstr Mar Coastal Eutrophication Symp, Bologna 1990

Smayda TS (1989) Primary production and the global epidemic of phytoplankton blooms in the sea: a linkage. In: Cosper EM, Bricelj VM, Charpenter EI(eds) Novel Phyoplankton blooms. Coastal marine studies, 35. Springer, Berlin, Heidelberg New York, pp 449–483

Smayda TS (1990) Novel and nuisance phytoplankton blooms in the sea: evidence for a global epidemic. In: Granéli E, Wallström K, Larsson U, Granéli W (eds) Toxic marine phytoplankton. Elsevier, Amsterdam, pp 29–41

Sournia A, Erard-Le Denn L,Grzebyk D, Lassus P, Partensky F (1990) Plancton nuisible sur les côtes de France. Pour la Science 153 (juillet): 60–66

Tatara K (1981) Relation between the primary production and the commercial fishery production by the boat fishery: Production in the fishery ground – Utilization of the primary production. Bull Nansei Reg Fish Res (a) 13: 111–113

Underdal B, Skulberg OM, Dahl E, Aune T (1989) Disastrous bloom of *Chrysochromulina polylepis* (Prymnesiophyceae) in Norwegian coastal waters 1988 – mortality in marine biota. Ambio 18, 5: 263–270

Van Bennekom AJ, WWC, Tijssen SB (1975) Eutrophication of Dutch coastal waters. Proc R Soc London B 189: 359–374

Virnstein RW, Carbonara FA (1988) Seasonal abundance and distribution of drift algae and sea grasses in the Mid Indian River lagoon FL. Aquat Bot 23: 67–82

van Westernhagen H, Dethlefsen V (1983) North Sea oxygen deficiency 1982 and its effects on the bottom fauna. Ambio 12(5): 264–266

Waite T, Mitchell R (1972) The effect of nutrient fertilization on the benthic algae *Ulva lactuca*. Mar Bot 15: 151–156

3 Coastal Eutrophication and Marine Benthic Vegetation: A Model Analysis

I. De Vries[1], C.J.M. Philippart[2], E.G. DeGroodt[3], and M.W.M. van der Tol[1]

3.1 Introduction

The production of marine vegetation is mainly nutrient-limited in many coastal areas. The primary producers should therefore be easily affected by eutrophication, particularly in lagoons, semi-enclosed bays and larger port areas. This supposition is confirmed by experimental work where the addition of the limiting nutrient generally results in increased production and biomass of phytoplankton (Borum 1983, 1985; Granéli and Sundbäck 1985; Oviatt et al. 1986; Rudek et al. 1991), microphytobenthos (Granéli and Sundbäck 1985; Nilsson et al. 1991), epiphytes (Borum 1983, 1985; Silberstein et al. 1986), macroalgae (Harlin and Thorne-Miller 1981; Lapointe et al. 1987) and seagrasses (Orth 1977; Harlin and Thorne-Miller 1981).

Yet, in the field, eutrophication generally does not result in an overall increase of marine algal species already present, but rather in a shift of species composition and biomass. An often observed shift of algal species composition is the decline of seagrass in nutrient-enriched estuaries, e.g. in the bays on the southcoast of Long Island, USA (Kelly and Naguib 1984), in Lac Tunis, Tunisia (Kelly and Naguib 1984), in the Venice Lagoon, Italy (Sfriso et al. 1989b) and in Cockburn Sound, Australia (Shepherd et al. 1989). The macroalgae composition was found to shift from filamentous to foliose species. In the Venice Lagoon, the macroalgae *Cladophora prolifera* and *Laurencia pinnatifida* almost disappeared while *Ulva* spp. increased in distribution and biomass (Sfriso

[1]Rijkswaterstaat, National Institute for Coastal and Marine Management, P.O. Box 20907, 2500 EX The Hague, The Netherlands
[2]Institute for Forestry and Nature Research, P. O. Box 165, 1790 AD Den Burg, The Netherlands
[3]Rijkswaterstaat, National Institute for Coastal and Marine Management, P.O. Box 20907, 2500 EX The Hague, The Netherlands

Ecological Studies, Vol. 123
Schramm/Nienhuis (eds) Marine Benthic Vegetation
© Springer-Verlag Berlin Heidelberg 1996

1987). During the increasing eutrophication of the Peel-Harvey estuary, Australia, the macroalgal species composition suddenly shifted from *Cladophora montagneana* Kütz to *Chaetomorpha linum* Kütz, followed by gradual shift to *Enteromorpha* spp., mainly *E. intestinalis*, and *Ulva rigida* (Lavery et al. 1991). Nutrient enrichment of the Baltic Sea coincided with a shift from *Furcellaria lumbricalis* and *Fucus vesiculosus* to Ectocarpaceae (Zmudzinski 1992).

Blooms of macroalgae and phytoplankton represent extreme shifts in species composition and biomass of marine vegetation. Hypertrophic conditions in the Venice Lagoon resulted in an extraordinary growth of *Ulva* spp. with local biomass maxima up to 20 kg wet weight m^{-2} (Sfriso et al. 1989b). Coinciding with eutrophication, the phytoplankton primary production and biomass almost doubled in the Baltic (Zmudzinski 1992) and cell numbers increased tenfold in Kastela Bay, Yugoslavia (Kelly and Naguib 1984). Blooms can have dramatic ecological and economic consequences if they result in oxygen depletion and the associated mass mortality of the benthos (Kelly and Naguib 1984; Anonymous 1988; Sfriso et al. 1989b).

Both macroalgal and phytoplankton dominance can be considered as final algal succession stages in eutrophic shallow waters. The dominance may vary in space and time. Macroalgae and phytoplankton blooms occurred simultaneously in different parts of several lagoons along the coast of Adriatic Sea, e.g. the Po delta (Pugnetti et al. 1991) and the Venice Lagoon (Sfriso et al. 1991). In the Peel-Harvey estuary, Australia, the dense spring bloom of the blue-green alga *Nodularia* collapsed in summer, and was followed by a bloom of macroalgae (Lavery et al. 1991).

The mechanisms of the decline of seagrasses as the result of eutrophication are extensively reviewed in current literature (Shepherd et al. 1989) and will not be discussed in this chapter. However, the mechanism behind the succession from phytoplankton to macroalgae is still unknown. This succession seems contra-indicated from the point of view that a species with the highest affinity for the limiting resource has competitive advantage and will thus outcompete the other species. In general, phytoplankton species have a larger ratio of surface to volume than macroalgae, as do filamentous macroalgae when compared to foliose species. This ratio affects metabolic rates such as productivity and nutrient uptake, and hence growth, which is much faster for small organisms (Littler and Littler 1980). In agreement with these principles, the production (g C m^{-2} year^{-1}) over biomass (g C m^{-2}) ratio ($= P/B$ ratio year^{-1}) of macroalgae is 10–100 times lower than that of phytoplankton (Atkinson and Smith 1983; Sfriso et al. 1988). In nutrient-limited environments, phytoplankton is expected to dominate over macroalgae, and filamentous over foliose macroalgae. The success of foliaceous macroalgae over phytoplankton would therefore seem unfavourable under most environmental conditions, but may enable these algae to dominate a community undergoing increasing eutrophication. The key question addressed in this chapter is: What makes the foliated macroalgae *Ulva* spp. outcompete phytoplankton under eutrophic, but still nutrient controlled conditions?

To evaluate this question, we will pinpoint the controlling factors for the competition between benthic marine macroalgae and phytoplankton at different levels of eutrophication. For this purpose, three marine systems of different trophic status are investigated with respect to the dynamics of macroalgae and their competition with phytoplankton by means of model analysis. The analysis will not attempt a deterministic and prognostic simulation of population dynamics and seasonal evolution of the biomass of macroalgae and/or seagrass as, for example, was recently done for *Ulva* spp. along the French west coast (Menesguen 1992) and for *Zostera marina* in Danish waters (Bach et al. 1992; Bach 1993). Instead, a descriptive and comparative mass balance analysis is performed, resulting in annual carbon and nutrient budgets of the three lagoons. Several mechanisms explaining the enhanced competitive capabilities of macroalgae over phytoplankton are discussed with reference to the specific characteristic of the three lagoons.

3.2 Materials and Methods

3.2.1 Areas of Study

Three semi-enclosed polyhaline lagoons are considered in this comparative study. The hydrographical and hydrochemical characteristics of the three water systems are summarized in Table 3.1. The three lagoons cover the whole range of oligotrophic to hypertrophic conditions in terms of nutrient input and eutrophication symptoms.

Lake Grevelingen
Lake Grevelingen, a former estuary in the Delta region of the Netherlands (Fig. 3.1), was disconnected from the rivers Rhine and Meuse in 1964 and from the North Sea in 1971 by enclosure dams. The connection with the North Sea was partly re-established in 1978 by means of a small exchange sluice, which is opened during winter to

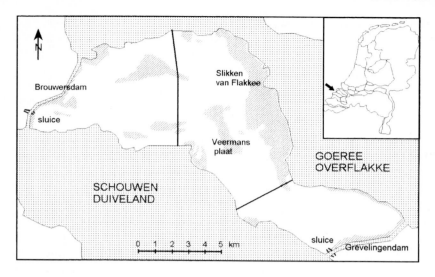

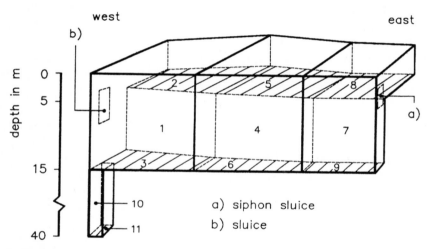

Fig. 3.1a,b. Lake Grevelingen. **a** Map with horizontal boundaries of model segments. **b** Model segments in vertical direction

prevent salinity stratification. The restricted flushing with sea water has reduced the flushing time, or water residence time, of the lake to its present value (Table 3.1). The bathymetry of the lake with large shallow areas and several deep gullies still reflects its estuarine origin. Approximately 60% of the area has a depth of less than 2.5 m.

The freshwater input from polders is very small compared to the lake volume, and to the other lagoons. As a consequence of the almost

Table 3.1. Hydrographical and hydrochemical properties of Lake Grevelingen, Lake Veere and the Venice Lagoon

	Lake Grevelingen	Lake Veere	Venice Lagoon
Area (km²)	108[a]	18–21[a]	400[b]
Volume (m³ × 10⁶)	575[a]	89[a]	625[b]
Average depth (m)	5.3[a]	4.5[a]	1.6[b]
Maximum depth (m)	48[a]	25[a]	11[b]
Flushing time (days)	270[a]	180[a]	10[e]
Salinity (‰)	30–33[a]	15–25[a]	22–34[b]
Fresh water load (m³ × 10⁶/year)	21[a]	83[a]	25[e]
Volume/fresh water load	27[a]	1[a]	25[e]
Average tidal amplitude (m)	–	–	0.35[b]
Common temperature, winter (°C)	1–2[c]	1–2[d]	6–7[b]
Common temperature, summer (°C)	18–22[c]	18–22[d]	24–26[b]
Extinction (m⁻¹)	0.2–0.5[a]	0.3–1.4[a]	0.5–1.5[b]
N–total input (g/m²/year)	4[c]	40[d]	20[f]
P–total input (g/m²/year)	0.4[c]	6[d]	4[f]
N/P ratio of input (by atoms)	21	14	11

[a]Nienhuis (1992).
[b]TECHNITAL (1992).
[c]De Vries et al. (1988).
[d]De Vries et al. (1990).
[e]E. Runca and L. Postma, (pers. comm).
[f]Sfriso et al. (1989a); Cossu et al. (1992).

complete hydrological isolation from the surrounding polders and from the river, the input of nutrients is small. The total input, depicted in Table 3.1, includes atmospheric deposition. The lake has developed into an oligotrophic ecosystem with a high transparency of the water. The phytoplankton concentration is very low (chlorophyll *a* rarely exceeds 10 mg m⁻³) and macrophytes, dominated by eelgrass (*Zostera marina*), cover the shallow areas.

Lake Veere

Lake Veere originated in 1960–1961 after enclosure from the adjacent Oosterschelde estuary and the North Sea by dams (Fig. 3.2). The water level is controlled by water exchange with the Oosterchelde estuary through the ship-lock in the eastern dam.

Compared to the other two lagoons, the lake is small and shallow. As in Lake Grevelingen, the former intertidal and shallow subtidal areas comprise at least 50% of the area of the lake. During winter, the lake receives excess water from the surrounding polders. To facilitate this function, the water level is artificially maintained at 0.7 m below mean sea level during winter, reducing the water surface area to 18 km². As a

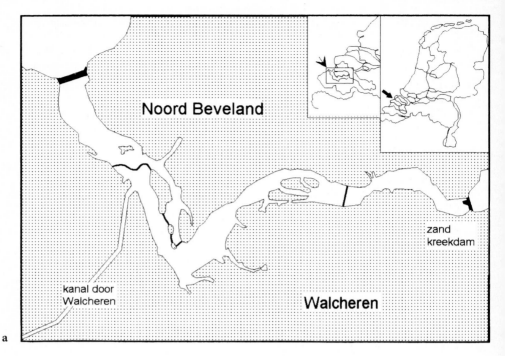

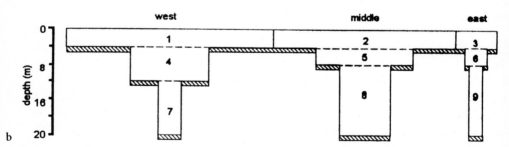

Fig. 3.2.a,b. Lake Veere. **a** Map with horizontal boundaries of model segments. **b** Models segments in vertical direction

consequence, the hydrology of the lake is dominated by the throughput of fresh water, which causes a strong and almost permanent salinity stratification in parts of the lake. The fresh-water input also results in a considerable input of nutrients due to the agricultural use of the polderland. The conditions in the lake are eutrophic, with dense phytoplankton blooms and prolonged oxygen depletion in the bottom water layer below 6–10 m.

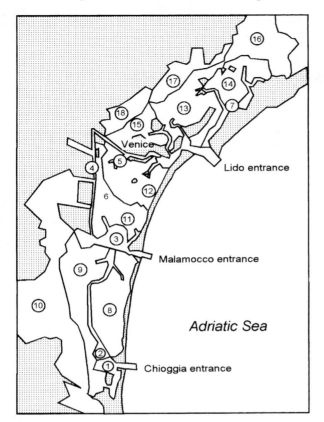

Fig. 3.3. Venice Lagoon. Map with horizontal boundaries of model segments.

Venice Lagoon

The Venice Lagoon is a shallow marine embayment, a large part of the lagoon has a depth of less than 0.5 m. The mean tidal amplitude is about 35 cm (Fig. 3.3).

The lagoon is connected to the Adriatic Sea by three entrances. The average effective flushing time of the lagoon by tidal exchange is approximately 10 days. (E. Runca, L. Postma, pers. comm). However, large differences exist between different parts of the lagoon. The central part around the city of Venice, bordered by the Malamocco–Marghera Canal in the west and the south and by the tidal lands extending from the airport to the island of Burano in the north, has a water residence time in the order of days. The residence times of the inner regions are much longer, in the order of months.

The freshwater input due to land run-off causes horizontal salinity gradients, from 22–26 PSU in the vicinity of the freshwater discharges to 31–34 PSU at the entrances to the Adriatic Sea. The lagoon receives considerable amounts of nutrients from domestic (resident and tourist population), industrial and agricultural sources, directly due to run-off and discharges and indirectly by atmospheric deposition. The conditions in parts of the lagoon are characterized as hypertrophic, with massive *Ulva* spp. blooms and subsequent anoxic conditions with mass mortality of aerobic organisms as the most striking features (Sfriso et al. 1989b).

3.2.2 Mass Balance Analysis

The three lagoons were compared by means of a descriptive mass balance analysis, results of which showed annual carbon and nutrient budgets which include the quantification of the role of the dominant macrophytes.

Existing applications of an ecological model have been used for this purpose (De Vries et al. 1988; Anonymous 1989; De Vries et al. 1990). The model consists of two main parts with distinct functions. The transport module deals with the transport of dissolved and suspended substances between the computational elements (model segments) in one, two or three directions (Postma 1988). Within each computational element, the ecological module describes the most important processes between substances and functional groups of organisms.

The schematization of Lake Grevelingen resulted in 11 model segments in horizontal and vertical directions. Four water segments and seven benthic segments were distinguished (Fig. 3.1). All mass transport by water movement was formulated as dispersive transport because of the wind driven circulation and a virtual absence of stratification.

The schematization of Lake Veere resulted in 18 segments in horizontal and vertical directions, including nine water segments which allowed a distinction to be made between shallow and deeper parts and water layers above and below the halocline. Each water segment had an adjacent bottom segment (Fig. 3.2). Information on the hydrodynamical environment was obtained from a detailed stratification model (Bollebakker and van de Kamer 1989). The output consisted of a time series of daily values for horizontal and vertical advective flows, entrainment, and vertical dispersion. For the Venice Lagoon, a two-dimensional horizontal schematization was used with 18 water segments and 18 adjacent bottom segments (Fig. 3.3), allowing the representation of differences due to local bathymetry as well as gradients imposed by run-off from land and exchange with the Adriatic Sea.

The choice of the state variables in the ecological model was based on their estimated involvement in the mass cycling. Only those functional groups of organisms which contributed more than 10% to the total mass flow or total pool size of organic carbon and nutrients were distinguished in the model. As a consequence, organisms at higher trophic levels such as birds and fishes, and specific functional groups of organisms with a low biomass such as benthic meiofauna and epibenthic crustaceans, were not incorporated in this model analysis. Following these rules, the state variables considered necessary to be introduced in the ecological model are shown in Fig. 3.4.

Some of the variables were not incorporated and treated as prognostic state variables but imposed as forcing functions derived from empirical data. This was the case for the primary consumers as well as for the eelgrass and *Ulva* spp. in Lake Grevelingen and Lake Veere. The forcing function for eelgrass in Lake Grevelingen was derived from a separate eelgrass model which was specifically developed for, and applied to, Lake Grevelingen (Verhagen and Nienhuis 1983). *Ulva* spp. in the Venice Lagoon were described in an intermediate way: the potential biomass development in each of the segments was derived from empirical data while the calculated biomass development could deviate from this potential biomass curve due to nutrient constraints.

Table 3.2 lists the simplified budget equations for the state variables. Processes such as denitrification, which directly influences mass balances by taking material from the system, have been emphasized in the model development and application. In addition to the state variables and processes depicted in Fig. 3.4 and Table 3.2, an oxygen balance was included in the model. Further details and parameter values are given in the references for the applications, and in Anonymous (1990).

We adopted the ratios (by weight) used by Sfriso et al. (1989b) for the stoichiometry of *Ulva* biomass:WW:DW:C:N:P $\sim$ 30:3.9:1:0.11:0.01.

3.3 Results

3.3.1 Model Calibration

Model calibration was performed for estimated average years by mixing existing data from different years, and by using long-term averages for meteorological forcing and other boundary conditions, since only a few complete sets of data for specific years were available. As can be seen

Table 3.2. Simplified equations of the state variables of the ecological model

State Variables	Equations[a]
Primary producers:	
phytoplankton	$dX/dt = prd - exc - mor - gra - sed$
microphytobenthos	$dX/dt = prd - mor - gra - res - bur$
macrophytes	$dX/dt = prd - mor$
Consumers:	
zooplankton	$dX/dt = gra - exc - fae - mor$
suspension feeders	$dX/dt = gra - exc - fae - mor$
deposit feeders	$dX/dt = gra - exc - fae - mor$
Detritus:	
suspended detritus	$dX/dt = mor + fae + res - gra - min - sed$
labile detritus sediment surface	$dX/dt = mor + fae + sed - gra - min - res - bur$
bottom detritus	$dX/dt = bur - gra - min$
Dissolved nutrients – water	
nitrate	$dX/dt = inp + nit - den - prd \pm exh$
ammonium	$dX/dt = inp - nit + min + exc - prd \pm exh$
phosphate	$dX/dt = inp + min + exc - prd \pm exh$
silicate	$dX/dt = inp + min - prd \pm exh$
Dissolved nutrients – bottom	
nitrate	$dX/dt = nit - den \pm exh$
ammonium	$dx/dt = min - nit \pm exh$
phosphate	$dX/dt = min \pm exh$
silicate	$dX/dt = min \pm exh$

[a]Legend: prd = primary production; exc = excretion (producers and consumers); mor = mortality; gra = grazing; fae = faeces and pseudofaeces biodeposition; min = mineralization; sed = sedimentation; res = resuspension; bur = burial; inp = input of nutrients (loading); nit = nitrification; den = denitrification; exh = sediment/water exchange.

from the figures, simulation runs for annual periods were repeated until dynamical stability from one year to the next was reached.

Figures 3.5, 3.6 and 3.7 show satisfactory agreement between measured and calculated values for Lake Grevelingen and Lake Veere, including the large difference of almost an order of magnitude in trophic status of the two lagoons. The model consistently describes the combination of nutrient dynamics, phytoplankton biomass and oxygen depletion resulting from organic loading of the sediment under stratified conditions.

The model describes the spatial gradients of nitrates and chlorophyll in the Venice Lagoon, with highest concentrations in the vicinity of the nutrient sources, in good agreement with measured values as shown in Figs. 3.8 and 3.9. The seasonal variation of nitrate is also simulated satisfactorily, but chlorophyll peaks do not coincide with measured maxima.

All three lagoons are nitrogen-limited during summer. Although the DIP concentration (dissolved inorganic phosphorus) in the Venice

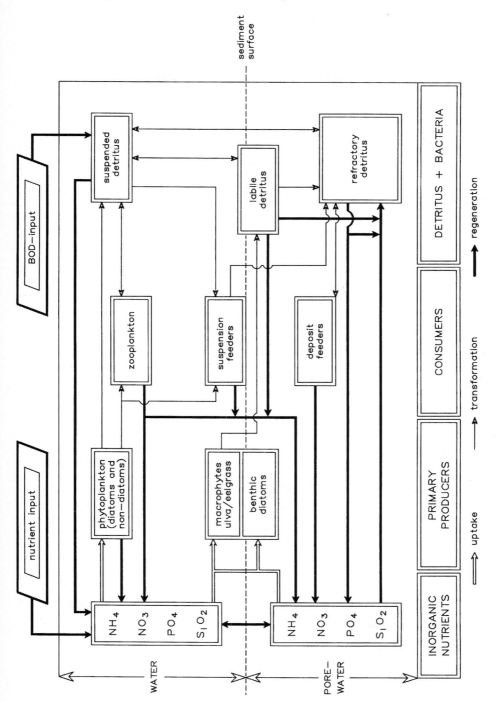

Fig. 3.4. State variables and processes in the ecological model

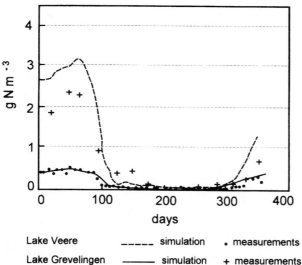

Fig. 3.5. Measured and simulated nitrate concentrations in Lake Grevelingen (model segment 4, cf. Fig 3.1) and Lake Veere (model segment 2, cf. Fig 3.2)

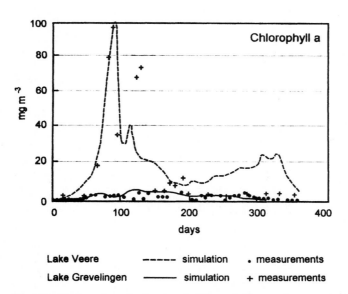

Fig. 3.6. Measured and simulated chlorophyll *a* concentrations in Lake Grevelingen (model segment 4, cf. Fig. 3.1) and Lake Veere (model segment 2, cf. Fig. 3.2)

Lagoon may become limiting too, Sfriso and co-workers (1989b) have proven that *Ulva* spp. in the lagoon are nitrogen-limited due to: (1) the nutrient input into the lagoon which is relatively poor in nitrogen (N/P ratio by atoms about 11 compared to the Redfield ratio of 16; see

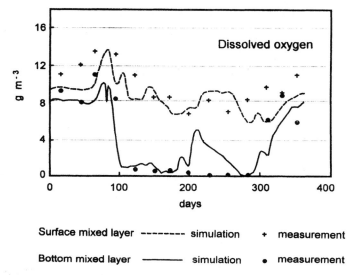

Fig. 3.7. Measured and simulated oxygen concentrations in the surface mixed layer (SML) and the bottom mixed layer (BML) of Lake Veere (model segments 2 and 8, cf. Fig. 3.2)

Table 3.1); and (2) the nitrophily of green macroalgae (N/P ratio about 25; Sfriso 1987).

The nitrogen limitation in the other two lagoons is even more obvious. Even though the N/P ratio of the nutrient input was close to or even higher than the Redfield ratio (Table 3.1), the DIN/DIP ratio (DIN = dissolved inorganic nitrogen in the form of ammonium and nitrate) of the winter concentrations was 2–4 in Lake Grevelingen and 7–8 in Lake Veere. This nitrogen limitation is partly explained by the remobilization of phosphorus from the sediment, which had been previously accumulated from the river input during the estuarine history of the lagoons (Kelderman 1985). Additionally, nitrogen removal by denitrification has been identified as a key process in the nitrogen cycle of shallow-water marine ecosystems. This process is responsible for the natural tendency towards N-limitation, and is particularly effective in water systems with long water residence times (De Vries and Hopstaken 1984; Seitzinger 1987; Smith 1991).

The mass balance analysis therefore focuses on the nitrogen cycle and evaluation of the phosphorus cycle is omitted.

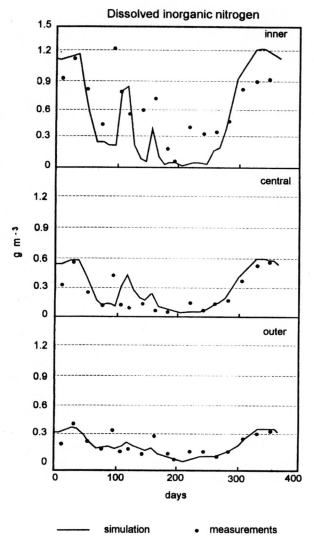

Fig. 3.8. Measured and simulated DIN concentrations
in the Venice Lagoon. Inner
Segments 17 and 18; central
segments 11, 13, and 15;
outer segment 6 (cf. Fig.3.3).
Measurements after TECH-
NITAL (unpubl. results)

3.3.2 Carbon and Nitrogen Budgets

The annual carbon budgets of the three lagoons derived from the model calculations are given in Fig. 3.10. The budgets are averaged over each entire lagoon, i.e. including the deeper parts (gullies) without macrophytes. The individual fluxes are rounded to 5 g C m^{-2} year^{-1}.

Only some of the calculated fluxes incorporated in the budgets can be directly compared to the measurements. Calculated phytoplankton

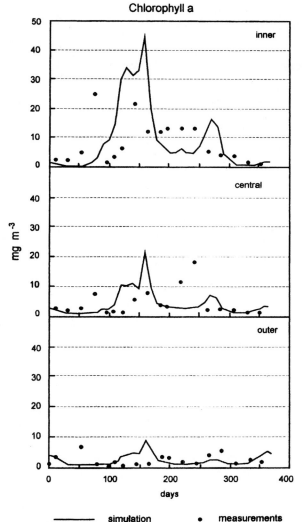

Fig. 3.9. Measured and simulated chlorophyll *a* concentrations in the Venice Lagoon. Inner Segments 17 and 18; central segments 11, 13 and 15; outer segment 6 (cf. fig.3.3). Measurements after TECHNITAL (unpubl. results)

production agrees very well with measured values for Lake Grevelingen (150–225 g C m^{-2} year^{-1} in 1979–1981; Vegter and De Visscher 1984) and for Lake Veere (300 g C m^{-2} year^{-1} in 1983; Stronkhorst et al. 1985). Most of the other terms in the budgets are more or less hypothetical, since empirical data for direct comparison are lacking. However, there is circumstantial evidence for the adequacy of the calculated budgets. They are consistent with other parts of the model output, e.g. phytoplankton

LAKE GREVELINGEN   LAKE VEERE   VENICE LAGOON

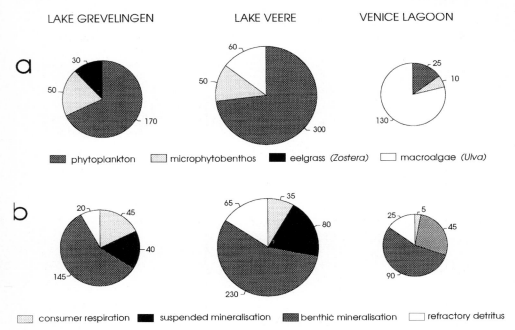

Fig. 3.10. Calculated organic carbon budgets of three lagoons (g C m^{-2} year^{-1}). a Net primary production. b Community respiration and sink of organic carbon in the sediment

and nutrient concentrations, showing satisfactory agreement with measurements mentioned in the previous section.

This means, for instance, that the calculated production of macrophytes may be considered as the level of production which can be sustained by the available nutrients under the constraints that measured biomass, and nutrient concentrations must be reproduced by the model.

The measured maximum biomass of *Zostera marina* in lake Grevelingen, approximately 10 g C m^{-2}, is reproduced by the model with a ratio between the annual production and the maximum standing crop (P/B) of 3. This is in agreement with the eelgrass model of Verhagen and Nienhuis (1983). Nienhuis (1992) roughly estimated the *Ulva* spp. production in Lake Veere at 120 g C m^{-2} year^{-1}. Our estimate is half as much, but is consistent with the range of maximum biomass in 1987 and 1989 (8.2 and 4.5 g C m^{-2}, Hannewijk 1988; Apon 1990) and an annual P/B ratio of 10 (Nienhuis, pers. comm.). Sfriso and coworkers (1989b) estimated the *Ulva* spp. production in the central part of the Venice Lagoon which has a water surface of 132 km^2, at 253 g C m^{-2} year^{-1}, based on an estimated maximum standing crop of 134 g C m^{-2} and a

P/B ratio of 1.2–1.6. Our estimate, depicted in Fig. 3.10, is based on a lower maximum standing crop of 86 g C m^{-2} in the coastal area, and 50 g C m^{-2} averaged over the total water surface of the lagoon (TECHNITAL, unpubl. data). The annual P/B ratio resulting from the simulation was approximately 2.5. The difference in P/B ratio of *Ulva* spp. in Lake Veere and in the Venice Lagoon is most probably caused by the difference in viability. In Lake Veere, the total standing crop remains viable throughout the growing season, resulting in a high turnover of the biomass (Nienhuis pers. comm.). In the Venice Lagoon, thick mats of *Ulva* spp. are formed during the growing season, with only the upper layer remaining viable and productive. Recently, the P/B ratio of *Ulva* spp. in the Venice Lagoon was found to be inversely related to the biomass density, ranging from 1.5 to 4.5 (Sfriso et al. 1993).

The carbon budgets of the three lagoons show a remarkable similarity with differences of less than a factor three between total as well as individual fluxes, except for phytoplankton and *Ulva* in the Venice Lagoon. This similarity contrasts with differences of one order of magnitude in nutrient input, winter nutrient concentrations and spring maximum of chlorophyll. Related to this apparent robustness of the carbon budget, the internal cycling, i.e. the flux of carbon through the biological cycle divided by the size of the biological components, is very fast in Lake Grevelingen, slower in Lake Veere and very slow in the Venice Lagoon. The consumer foodchain is dominant in Lake Grevelingen, with more than 50% of the total net primary production being directly consumed by zooplankton and benthic suspension feeders. This results in a strong grazing control and rapid turnover of the phytoplankton biomass. Direct consumption of the total net primary production in Lake Veere is estimated at 40%. The consumer food chain in the Venice Lagoon is not important. Phytoplankton, the only primary producer which would be subject to substantial grazing, only accounts for 15% of total primary production. The detrital chain is dominant in Venice Lagoon, resulting in a slow internal cycling, in an accumulation of organic material, and in consequent anoxia.

The overall characteristics of nitrogen cycling in the three lagoons are illustrated by the annual nitrogen budgets in Table 3.3, which compares external balances and internal cycling.

The external input of nitrogen into the three lagoons per unit area or volume is different by about one order of magnitude. Lake Veere receives the highest input per unit area, while the Venice Lagoon, due to its smaller depth, has the highest input per unit volume.

A large amout of the available nitrogen in Lake Grevelingen is removed by denitrification and by retention in refractory detritus. This

Table 3.3. Nitrogen budgets of three lagoons (units gN/m^2/year, unless indicated otherwise)

	Grevelingen	Lake Veere	Venice Lagoon
External balance:			
external input	3.5	35	20
external input (gN/m^3/year)	0.7	8	13
net export	−0.5	18	7
denitrification + retention	4	17	13
Internal cycling:			
uptake primary producers	32	72	22
cycling index (−)	9.1	2.1	1.1

results in a lower DIN concentration than in the adjacent coastal zone of the North Sea, which in turn results in a small net import of nitrogen by the water exchange with the North Sea. The nitrogen removal calculated for the other two lagoons is less efficient and compensates for only part of the external input. As a consequence, the average DIN concentrations of these lagoons are higher than in the adjacent water systems, resulting in an export of a significant part of the input of nitrogen to the North Sea and the Adriatic Sea, respectively.

The total annual uptake of nitrogen by primary production as calculated by the model illustrates the intensity of internal cycling through the biological components. The differences in cycling intensity between the three lagoons are smaller than the differences in external input.

The concept of new and regenerated production, originally formulated defined by Dugdale and Goering (1967) for pelagic oceanic systems, is applicable to this cycling intensity.

Dugdale and Goering's definition is based on the chemical distinction between nitrogen sources:

– new production: production based on nitrate as nitrogen source.
– regenerated or old production: production based on ammonium as nitrogen source.

For the pelagic oceanic ecosystem, the ratio of old versus new production can be interpreted as an index of the number of times nitrogen cycles before being exported from the upper water column as particulate nitrogen. On the shelf at the middle Atlantic Bight, nitrogen cycles on average two times before sinking (Harrison et al. 1983).

This original definition, however, cannot be applied to shallow coastal ecosystems such as lagoons and estuaries because: (1) external input from land run-off and rivers consists partly of ammonium (Wassmann 1986); and (2) the pelagic subsystem cannot be considered independent of the benthic subsystem, due to the tight and continuous sediment–water interaction. We therefore propose the following definition: (1) new production: production based on the external input of nitrogen; (2) old (regenerated) production: production based on pelagic and benthic regeneration of nitrogen, i.e. internal input or loading.

The cycling index given in Table 3.3 is based on this definition, and is defined as the ratio of uptake by primary producers and external input. When comparing the three water systems, this cycling index shows a strong inverse relation with the external input per unit volume.

3.3.3   Contribution of Macrophytes to Nitrogen Cycling

Figure 3.11 illustrates the differences between the three lagoons with respect to the dominance of macrophytes and phytoplankton, by comparing the contribution of primary producers to nitrogen cycling as well as to the accumulation of nitrogen in their biomass.

The difference of one order of magnitude in nutrient input between Lake Grevelingen and Lake Veere only causes quantitative and gradual differences in nitrogen cycling and storage. Phytoplankton is the dominant producer in both lagoons. The production by macrophytes is even smaller than the calculated production by benthic diatoms on the shallow sediment surface. Due to the high P/B ratio of phytoplankton and microphyto-benthos, the accumulation of nitrogen in their biomass is small. Only 15–20% of the winter accumulated DIN is stored in the biomass of these producers during spring and summer. Eelgrass (*Zostera marina*), the dominant macrophyte in Lake Grevelingen, is replaced by approximately the same amount of *Ulva* spp. in Lake Veere, which however represents a significantly larger nitrogen uptake due to its nitrophily and higher P/B ratio.

The Venice Lagoon shows qualitative differences with the other two lagoons. *Ulva* spp. are the dominant producers in the Venice Lagoon due to their higher production compared to the other lagoons but also due to the low production of phytoplankton and microphytobenthos. The calculated total production and nitrogen uptake in the Venice Lagoon is even smaller than in Lake Grevelingen, due to this shift in dominance from phytoplankton to *Ulva*.

The most significant difference between the Venice Lagoon and the other two lagoons is the accumulation of *Ulva* biomass and coinciding storage of nitrogen. The maximum biomass accumulating in the Venice Lagoon is almost ten times the maximum macrophyte biomass estimated

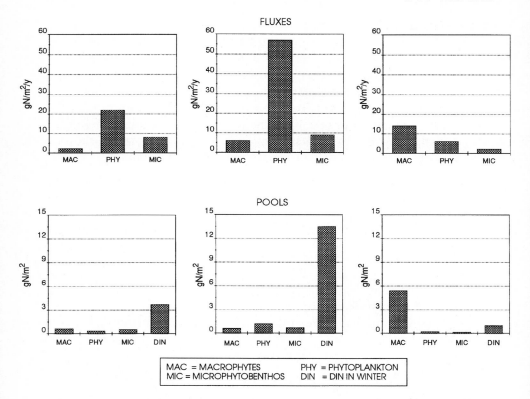

Fig. 3.11. Nitrogen cycling by primary production (fluxes: g N m^{-2} year^{-1}) and nitrogen storage in primary producers and winter–accumulated DIN (pools: g N m^{-2}) in three lagoons

for the other lagoons. The maximum storage of nitrogen in *Ulva* biomass is more than five times larger than the DIN accumulated in the water column of the Venice Lagoon in winter.

The comparison with the other two lagoons illustrates that this algal biomass accumulation feature, responsible for the biggest *Ulva* spp. mass bloom in the world ever reported in the literature, cannot be explained by the nitrogen input and resulting DIN concentration as such, nor by the productivity of *Ulva* itself.

3.4 Discussion

Comparison of the three lagoons reveals the following main findings:

1. *Rubustness of Overall Carbon Budgets and Nitrogen Cycling.* Differences of one order of magnitude in nitrogen input by area and by volume leads to or is accompanied by, differences of less than a factor three between total as well as individual fluxes of carbon and nitrogen through the biological components, with only a few exceptions. The cycling index or ratio of old versus new production is inversely related to the nutrient input.
2. *Differences in Dominance Between Primary Producers at Comparable High Levels of Nitrogen Input.* The high input of nitrogen into Lake Veere coincides with a sustained dominance of phytoplankton over macroalgae, whereas a comparable input of nitrogen into the Venice Lagoon (a smaller input by area but a larger input by volume) coincides with a dominance of *Ulva* spp. over phytoplankton in terms of production and nitrogen uptake.
3. *Differences in Biomass Accumulation and P*B Ratio of Ulva* spp. *in Different Lagoons.* The storage of nitrogen in accumulating *Ulva* spp. biomass in the Venice Lagoon is a significant fraction ($> 25\%$) of the annual nitrogen input, and is more than five times larger than the accumulated DIN in the water column in winter. The storage of nitrogen in *Ulva* spp. biomass in Lake Veere is insignificant when compared to the input (about 2%) as well as to the DIN pool in the water column during winter (less than 5%). Concomitantly, the P/B ratio of *Ulva* in the Venice Lagoon is much lower (at least a factor 4) than the potential P/B ratio of a viable population.

These findings illustrate that eutrophication symptoms such as dominance and mass accumulation of macroalgae, specifically *Ulva* spp. are not unequivocally related to the causes of eutrophication viz. high nutrient loadings and resulting concentrations of nutrients. Other factors must play a decisive role in determining the specific eutrophication symptoms which emerge under generally eutrophic conditions. The following factors are identified and discussed: (1) light climate, (2) storage capacity for nutrients, (3) nutrient supply to macroalgae by flushing and (4) control of phytoplankton by grazing and flushing.

3.4.1 Light Climate

Primary producers occupy different positions in the water column. Microphytobenthos and macrophytes have a fixed position in the eutrophic zone, being more or less attached to the bottom surface. Phytoplankton is permanently mixed over the entire water column, as long as no stratification occurs and enough turbulence is present to prevent buoyancy control. The ability of the different primary producers to compete for light is therefore determined by the light extinction in the water column and the morphological and hydrographic characteristics of the system.

These aspects of the light climate can be quantified with the following equations (see also Fig. 3.12). The light extinction in the water column can be described according to the law of Lambert-Beer:

$$I(z) = I(o)e^{-kz} \tag{1}$$

where $I(o)$ = surface irradiance ($\mathrm{Wm^{-2}}$)
$\quad\quad\quad I(z)$ = irradiance at depth z($\mathrm{Wm^{-2}}$)
$\quad\quad\quad k$ = light extinction coefficient ($\mathrm{m^{-1}}$)
$\quad\quad\quad z$ = depth (m).

The relation between area and depth in lagoons and estuaries with shallow areas alternating with gullies can be approximated accordingly:

$$A(z) = A(o)e^{-az} \tag{2}$$

where $A(o)$ = area at $z = 0$ ($\mathrm{m^2}$)
$\quad\quad\quad A(z)$ = area at depth z($\mathrm{m^2}$)
$\quad\quad\quad a$ = coefficient relating area to depth ($\mathrm{m^{-1}}$).

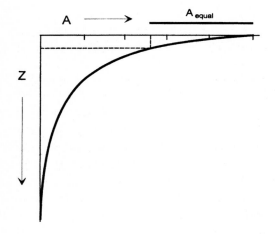

Fig. 3.12. Illustration of the terms used in Eqs (1–4)

The value of the coefficient relating area to depth (a) was quantified as the inverse value of the mean depth ($1/z_{nean}$) according to Klepper (1990).

The average irradiance integrated over the water column I_{col} (W m^{-2}) can be derived from Eqs. (1) and (2) accordingly:

$$I_{col} = \frac{a/(a+k)\, I(o)(1 - e^{-(a+k)z})}{(1 - e^{-az})} \tag{3}$$

The average irradiance in the water column of the stratified Lake Veere was calculated for the surface mixed layer of 8 m instead of the maximum depth of 25 m.

It may be hypothesized that benthic primary producers can successfully compete with phytoplankton for light on that part of the bottom area where the irradiance $I(z)$ is equal to or higher than the average irradiance in the water column I_{col}.

The bottom area defined in this way represents the habitat suitable for benthic primary producers considering the competition for light with phytoplankton, when the same affinity for light between the different primary producers is assumed. This bottom area A_{equal} can be calculated accordingly (Fig. 3.12):

$$A_{equal} = 1 - \left(\frac{I_{col}}{I(o)}\right)^{a/k} \tag{4}$$

The ability of benthic primary producers to compete for light with phytoplankton will be larger if a water system contains a larger area A_{equal}, where $I(z) > I_{col}$. Since this area depends on the ratio a/k [Eq. (4)], this ratio can be used as an index for the competition for light between benthic and planktonic primary producers. High values for this index, caused by shallowness of the water system (large a) or by a high transparency of the water (small k), indicate a high competitive ability of benthic primary producers.

The aspects of the light climate in the three water systems calculated with Eqs. (1–4) using the characteristics depicted in Table 3.1, are given in Table 3.4.

The average irradiance in the water column is equally high in Lake Grevelingen and the Venice Lagoon at approximately 30–50% of the surface irradiance (Table 3.4). This is due to the high transparency of the water in Lake Grevelingen and due to the shallowness of the Venice Lagoon. As a consequence, a considerable fraction of the bottom area, approximately 40–50%, is identified as suitable habitat for benthic

Table 3.4. Characteristics of the light climate in three lagoons for the range of extinction values depicted in Table 1 [units: irradiance as fraction of $I(o)$, area as fraction of $A(o)$ and depths in m]

	Lake Grevelingen	Lake Veere	Venice Lagoon
a	0.189	0.222	0.625
I_{col}	0.27–0.49	0.16–0.50[a]	0.29–0.56
z_{equal}	2.6–3.6	1.3–2.3	0.8–1.2
$A(z_{equal})$	0.39–0.49	0.25–0.40	0.40–0.52
a/k_{mean}	0.54	0.26	0.63

[a]The average irradiance in the water column of stratified Lake Veere is calculated for the surface mixed layer of 8 m depth.

primary producers (Table 3.4). Moreover, the irradiance at the lower depth limit (z_{equal}) of this area is high. This means that also deeper bottom areas than indicated in Table 4 can be colonized by benthic primary producers, if phytoplankton development is limited by factors other than light.

The light climate is probably not an important regulating factor in Lake Grevelingen, due to the strong nitrogen limitation during almost the entire growing season. In the Venice Lagoon however, primary production is less nitrogen-limited. The shallow bottom area suitable for macroalgal growth is most probably an important factor causing the high competitive ability of macroalgae in the Venice Lagoon.

The situation with respect to the light climate is quite different in Lake Veere. This difference is indicated by the competition index, the value of which is twice as low for Lake Veere compared to the other two lagoons. The fraction of the bottom area which is suitable for macroalgae is estimated at 25–40% (Table 3.4). Not only is the suitable area smaller than in the other two lagoons, but the irradiance at the lower depth limit of this bottom area is also lower. This means that there is less possibility for macroalgae to occupy deeper bottom areas when phytoplankton is limited by factors other than light. The low competitive ability of macroalgae in Lake Veere is consistent with these features of the Lake Veere light climate.

3.4.2 Storage Capacity for Nutrients

Macroalgae are able to store reserve material, e.g. soluble nitrogen, carbohydrates and starch (Rosenberg and Ramus 1982). This characteristic enables them to survive growth-limiting conditions for longer periods than phytoplankton. In addition, the survival of macroalgae is favoured by their adaptive nature. For example, *Ulva lactuca* was found to adapt to very low nutrient concentrations (Cohen and Neori 1991) and light conditions (Vermaat and Sand-Jensen 1987).

The growth of primary producers can be limited by nutrients or by other factors. If the growth of macroalgae is limited by nitrogen, the frequency at which pulses of dissolved nitrogen occur in the surface water may affect the species composition of a mixed algal assemblage containing species which differ in uptake rate and storage capacity (Fujita 1985). The nitrogen storage capacity of macroalgae is expected to favour its competitive abilities over phytoplankton in environments with episodic nitrogen availability, compared to environments with a more regular nitrogen supply.

During winter in temperate regions, growth is often limited by climatic conditions, e.g. light and temperature. The rapid spring growth of macroalgae could be enhanced by the presence of macroalgal biomass that survived this unfavourable period. It is hypothesized that this survival during winter may give macroalgae a particularly strong competitive advantage over phytoplankton in areas with a strongly seasonal light regime (Mann and Chapman 1975).

These principles are in agreement with findings on the timing and character of the blooms of the main primary producers in the three lagoons under consideration (Fig. 3.13). All systems receive additional nitrogen originating from external sources during winter and early spring. These nutrients cannot immediately be used for new production due to restricted climatic conditions.

No substantial biomass of phytoplankton nor aboveground biomass of macrophytes is present during the winter in Lake Grevelingen. The nutrients which were accumulated during winter are used to support the spring bloom of phytoplankton, due to their competitive advantage over macrophytes by their high P/B ratio. Microphytobenthos also contributes significantly to nutrient depletion, especially silicate, during spring (De Vries and Hopstaken 1984). This new production of phytoplankton and microphytobenthos in spring is followed by a high regenerated production of phytoplankton with a low biomass in summer and autumn. Aboveground biomass of eelgrass develops after the depletion of nutrients in the water column by the spring bloom. Initial eelgrass growth is supported by reserve material stored in overwintering underground biomass. Further eelgrass growth depends mainly on nutrients from the pore water (Verhagen and Nienhuis 1983).

Macroalgae in Lake Veere are not able to survive the winter as a result of the restrictive climatic conditions at this latitude. The relatively high P/B ratio of phytoplankton and the absence of macroalgae biomass in spring, results in nutrients accumulated during winter being used for the

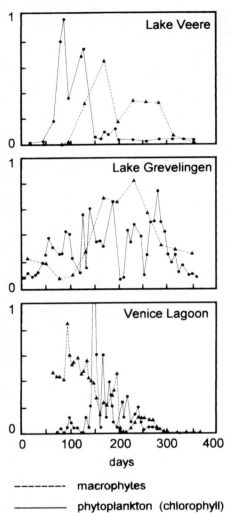

Fig. 3.13. Biomass development of phytoplankton and macrophytes in three lagoons. Units are relative, normalized between 0–1; absolute levels are indicated in Figs. 3.6, 3.9 and 3.11. Data for Lake Grevelingen and Lake Veere derived from Nienhuis (1992); data for Venice Lagoon derived from Sfriso et al. (1988)

- - - - - - - macrophytes

——————— phytoplankton (chlorophyll)

spring bloom of phytoplankton. Macroalgae only occur when the nutrients, transferred to detritus by phytoplankton production and mortality, are regenerated again. The new production in Lake Veere consists of phytoplankton, whereas there is only regenerated production of macroalgae. The macroalgal biomass persists during the rest of the year until it dies off at the end of autumn due to unfavourable environmental conditions.

The climatic conditions in the Venice Lagoon allow a certain amount of macroalgae to survive the winter period, and this amount is probably variable from one year to another. The initial biomass is, at least is some years, sufficiently high to enhance the initially low competitive ability caused by the low P/B ratio, and subsequently supports a rapid spring growth. The growth of *Ulva* spp. occurs through vegetative increase of pre-existing frond fragments, or by sexual reproduction on solid supports, e.g. shells and crab carapaces (Sfriso et al. 1991). The high new production of macroalgae depletes the DIN accumulated in the water column during winter as well as the considerable nutrient inputs which occurred in winter and early spring. The collapse of the macroalgae bloom after nutrient depletion is followed by partial decomposition of the produced material. The resulting sudden availability of nutrients thus released into the water triggers a subsequent low regenerated production of phytoplankton (Sfriso et al. 1988). The biomass of the macroalgae fluctuates throughout the summer and autumn, until the climatic conditions again become limiting but not lethal to the macroalgae.

The principal difference between the three water systems is that new production consists of phytoplankton in Lake Grevelingen and Lake Veere, whereas the new production in the Venice Lagoon consists of macroalgae (Fig. 3.13). We suggest that the causal factor for this difference is climatic. Low winter temperatures prevent the survival of macroalgae in temperate regions. Mild winters at lower latitudes allow the survival of macroalgae, representing initial storage capacity for nutrients, and enabling them to out-compete phytoplankton at the onset of the growing season.

3.4.3 Nutrient Supply to Macroalgae by Flushing

A poor water exchange is generally considered to be beneficial for macroalgae development and accumulation. Stagnancy of the water is even suggested as the main factor responsible for the mass blooming of *Ulva* spp. in the Venice Lagoon under the present conditions of nutrient availability (Sfriso et al. 1989b).

However, if a distinction is made between the partly overlapping phases in the seasonal cycle of (1) nutrient uptake and initial biomass production, and (2) biomass accumulation and decay, it may be argued that a certain amount of water exchange or flushing is beneficial or even necessary for macroalgae growth during the initial production phase.

It has been found for *Porphyra tenera* (Rhodophyta), an algal species morphologically similar to *Ulva*. that increasing water exchange enhances growth at sites with comparatively low nutrient concentrations (Matsumoto 1959). In general, a current velocity of at least 15 cm s^{-1} was needed for optimal growth, while in nutrient-deficient waters 30 cm s^{-1}

was required. Menesguen (1992) explains the excessive proliferation of *Ulva* spp. on very open sites along the French Atlantic coast by the dynamic trapping of water masses in bays where the tidal residual drift vanishes.

The following calculation illustrates the amount of flushing necessary for the initial *Ulva* production in the Venice Lagoon. A typical maximum *Ulva* biomass, averaged over the area occupied by *Ulva*, is 2.5 kg m^{-2} wet weight. This biomass develops in approximately 1.5 months (March–April) from the winter wet weight biomass of 0.5 kg m^{-2}. For such an increase in biomass 7.2 g Nm^{-2} would be needed. The supply of nitrogen released from the sediment may be assumed to be negligible, since the lowest concentration of nitrogen in the sediment was found in this period (February–April; Sfriso et al. 1988). The typical winter concentration of DIN in the inner lagoon is 0.7 gN m^{-3}. Given a mean water depth of 1 m in the central area, the water volume of the area occupied by *Ulva* would have to be exchanged ten times in a period of 1.5 months to supply the needed amount of nitrogen. It therefore seems probable that the effective flushing time of the central area of the lagoon of Venice occupied by Ulva during March–April is less than 1 week. The flushing may be over estimated in this manner, since storage of nitrogen in *Ulva* tissue during winter may sustain part of the growth in the initial months of the growing season (Rosenberg and Ramus 1982). On the other hand, the calculation is based on the average maximum biomass in the central area of the lagoon. Maximum standing crop can locally exceed 20 kg m^{-2} wet weight, developing from a very low winter biomass (Sfriso et al. 1988, 1989b).

3.4.4 Control of Phytoplankton by grazing and Flushing

Grazing, particularly by benthic suspension feeders, can effectively reduce phytoplankton biomass, thus acting as a natural control on effects induced by eutrophication. Grazing by suspension feeders is more effective than grazing by zooplankton, since the overwintering standing stock of suspension feeders represents a grazing capacity already present at the onset of phytoplankton development. Phytoplankton biomass is controlled by grazing when the water recycling time for the suspension feeding community is equal to the time constant of phytoplankton growth (Cloern 1982; Officer et al. 1982). This comparison of time constants is appropriate, since it takes account of the increase of the production and P/B ratio of phytoplankton caused by the acceleration of nutrient cycling by suspension feeding activity (DeVries and Hopstaken 1984).

Flushing may also reduce phytoplankton biomass by exporting it out of the water system. The same comparison of time constants can be applied for flushing. If the water residence time is shorter or equal to the time constant of phytoplankton growth, flushing will control phytoplankton biomass and prevent high chlorophyll concentrations.

If phytoplankton biomass is controlled by grazing and/or flushing, the ability of phytoplankton to compete with other primary producers not subject to grazing or flushing is affected. Grazing control and/or flushing may thus enhance the dominance of *Ulva* over phytoplankton.

An effective grazing control exists in Lake Grevelingen. The calculated annual average time constant for phytoplankton growth (the B/P ratio) is approximately 3 days, and the benthic suspension feeders are able to filter the whole volume of the lake in several days (Verhagen 1983). The phytoplankton biomass remains very low due to grazing, notwithstanding the considerable DIN concentration during winter. A large part of the winter pool of DIN is transferred and reversibly stored in the detritus pool on the sediment surface during summer. This is done mainly by biodeposition, thus increasing nitrogen limitation.

An East-West gradient in grazing control of phytoplankton exists in lake Veere, resulting in high phytoplankton concentrations in the eastern part of the lake and low concentrations in the western part. However, the highest *Ulva* coverage is also found in the eastern part of the lake. Grazing control of phytoplankton in the western part obviously does not lead to an increase of macroalgal production. This is probably due to a restricted habitat suitability for macroalgae. The western part of Lake Veere is more wind exposed and *Ulva* thalli may be torn away by shearing forces of waves and exported into the gullies.

Because of limited information on macrozoobenthic standing stock and species composition, i.e. the share of suspension feeders, no conclusive evidence exists with respect to grazing control in the Venice Lagoon. A recent survey over six transects each containing an inner, central and outer station, revealed significant macrobenthic biomass with an average value of 200 g m^{-2} wet weight (Consorzio Venezia Nuova unpub.data), which equals approximately 8 g C m^{-2} (dry wt/wet wt = 0.1 and C/DW = 0.4, approximately). A bivalve standing stock of 7 g m^{-2} ash-free dry weight, averaged over four stations in the central lagoon and over five sampling periods was measured in June 1991–September 1992 (C.J.M. Philippart, unpubl. data). Both data sets indicate a substantial standing stock of benthic suspension feeders, of approximately 3 g C m^{-2}. Assuming an effective clearance rate of 0.10 m^3 g C^{-1} day^{-1} (Jorgensen 1990), the water recycling time by suspension feeders in the Venice Lagoon is 5–6 days, and 3–4 days in the central area of the lagoon. The flushing time of the central area is also in the order of day (Sects. 3.2.1 and 3.4.3). The calculated annual average time constant for phytoplankton growth is 5–6 days.

Thus, control of phytoplankton biomass by the combined action of flushing and grazing is probably very effective in the Venice Lagoon, especially in the central part.

3.4.5 Conclusions

Eutrophication of marine coastal areas generally results in a decline of seagrasses and in an increasing dominance of macroalgae or phytoplankton. These shifts in species composition and biomass towards macroalgal and phytoplankton dominance as final algal succession stages under eutrophic conditions are also apparent in the three (semi-) enclosed polyhaline lagoons considered in this comparative study. The carbon and nitrogen mass balances of the three water systems indicate, however, that the causes of eutrophication, viz. high nutrient inputs and resulting concentrations, do not determine the specific symptoms that will emerge. The mass bloom and accumulation of *Ulva* spp. in the Venice Lagoon arises from similar nitrogen loadings and resulting DIN concentrations as in Lake Veere, where phytoplankton is dominant. Thus, factors other than nutrient loading must play a decisive role in determining the specific eutrophication symptoms which emerge under generally eutrophic conditions.

Four factors which increase the competitive capabilities of foliose macroalgae over phytoplankton are identified. Although the evidence is by no means unambiguous, all four factors appear capable of explaining the dominance hierarchy among the primary producers at comparable high levels of nitrogen input. However, it is difficult to determine which factor plays the most important role in Lake Veere and in the Venice Lagoon, as the factors involved operate simultaneously and may also interact with each other.

Climatic conditions, especially low winter temperatures which prevent the survival of macroalgal biomass in temperate regions, are suggested as causal factors for the absence of new macroalgal production in eutrophic Lake Veere. With no initial biomass present, phytoplankton and microphytobenthos take advantage of their high P/B ratio and outcompete macroalgae in the initial stage of the growing season. The sustained dominance of phytoplankton over macroalgae in Lake Veere in the regeneration production phase is consistent with the light climate, determined by the bathymetry of the lake and light extinction. Due to the comparatively large depth and episodic high turbidity, the maximum depth and bottom area where macroalgae can successfully compete with phytoplankton for light is restricted, and are estimated at 1.3–2.3 m and 25–40% of the total area, respectively. A further restriction for macroalgal development in the shallow areas is formed by the shearing force of wind–induced waves in the western part of the lake. It is suggested that export of *Ulva* biomass into the gullies by wave action is the cause of macroalgal production not being increased by the grazing control of phytoplankton by suspension feeders.

Milder climatic conditions at lower latitudes enable the survival of macroalgal biomass. The coherent storage capacity for nutrients compensating the low P/B ratio may explain the dominance of *Ulva* spp. over phytoplankton in the new production phase in the Venice Lagoon. The dominance of *Ulva* is further facilitated by the shallowness of the Venice Lagoon. Notwithstanding the high turbidity, the maximum depth and area where macroalgae can successfully compete for light with phytoplankton are estimated at 0.8–1.2 m and 40–50% of the total area, respectively. Moreover, the irradiance at larger depths is sufficiently high to allow colonization of deeper bottom areas when phytoplankton growth is impeded by factors other than light.

It is argued that the effective flushing of the central area of the lagoon is in two ways beneficial for macroalgae development. We estimate that the effective flushing time of the central area of the lagoon must be less than 1 week to supply sufficient nutrients allowing the observed initial biomass production during March–April. The same amount of flushing may effectively reduce the phytoplankton concentration by exporting phytoplankton out of the lagoon at a rate equal to or higher than the production rate. Grazing by suspension feeders probably intensifies the control of the phytoplankton biomass. With regard to the hydrodynamic aspects, flushing or water exchange ($m^3 \ s^{-1}$) which is beneficial or even necessary for nutrient supply in the production phase has to be distinguished from the absence of high shearing forces resulting from high flow velocities ($m \ s^{-1}$) and wave action in the biomass accumulation phase.

In summary, the mass blooming of *Ulva* in the Venice Lagoon and its dominance over phytoplankton at the present level of nutrient input can be explained by the combined influence of mild winters, shallowness and effective flushing of the lagoon, and grazing of phytoplankton by suspension feeders.

Acknowledgments. The authors especially wish to thank L. Postma and E. Runca for their valuable contribution to this study, and A. van Leeuwen and J. Joordens for linguistic advice and corrections. We acknowledge Consorzio Venezia Nuova, TECHNICAL, A. Sfriso and P.H. Nienhuis for their permission to use unpublished data. The figures in this chapter have been prepared by C.P.L. Haenen and E. Vennix.

References

Anonymous (1988) Eutrophication in the Mediterranean Sea: receiving capacity and monitoring of long-term effects. UNESCO Rep Mar Sci 49. UNESCO Paris

Anonymous (1989) Preliminary assessment of water quality and *Ulva* in the Venice Lagoon. A model simulation study. Delft Hydraul. Rep T604 Delft, 43pp

Anonymous (1990) ECOLUMN: Ecological column model for marine system nutrient dynamics. Delft Hydraul Rep T650, Delft, 90 pp

Apon LP (1990) Verspreiding en biomassa van het macrophytobenthos in het Veerse Meer in 1989. Delta Institute for Hydrobiological Research, Rep. 1990-92, Yerseke

Atkinson MJ, Smith Sv (1983) C:N:P ratios of benthic marine plants. Limnol Oceanogr 28: 568-574

Bach HK (1993) A dynamical model describing seasonal variation in growth and distribution of eelgrass (*Zostera marina*). I. Model theory. Ecol Model 65: 31-50

Bach HK, Malmgren-Hansen A, Birklund J (1992) Modelling of eutrophication effects on coastal ecosystems with eelgrass as the dominating macrophyte. In: Vollenweider RA, Marchetti R, Viviani R (eds) Marine coastal eutrophication. Proc Int Conf, Bologna, Italy 21-24 March 1990. Elsevier, Amsterdam, pp 993-999

Bollebakker GP, van de Kamer JPG (1989) IJking en validatie van het stratificatie-model STRESS-Veerse Meer en toepassing van het model voor de beleidsanalyse Veerse Meer. Rijkswaterstaat Tidal Waters Division, Report GWWS-88.411, Rijkswaterstaat, Middelburg

Borum J (1983) The quantitative role of macrophytes, epiphytes, and phytoplankton under different nutrient conditions in Roskilde Fjords, Denmark. Proc Int. Symp Aquat Macrophytes, 18-23 Sept, Nijmegen: pp 35-40

Borum J (1985) Development of epiphyte communities on eelgrass (*Zostera marina*) along a nutrient gradient in a Danish estuary. Mar Biol 87: 211-218

Cloern JE (1982) Does the benthos control phytoplankton biomass in South San Franciso Bay? Mar Ecol Prog Ser 9: 191-202

Cohen I, Neori A (1991) *Ulva Lactuta* biofilters for marine fishpond effluents. I. Ammonia uptake kinetics and nitrogen content. Bot Mar 34: 475-482

Cossu R, Andreottola G, Ragazzi M, Casu G (1992) Direct and indirect domestic nutrient load evaluation by mathematical model: the Venice Lagoon case study. In: Vollenweider RA, Marchetti R, Viviani R (eds) Marine coastal eutrophication. Proc Int Conf, Bologna, Italy, 21-24 March 1990. Elsevier, Amsterdam, pp 383-392

De Vries I, Hopstaken CF (1984) Nutrient cycling and ecosystem behaviour in a salt-water lake. Neth J Sea Res 18: 221-245

De Vries I, Hopstaken F, Goossens H, De Vries M, De Vries H, Heringa J (1998) GREWAQ: an ecological model for Lake Grevelingen. Delft Hydraul Rep. T215-03, Delft, 159 pp

De Vries I, De Vries M, Goossens H (1990) Ontwikkeling en toepassing VEERWAQ ten behoeve van beleidsanalyse Veerse Meer. Delft Hydraul Rep T430, Delft, 87pp

Dugdale RC, Goering JJ (1967) Uptake of new and regenerated forms of nitrogen in primary productivity. Limnol Oceanogr 12: 196–206

Fujita RM (1985) The role of nitrogen status in regulating transient ammonium uptake and nitrogen storage by macroalgae. J Exp Mar Biol Ecol 92: 283–301

Graneli E, Sundbäck K (1985) The response of planktonic and microbenthic algal assemblages to nutrient enrichment in shallow coastal waters, southwest Sweden. J Exp Mar Biol Ecol 85: 253–268

Hannewijk A (1988) De verspreiding en biomassa van macrofyten in het Veerse Meer, 1987. Delta Institute for Hydrobiological Research, Rep 1988–02, Yerseke

Harlin MM, Thorne-Miller B (1981) Nutrient enrichment of seagrass beds in a Rhode Island coastal lagoon. Mar Biol 65: 221–229

Harrison WG, Douglas D, Falkowski P, Rowe G, Vidal J (1983) Summer nutrient dynamics of the middle Atlantic Bight: nitrogen uptake and regeneration. J Plankton Res 5: 539–556

Jorgensen CB (1990) Bivalve filter feeding: hydrodynamics, bioenergetics, physiology and ecology. Olsen and Olsen, Fredensborg, Denmark, 140 pp

Kelderman P (1985) Nutrient dynamics in the sediment of Lake Grevelingen (SW Netherlands). Thesis, Univ Groningen, Delft University Press, Delft

Kelly M, Maguib M (1984) Eutrophication in coastal marine areas and lagoons: a case study of 'Lac de Tunis', UNESCO Rep Mar Sci 29. UNESCO, Paris

Klepper O (1990) A model of carbon flows in relation to macrobenthic food supply in the Oosterschelde estuary (S.W. Netherlands). Thesis Univ Wageningen, Wageningen

Lapointe BE, Littler MM, Littler DS (1987) A comparison of nutrient-limited productivity in macroalgae from a Caribbean barrier reef and from a mangrove ecosystem. Aquat Bot 28: 243–255

Lavery PS, Lukatelich RJ, McComb AJ (1991) Changes in the biomass and species composition of macroalgae in a eutrophic estuary. Est Coast Shelf Sci 33: 1–22

Littler MM, Littler DS (1980) The evolution of thallus form and survival strategies in benthic marine macroalgae: field and laboratory tests of a functional form model. Am Nat 116: 25–44

Mann KH, Chapman RO (1975) Primary production of marine macrophytes. In: Cooper JP (ed) Photosynthesis and productivity in different environments. Cambridge University Press, Cambridge, pp 207–223

Matsumoto F (1959) Studies on effects of environmental factors on the growth of 'nori' (*Porphyra tenera* Kjellm.) with special reference to water current. J Fac Anim Husb, Hiroshima Univ 2: 249–333

Menesguen A (1992) Modelling coastal eutrophication: the case of the French *Ulva* blooms. In: Vollenweider RA, Marchetti R, Viviani R (eds) Marine coastal eutrophication. Proc Int Conf, Bologna, Italy, 21–24 March 1990. Elsevier, Amsterdam, pp 979–992

Nienhuis PH (1992) Ecology of coastal lagoons in The Netherlands (Veerse Meer and Grevelingen). Vie Milieu 42: 59–72

Nilsson P, Jönsson B, Lindström Swamberg I, Sundbäck K (1991) Response of a marine shallow-water sediment system to an increased load of inorganic nutrients. Mar Ecol Prog Ser 71: 275–290

Officer CB, Smayda JH, Mann R (1982) Benthic filter feeding: a natural eutrophication control. Mar Ecol Prog Ser 9: 203–210

Orth RJ (1977) Effect of nutrient enrichment on growth of the eelgrass *Zostera marina* in the Chesapeake Bay, Virginia, USA. Mar Biol 44:187–194

Oviatt CA, Keller AA Sampou PA, Beatty LL (1986) Patterns of productivity during eutrophication: a mesocosm experiment. Mar Ecol Prog Ser 28: 69–80

Postma L (1988) DELWAQ. Users manual and technical reference, version 4.2. Delft Hydraul, Delft

Pugnetti A, Corazza C, Ceccherelli VU (1991) Dystrophic events in a lagoon of the northern Adriatic Sea: causes and recovery processes. In: Ravera O (ed) Terrestrial and aquatic ecosystems: perturbation and recovery. Ellis Horwood Series in Environmental Management, Science and Technology, New York pp 402–409

Rosenberg G, Ramus J (1982) Ecological growth strategies in the seaweeds *Gracilaria foliifera* (Rhodophyceae) and *Ulva* sp. (Chlorophyceae): soluble nitrogen and reserve carbohydrates. Mar Biol 66: 251–259

Rudek J, Paeri HW, Mallin MA, Bates PW (1991) Seasonal and hydrological control of phytoplankton nutrient limitation in the lower Neuse River estuary, North Carolina. Mar Ecol Prog Ser 75: 133–142

Seitzinger SP (1987) Nitrogen biogeochemistry in an unpolluted estuary: the importance of benthic denitrification. Mar Ecol Prog Ser 41: 177–186

Sfriso A (1987) Flora and vertical distribution of macroalgae in the lagoon of Venice: a comparison with previous studies. G Bot Ital 121: 69–85

Sfriso A, Pavoni B, Marcomini A, Orio AA (1988) annual variations of nutrients in the Lagoon of Venice. Mar Pollut Bull 19(2): 54–60

Sfriso A, Marcomini A, Pavoni B, Orio AA (1989a) Macroalgal production and nutrient recycling in the Lagoon of Venice. Ing San 5: 256–266

Sfriso A, Pavoni B, Marcomini A (1989b) Macroalgae and phytoplankton standing crops in the central Venice Lagoon: primary production and nutrient balance. Sci Total Environ 80: 139–159

Sfriso A, Raccanelli S, Pavoni B, Marcomini A (1991) Sampling strategies for measuring macroalgal biomass in the shallow waters of the Venice Lagoon. Environ Techn 12: 263–269

Sfriso A, Marcomini A, Pavoni B, Orio AA (1993) Species composition, biomass and net primary production in shallow coastal waters: The Venice Lagoon. Biores Technol 44 (3): 235–250

Shepherd SA, McComb AJ, Bulthuis DA, Neverauskas V, Steffensen DA, West R (1989) Decline of seagrasses. In: Larkum AWD, McComb AJ, Shepherd SA (eds) Biology of seagrasss. Elsevier, Amsterdam, pp 346–393

Silberstein K, Chiffings AW, McComb AJ (1986) The loss of seagrass in Cockburn Sound, western Australia. III. The effects of epiphytes on productivity of *Posidonia australis* Hook. F.. Aquat Bot 24: 355–371

Smith SV (1991) Stoichiometry of C:N:P fluxes in shallow–water marine ecosystems. In: Cole J, Lovett G, Findlay S (eds) Comparative analyses of ecosystems. Patterns, mechanisms and theories. Springer, Berlin Heidelberg New York, pp 259–286

Stronkhorst J, Duin R, Hass H (1985) Primaire produktie onderzoek in het Veerse Meer (1982–1983). Report Tidal Waters Division, DDMI-85.10, Rijkswaterstaat, Middelburg

Technital (1992) Progetto generale degli interveni sulla morfologia. VE3910, unpubl

Vegter F, De Visscher PRM (1984) Seasonal periodicity in phytoplankton primary production in brackish Lake Grevelingen (SW Netherlands) during 1976–1981. Neth J. Sea Res 18: 246–259

Verhagen JHG (1983) A distribution and population model of the mussel *Mytilus edulis* in Lake Grevelingen. In: Lauenroth WK, Skogerboe GV, Flug M (eds) Analysis of ecological systems: State-of-the-art in ecological modelling. Elsevier Amsterdam, pp 373–383

Verhagen JHG, Nienhuis PH (1983) A simulation model of production, seasonal changes in biomass and distribution of eelgrass (*Zostera marina. L.*) in Lake Grevelingen. Mar Ecol Prog Ser 10: 187–195

Vermaat JE Sand-Jensen K (1987) Survival, metabolism and growth of *Ulva Lactuta* under winter conditions: a laboratory study of bottlenecks in the life cycle. Bot Mar 95: 55–61

Wassmann P (1986) Benthic nutrient regeneration as related to primary productivity in the west Norwegian coastal zone. Ophelia 26: 443–456

Zmudzinski L (1992) Present state of the Baltic Sea ecosystem. Biol Mor Inst Ryb/Bull Sea Fish Inst 1: 53–62

4 Aquaculture Methods for Use in Managing Eutrophicated Waters

J. E. Merrill

4.1 Introduction

Macroalgae (seaweeds) are important components of nearshore marine and brackish water ecosystems throughout the world. Many macroalgae are highly productive, effectively utilizing solar energy to fix carbon into biomass. In this process, they absorb nutrients and other substances from the water, incorporating them into the thallus biomass. This biomass can serve as habitat or food for fish, shellfish, or other organisms, and if harvested can provide raw material for energy production, human food, animal fodder and industrial feedstocks for chemical production (Chapman and Chapman, 1980; Waaland 1981; Guiry and Blunden 1991). A considerable amount of recent attention has been focused on the problems that can occur when blooms of these macroalgae result from eutrophication caused by wastewater discharge and agricultural runoff. In such circumstances, the macroalgal biomass can represent a significant nuisance, impeding navigation, fouling recreational areas, and upsetting the ecological/biochemical balance in the marine system. Such problems, serious now, are sure to become even worse in the future as populations expand and industrialization spreads (Maurits la Rivière 1989). In spite of these negative effects, the environmental benefits of macroalgae, particularly their capacity to improve water quality, should be more effectively managed for beneficial purposes.

The idea of using aquatic plants for wastewater treatment in conjunction with traditional primary or secondary sewage treatment facilities is well established, and many projects are in place that utilize this concept on an operational or experimental basis (e.g. Reddy and Smith 1987; Brix and Schierup 1989). These systems rely on fresh water aquatic macrophytes

Department of Microbiology, 178 Giltner Hall, Michigan State University, East Lansing, MI 48824, USA

Ecological Studies, Vol. 123
Schramm/Nienhuis (eds) Marine Benthic Vegetation
© Springer-Verlag Berlin Heidelberg 1996

such as water hyacinth (*Eichhornia* sp.), and duckweed (*Lemma* sp.); emergent aquatic plants, such as reed (*Phragmites* sp.) or cattail (*Typha* sp; Wolverton 1987); or fresh water microalgae such as *Chlorella, Spirulina, Scenedesmus, Euglena,* etc. (Abeliovich 1986; Oswald 1988; de la Noüe et al. 1992). Similar considerations with marine plants are rare and the scientific background is correspondingly less well developed (Ryther et al. 1979; Ryther 1983; Schramm 1991; Subander et al. 1993). To a large extent, a seaweed bed can be functionally similar to a wetland with regard to the beneficial characteristics of nutrient removal and water quality improvement. There is no reason why discussions of "artificial wetlands" should not encompass artificial marine wetlands composed of macroalgal species and their associated biota. This may be especially important where it is not possible nor practical to intercept wastewater streams before they enter the marine environment (i.e., non-point sources). In such circumstances, artificial plantings of macroalgae could constitute effective in situ phytodepuration systems.

Much of the emphasis applied to finding solutions to the problems of excess macroalgae growth has been focused on harvest and disposal or utilization of the algal biomass causing the problem (Briand 1991; Morand et al. 1991). In this report I will argue for application of an "agricultural imperative" to the consideration of macroalgal blooms in eutrophicated marine systems. By this I mean that the essential criterion of a eutrophicated system, enrichment with minerals that are plant nutrients, should be viewed as a potentially valuable resource. A terrestrial farmer, when confronted with the rampant growth of weeds or other vegetation, would probably not ask: "What value can I get from these weeds?", but rather: "What valuable crop could I plant here to take advantage of the obviously fertile conditions?" If such an approach can be adopted, with significant development of macroalgal seafarms in affected areas, water quality can be expected to improve through the harvest of useful biomass, and this at potentially no cost to governments. This would mean a great benefit compared to the costs (for harvest and disposal) or losses (from reduced tourism and fisheries revenues) currently incurred from some of the most severe examples of marine macroalgal blooms.

4.2   The "Green Tide" Phenomenon

A frequent effect of eutrophication in coastal marine ecosystems is the proliferous growth of macroalgae. Such a response to nutrient enrichment should normally be considered a natural and healthy response. However, in the more extreme cases, the resulting algal biomass can create ecological imbalances that can result in significant problems, including direct fouling of recreational beaches and navigation channels, and the indirect impacts of hypoxia and release of noxious gases (principally hydrogen sulfide) during their eventual death and anaerobic decay (Briand 1987; Merrill and Fletcher 1991). A notable form in which the proliferous biomass is sometimes observed is exemplified along the Brittany coast of France where *Ulva* biomass accumulates in extensive windrows along the beaches, leading to the use of the descriptive term "green tide". This name might well be applied to other excess macroalgae situations in order to bring attention to their similarities as well as the commonalities of cause, even though such phenomena are not always the result of green algae, nor are they always associated with tidal events (a model adapted from the concept of "red tides" associated with dinofagellate blooms; Merrill and Fletcher 1991).

The conditions which lead to green tides typically include: (1) high inorganic nutrient fluxes, particularly nitrogen; and (2) relatively shallow water, without strong currents or waves (Sfriso et al. 1987). These conditions are found throughout the world wherever bays or lagoons with suitable physical characters are impacted by the eutrophicating affects of agricultural runoff or human sewage discharge.

The most common macroalage leading to green tides are species of the green alga *Ulva*, although significant additional biomass may come from other green algae such as *Enteromorpha* or *Monostroma*. It is a somewhat unique characteristic of *Ulva* that it can thrive without any attachment to the substratum whatsoever. This is a particularly important point since the types of locations where green tides are most significant tend to have mud, silt, or sand bottoms as a result of the relative lack of water movement. The ability to thrive unattached, coupled with a strong ability for nitrogen uptake from the seawater, contribute to its domination in the habitats previously mentioned. Specific inhibitory effects of *Ulva* upon other algae may also contribute to its dominance (Svirski et al. 1993).

In many places direct action has been necessary to reduce or eliminate the damaging impacts of green tides. Such action typically consists of harvesting or otherwise removing the offending algal mass (Fig. 4.1). This type of harvesting has reached major proportions in some places, for example, Venice Lagoon where amounts in excess of 200 tons/day are removed over extended periods through spring, summer and fall seasons (Sfriso et al. 1987), and the Brittany coast of France, where 98 000 m^3 were removed in 1991, involving 25% of the coastal communitites in the area (Dion and Le Bozec, Chap. 9, this vol.). Intervention on such a massive scale is achieved only at tremendous expense. The harvest of *Ulva* from Venice

Fig. 4.1. Harvest of problem *Ulva* biomass in Goro, Po River Delta area, Italy, June 1990. Small harvest boat in foreground offloading harvested biomass to transport barge

Lagoon in 1989 involved approximately 50 workers operating 39 boats (14 harvesters, 25 transporters) at a cost of some US $5.6 million (Fletcher 1990). Costs for removal of green algal biomass along the Brittany coast reached FF 3.2 million in 1991 (P. Dion, pers. comm.). It is thus imperative that ways be found to reduce the costs associated with the treatment of green tides.

In is unfortunate that the algae typically associated with green tides are ones for which there is no significant market potential at present, for if these seaweeds had intrinsic value, their removal and disposal could be the basis of a viable maritime industry. Although schemes for processing green tide biomass into valuable commodities, such as biogas (Croatto 1983), or compost (Briand 1987), have been proposed, in general these do not achieve sufficient value to justify the costs of development. The agricultural imperative would have us look not for ways to create value from the naturally occurring "weeds", but rather look for valuable agricultural crops, suited to climatic and market conditions, that could be planted and managed to the farmers' economic benefit. In most circumstances it should be possible to cultivate more valuable species in these problem areas through the use of aquaculture methods.

4.3 Macroalgal Cultivation

Macroalgae for commercial applications have traditionally been obtained from two sources: harvests from naturally occurring wild beds or aquacultural production. While these categories provide a convenient framework into which to categorize raw material sources, they might justifiably be considered to be just the two extremes in a continuum spanning a wide range of intermediate levels of management complexity (Fig. 4.2). Introduction of stones or other solid substrate into habitats otherwise dominated by sand or mud bottoms is used as a means of increasing the extent of macroalgal beds and might be considered a small step beyond simple wild harvest. A somewhat higher level of effort might include pre-inoculation of such substrate with desired spores prior to distribution in the sea. Collection of naturally occurring spores during appropriate seasons and transport of the resulting sporelings to other areas was widely practiced during the early history of *Porphyra* cultivation in Asia. Complete management over all life history stages and manipulation of environmental factors to control sporulation, settlement, growth, etc., such as applied in the modern methods for many cultivated macroalgae are examples of what might be classed "full cultivation". Still further levels of sophistication, however, are represented by such systems as onshore tank or pond cultivation methods, or even the still mostly experimental cases of in vitro cell tissue culture, immobilized cells or protoplasts. In this regard, when evaluating the potential for macroalgal production at a given site,

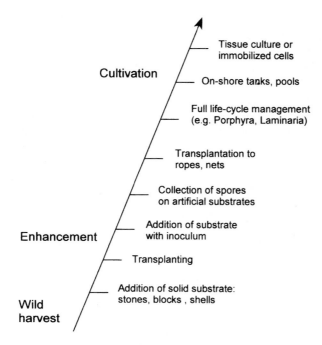

Cultivation
— Tissue culture or immobilized cells
— On-shore tanks, pools
— Full life-cycle management (e.g. Porphyra, Laminaria)
— Transplantation to ropes, nets
— Collection of spores on artificial substrates
Enhancement
— Addition of substrate with inoculum
— Transplanting
— Addition of solid substrate: stones, blocks, shells
Wild harvest

Fig. 4.2. Conceptual schematic for range of intervention levels based on management of macroalgal populations. Intervention at a specific site should consider applicability and feasibility of various techniques along this continuum

consideration should be given to the full range of available options along this continuum.

On a worldwide basis, many different species of macroalgae are cultivated (Waaland 1981; Ryther 1985; Perez et al. 1992). Most of the types cultivated successfully on a commercial scale are those used for human food, particularly in Asia (Abbott 1988). The best known examples include the red alga *Porphyra* ("nori"), and the brown algae *Undaria* ("wakame") and *Laminaria* ("kombu"); (Miura 1980; Waaland 1981; Arasaki and Arasaki 1983). Other species have been cultivated as sources for valuable extractive products, e.g. the red algae *Eucheuma* for caragheenan and *Gracilaria* for agar (Jansen 1979; Hanisak and Ryther 1984; Bird and Benson 1987; Lewis et al. 1988).

Ideally, the species selected for cultivation would be one for which commercial methods were already established, otherwise a potentially long and costly "domestication" process might be required prior to commercial scale cultivation. In most cases utilization of a locally occurring species will be preferable to introduction of a nonnative species due to the current intenstity of opposition to species introductions. Fortunately, many of the most productive cultivated genera have very wide geographic distribution, and it is therefore likely that for a given intervention target area, there will be a useful local representative that can be brought into production with a relatively modest research and development effort.

A key consideration in every example of seaweed farming is the selection and optimization of a suitable substrate. The sea farm grounds are often in areas where the same species would not otherwise occur because of the lack of suitable substrate materials at the appropriate water depth. The exception to this principle is in tank or pond cultivation where vigorous water mixing by aeration or other means allows free floating culture.

If we consider the conditions present in a typical bay or lagoon that is subject to repeated green tide occurrences we know that, by definition, the overall annual nutrient budget of the water system is excellent for support of algal growth. These nutrients come either from current sources of runoff and discharge or from mineral cycling processes within the sediments (cf. Bianchi et al. 1988). In many cases, by providing a suitable substratum, and by providing adequate inoculum, we can cause desirable species of seaweed to dominate. As with any agricultural crop, however, it is essential to understand the particular biological requirements of the target species in order to successfully manage it under cultivation, especially if one of the major objectives is to out-compete the green tide.

Some examples will serve to demonstrate typical methods of seaweed cultivation and how these might be applied in eutrophicated bays and lagoons.

4.3.1 *Porphyra* ("nori") Cultivation

The most valuable cultivated seaweed on a worldwide basis is *Porphyra* (Chapman and Chapman 1980; Perez et al. 1992; Merrill 1993). Selected strains of *P. yezoensis* and *P. tenera* are farmed on a very large scale in Japan as well as in Korea and China for the production of the edible product "nori" (Miura 1975; Mumford and Miura 1988). Nori is a key ingredient in the Japanese dish called "sushi", in which nori is used as a flavorful wrapper around a filling of vinegared rice and various vegetables or fish (Arasaki and Arasaki 1983).

Nori is grown on nets (Fig. 4.3) that are suspended just at the water surface (Miura 1975; Mumford and Miura 1988; Merrill 1989; Perez et al. 1992). These nets can be supported on poles driven into shallow mud or sand bottoms, or in floating gridworks of rope in deeper sites. Key environmental requirements for the cultivation of currently valuable strains include temperatures in the range of 6–16°C and salinitites slightly below full-strength seawater, 25–32 ppt. An essential aspect of the techniques for nori farming is the need to provide periodic drying to the juvenile plants as a means of eliminating competing species. This is achieved either by the normal exposure during low tides in the case of pole culture, or by controlled lifting of the nets in the case of floating culture.

Nori is a very fast growing plant, requiring only 45 days from spore settlement to first harvest. Subsequent harvests can be taken at roughly

Fig. 4.3. *Porphyra* (nori) cultivation on artifical nets. Shown are nets at ca. 30 days after seeding. Nets are normally deployed at water surface; here being lifted for drying treatment on floating nursery frameworks. October, 1987; Puget Sound, Washington State, USA

10-day intervals. It has a very high protein (hence nitrogen) content, up to 45% by dry weight. The productivity and nitrogen uptake of this plant are sufficiently great that it is an excellent choice for eutrophication abatement where other conditions are suitable.

4.3.2   Kelp Species, e.g. *Laminaria, Undaria*

Laminaria ("kombu") and *Undaria* ("wakame") are kelps cultivated principally in Asian countries (Druehl 1988). These are used primarily in preparing soups and as flavoring in other dishes (Arasaki and Arasaki 1983). They have markets that are almost as large as that for nori. These and other kelp species may also have potential value for extraction of alginate, although alginate is currently obtained entirely from harvests of wild stocks.

Kelps are typically grown on large diameter (15–20 mm) rope long-lines (Fig. 4.4) which are suspended at suitable depths below the surface (Kawashima 1984; Druehl et al. 1988; Merrill and Gillingham 1991a). Inocculation of the long-lines is achieved by careful preparation of seed-stock inoculum in on-shore facilities. Although the time required from seeding to harvest may be longer than for nori, the final biomass may be equivalent or greater.

Fig. 4.4. Typical kelp cultivation by long-line method. Pictured plants (*Nereocystis luet-keana*) are ca. 90 days from zoospore settlement on seedling twine. Long-line, typically depolyed below water surface, here being litted for evaluation. November, 1990; Puget Sound, Washington state, USA

Protein content is lower in the kelps, typically in the range of 10–20% by dry weight. Their value in pollution abatement is augmented by their value as habitat for fish and shellfish (Merrill and Gillingham 1991b).

4.3.3 *Gracilaria* and Other Agarophytes

Gracilaria is one of the principal sources of agar and is consequently one the most commercially important of all macroalgae other than the main food sources. Cultivation efforts in Chile and Taiwan are particularly notable as examples of the technology applied (Santelices and Doty 1989). Extensive beds of *Gracilaria* have been planted in Chile using sand-filled plastic tubing into which have been inserted small vegetative plants. Areas thus planted may support repeated croppings (Fig. 4.5).

In Taiwan, shallow tidal ponds, originally constructed for the aquaculture production of fish or shellfish, have been converted to *Gracilaria* production (Shang 1976; Chiang 1981). Vegetative fragments are used as stocking material with harvests at 10–30 day intervals after initial development. Nutrient enrichment of the ponds is frequently supplied by fermented animal manures or effluent from adjacent fish or shellfish ponds. This highlights one of the key advantages of pond or tank cultivation as well as its principal limitation: this method is best suited to the treatment of point-source discharges of wastewater before they enter the open environment, rather than those of non-point origin as often associated with agricultural runoff.

Fig. 4.5. *Gracilaria* harvest by "hookah" divers. This harvest is from artificially planted beds at Maullin River, near Puerto Montt, Chile; July 1993

An important advantage of *Gracilaria* for use in water quality improvement is its ability to tolerate a wide range of salinities and relatively low water motion; conditions that would be considered unfavorable for other species, but which are fairly common in coastal areas subject to damaging green tides. Many species of *Gracilaria* are also tolerant of relatively high temperatures (up to 30°C) and of wide temperature ranges. There are species of *Gracilaria* native to most of the world's coastlines and it is highly likely that strains with adequate commercial characteristics can be found in a given area.

4.3.4   Additional Alternatives

While the examples cited above represent cases of complete cultivation, there may be circumstances in which controlled cultivation is uneconomical, or is undesirable for other reasons. In such cases, it may be possible to apply simplified techniques toward "enhancement" of desired species. For example, in some shallow lagoons where unattached populations of *Ulva* dominate, the simple addition of shell fragments or pebbles, with or without pre-inoculation with spores, may be sufficient to stimulate a shift to populations of *Gracilaria* or *Gracilariopsis* in that these species seem to prefer at least some attachment point.

4.4   Nutrient Removal

Ryther (1983) has calculated that a one-hectare *Gracilaria* farm is capable of removing all of the nitrogen and much of the phosphorus from 350 m^3 of wastewater per day, equivalent to the output of 1000 people. The nutrient loading of fish cultivation and processing wastes have been effectively reduced by natural or cultivated macroalgal populations (Markovtsev and Krupnova 1988; Subandar et al. 1993).

A useful framework within which to consider the beneficial value of nutrient uptake and removal is that of Algal Biomass Potential (ABP; Oswald 1988). In the presence of sufficient quantities of other elements, a single nutrient may support a finite quantity of biomass production by a given species. For example, ABP can be defined for nitrogen as follows:

$$\text{ABP (algae mg dm}^{-3}) = \frac{\text{available N (mg dm}^{-3})}{\text{algal N content (\%)}}. \tag{1}$$

If we consider a system supporting growth of *Ulva* containing approximately 3.5% N with a nitrogen flux of 20 mg dm^{-3}, we can calculate the potential biomass ABP per unit volume as follows:

$$\frac{20 \text{ mg dm}^{-3} \text{N}}{3.5\% \text{ N}} = 571 \text{ mg dm}^{-3} \text{ ABP } Ulva. \tag{2}$$

Furthermore, in a shallow lagoon of 2 m depth, for example, we can calculate from the above value the ABP per unit area as 1.14 kg m^{-2}, or 11.4 t ha^{-1}. Interestingly, the *Ulva* biomass production of Venice Lagoon has been estimated as 18.5 t ha^{-1} (Sfriso et al. 1987).

It follows that any proportion of the ABP that can be utilized by a harvested crop will reduce that remaining for support of the growth of "problem" algae such as *Ulva*. To illustrate the nutrient uptake value of nori, we can make the following calculations for commercial nori production:

$$(450 \text{ g } Porphyra \text{ m}^{-2}) \ (0.07 \text{ S N g}^{-1} \ Porphyra) = 31.5 \text{ g N m}^{-2} \tag{3}$$

This production figure of $450 \text{ g m}^{-2} \text{ year}^{-1}$ dry weight is for a typical 5-month harvest season and is based on a system in which the actual net substrate only covers approximately 40% of the surface area occupied by the total cultivation system. Again, we can compare this nitrogen uptake value to those estimated for *Ulva* in Venice Lagoon of $50-70 \text{ g N m}^{-2}$ (Sfriso et al. 1987).

The removal of this quantity of nutrients from the water should have the immediate beneficial result of reducing the ABP of problem species. It is particularly important to note that even when complete replacement of problem species is not feasible, partial replacement may be sufficient to reduce the total algal biomass below the threshold of hypertrophic events. Furthermore, by continued harvests in successive years, it could be expected that significant amounts of nutrients could be removed from the sediment load.

4.5 Conclusion

There is a widespread need for active intervention in eutrophicated marine ecosystems in order to reduce the occurrence of green tides, hypoxia and other ill effects. A significant contribution to such efforts can be made by using aquaculture techniques to alter the species composition in the target area. By cultivating species with intrinsic economic value, the great costs associated with harvest and removal of undesirable species can be reduced or eliminated. If we adopt the agricultural imperative our focus changes from that of "what can we do with the seaweed that is here?" to "what valuable seaweed crop can we produce in these fertile waters?" Although the technologies for macroalgal seafarming are admittedly less well developed than are those for terrestrial plant crops, the systems that currently exist provide us with numerous options for either direct application or as starting points for development of new technologies for unique circumstances.

In the context of this report, combatting eutrophication and reducing macroalgal blooms are primary objectives. Nutrient uptake will place the cultivated species in direct nutrient competition with the naturally occurring bloom-forming species. The nutrients absorbed by the cultivated algae are removed from the ecosystem through harvest. When the method chosen for cultivation involves off-bottom suspension of nets or ropes as the cultivation substrate there will be an additional advantage in the competition for sunlight. The overall biomass of problem-causing species should decrease by these actions. In those situations where damaging phenomena occur only when the green tide biomass exceeds certain threshold levels (as in the case of hypoxia), the partial reduction in biomass resulting from competition with cultivated macroalgae may significantly alleviate the critical phenomena, even though a complete elimination of the causative species is not achieved. Furthermore, in the absence of any harvesting of biomass, the nutrients that contributed to the bloom are typically recycled back into the water column or the sediments when the algae die and decay. Such cycling of nutrients is a major contributor to successive blooms. Thus, any harvest of algal biomass will be beneficial with respect to decreasing the possibility of future blooms since such harvests will reduce that portion of the nutrient budget contributed by cycling.

A key element in any consideration of possible pollution abatement plans will be the overall costs associated with any specific action. I have summarized some of the more dramatic examples of green tides and the costly efforts applied to their abatement. Sea farming of macroalgae offers the possibility for water quality improvement to be accomplished at a net profit; in fact, with establishment of a permanent new industry in the affected communities. It is highly likely that government financial assistance will be required, especially during the planning and evaluation stage of any new endeavor, but these expenses should be a one-time investment, and more than justified in comparison to the recurrent costs of harvesting and disposal of low value green tide biomass.

References

Abbott IA (1988) Food and food products from seaweeds. In: Waaland JR, Lembi CA (eds) Algae and human affairs. Cambridge University Press, Cambridge, pp 331–338

Abeliovich S (1986) Algae in wastewater oxidation ponds. In: Richmond A (ed) Handbook of microalgal mass culture. CRC Press, Boca Raton, pp 331–338

Arasaki S, Arasaki T (1983) Vegetables from the sea. Japan Publications, Tokyo

Bianchi M, Van Wambeka F, Bianchi A (1988) Heterotrophic processes in eutrophicated aquatic marine environments. Prog Oceanog, 21: 159–166

Bird KT, Benson PH (eds) (1987) Seaweed cultivation for renewable resources. Elsevier, New York

Briand X (1987) Stranded algae and disposable biomass. In: Aquatic primary biomass (marine macroalgae): biomass conversion, removal and use of nutrients. Proc Ist Worksh COST 48 subgroup 3. Commission of the European Communities, Brussels, pp 13–16

Briand X (1988) Exploitation of seaweeds in Europe. In: Morand P, Schulte EH (eds) Aquatic primary biomass (marine macroalgae): biomass conversion, removal and use of nutrients. I. Proc 1st Worksh of the COST 48 Sub-Group 3. L'Hommeau, France, 12–14 Feb 1987. Commission of the European Communities, Brussels, pp 53–65

Briand X (1991) Seaweed harvesting in Europe. In: Guiry MD, Blunden G (eds) Seaweed resources in Europe: uses and potential. John Wiley Chichester, pp 259–308

Brix H, Schierup H-H (1989) The use of aquatic macrophytes in water pollution control. Ambio 18: 100–107

Chapman VJ, Chapman DJ (1980) Seaweeds and their uses, 3rd edn. Chapman & Hall, London

Chiang YM (1981) Cultivation of *Gracilaria* (Rhodophycophyta, Gigartinales) in Taiwan. Proc Int Seaweed Symp 10: 569–574

Croatto U (1983) Energy from macroalgae of the Venice Lagoon. In: Strub A, Chartier P, Schleser G (eds) Energy from biomass. 2nd EC Con, Commission of the European Communities, Brussels, pp 329–333

de la Noüe J, Laliberté G, Proulx D (1992) Algae and Waste water. J Appl Phycol 4: 247–254

Druehl LD (1988) Cultivated edible kelp. In: Waaland JR, Lembi CA (eds) Algae and human affairs. Cambridge University Press, Cambridge, pp 119–134

Druehl LD, Baird R, Lindwall A, Lloyd KE, Palula S (1988) Longline cultivation of some Laminariaceae in British Columbia, Canada. Aquacult Fish Manager 19: 253–263

Fletcher RL (1990) The "green tide" problem: a review. Proc 2nd Worksh Marine Biotechnology on Eutrophication and biotransformation in coastal waters. Sorrento, Italy, 18–24 Nov 1990, Ecolmare, pp 100–105

Guiry MD, Blunden G (eds) (1991) Seaweed resources in Europe: uses and potential. John Wiley, Chichester

Hanisak MD, Ryther JH (1984) The experimental cultivation of the red seaweed *Gracilaria tikvahiae* as an "energy crop": an overview. In: Barclay WR, Mclntosh RP (eds) Algal biomass technologies: an interdisciplinary perspective. Cramer, Berlin, pp 212–217

Jansen A (1979) Industrial utilization of seaweeds in the past, present and future. Proc Int Seaweed Symp 9: 17–34

Kawashima S (1984) Kombu cultivation in Japan for human foodstuff. Jpn J Phycol 32: 379–394

Lewis JG, Stanley NF, Guist GG (1988) Commercial production and applications of algal hydrocolloids. In: Waaland JR, Lembi CA (eds) Algae and human affairs, Cambridge University Press, Cambridge, pp 205–236

Markovtsev VG, Krupnova TN (1988) Biological principles of the cultivation of *Laminaria* algae for purification of fish processing plant effluent. Hydrobiol J 24: 95–99

Maurits la Rivière JW (1989) Threats to the world's water. Sci Am 261: 80–84

Merrill J (1993) Development of nori markets in the Western world. J Appl Phycol 5: 149–154

Merrill J, Fletcher R (1991) Green tides cause major economic burden in Venice Lagoon, Italy. Appl Phycol Forum 8: 1–3

Merrill JE (1989) Commercial nori (*Porphyra*) sea farming in Washington State. In: Kain JM, Andrews JW, McGregor BJ (eds) Aquatic primary biomass – marine macroalgae: outdoor seaweed cultivation. Proc COST 48 Worksh on Outdoor seaweed cultivation. EEC, Brussels, pp 90–100

Merrill JE, Gillingham DM (1991a) A handbook of bull kelp cultivation. National Coastal Resources Research and Development Institute, Newport, Oregon

Merrill JE, Gillingham DM (1991b) Seaweed management systems for use in habitat restoration, environmental management, and mitigation. In: Proc Puget Sound Research '91. Seattle, Jan 4–5, 1991, Puget Sound Water Quality Authority, Seattle, pp 354–363

Miura A (1975) Porphyra cultivation in Japan. In: Tokida J, Hirose H (eds) Advance of phycology in Japan. Junk, The Hague, pp 273–304

Miura A (1980) Seaweed cultivation: present practices and potentials. In: Borgese EM, Ginsberg N (eds) Ocean yearbook 2. University of Chicago Press, Chicago, pp 57–68

Morand P, Carpentier B, Charlier RH, Mazé J, Orlandini M, Plunkett BA, De Waart J (1991) Bioconversion of seaweeds. In: Guiry MD, Blunden G (eds) Seaweed resources in Europe. John Wiley, Chichester, pp 95–148

Mumford TF Jr, Miura A (1988) *Porphyra* as food; cultivation and economics. In: Waaland JR, Lembi CA (eds) Algae and human affairs. Cambridge University Press, Cambridge, pp 87–117

Oswald WJ (1988) Micro-algae and waste-water treatment. In: Borowitzka MA, Borowitzka LJ (eds) Micro-algal biotechnology. Cambridge University Press, Cambridge, pp 305–328

Perez R, Kaas R, Campello F, Arbault S, Barbaroux O (1992) La culture des algues marines dans le monde. L'Institut Français de Resherche Pour L'Exploitation de la Mer (IF-REMER), Brest

Reddy KR, Smith WH (eds) (1987) Aquatic plants for wastewater treatment: an overview. Magnolia Publishers, Orlando

Ryther JH (1983) The evolution of integrated aquaculture systems. J World Maricul Soc 14: 473–484

Ryther JH (1985) Technology for the commercial production of macroalgae. In Lowenstein MZ (ed) Energy applications of biomass. Elsevier, New York, pp 177–188

Ryther JH, DeBoer JA, Lapointe BE (1979) Cultivation of seaweeds for hydrocolloids, waste treatment and biomass for energy conservation. Proc Int Seaweed Symp IX 9: 1–17

Santelices B, Doty MS (1989) A review of *Gracilaria* farming. Aquaculture 78: 95–133

Schramm W (1991) Seaweeds for waste water treatment and recycling of nutrients. In: Guiry MD, Blunden G (eds) Seaweed resources in Europe: uses and potential. John Wiley Chichester, pp 149–168

Sfriso A, Marcomini A, Pavoni B (1987) Relationships between macroalgal biomass and nutrient concentrations in a hypertrophic area of the Venice Lagoon, Mar Environ Res 22: 297–312

Shang YC (1976) Economic aspects of *Gracilaria* culture in Taiwan. Aquaculture 8: 1–7

Subandar A, Petrell RJ, Harrison PJ (1993) *Laminaria* culture for reduction of dissolved inorganic nitrogen in salmon farm effluent. J Appl Phycol 5: 455–463

Svirski E, Beer S, Friedlander M (1993) *Gracilaria conferta* and its epiphytes. 2. Interrelationships between the red seaweed and *Ulva* cf. *lactuca*. Hydrobiologia 260/261: 391–396

Waaland JR (1981) Commercial utilization. In: Lobban CS, Wynne MJ (eds) The biology of seaweeds. University of California Press, Berkeley, pp 726–741

Wolverton BC (1987) Aquatic plants for wastewater treatment: an overview. In: Reddy KR, Smith WH (eds) Aquatic plants for water treatment and resource recovery. Magnolia Publishing, Orlando, pp 3–15

Part B
Regional Observations and Investigations

5   The Baltic Sea and Its Transition Zones

W. Schramm

5.1   Introduction

The Baltic Sea and its transition zones probably range among the most encumbered marine waters around the world.

Bordered by nine countries, most of them densely populated and highly industrialized, and not far away from the centres of air pollution in Europe, this brackish inland sea is particularly endangered and susceptible to man made eutrophication and pollution due to its specific hydrographical conditions and geomorphological features. During the past two decades, an increasing number of observations have been made, which indicate the response of the ecosystem Baltic to these man-made impacts.

The international concern of the Baltic states about the alarming signs and symptoms of pollution and eutrophication led to the Helsinki Convention on the protection of the Baltic environment (Baltic Marine Environment Protection Commission, HELCOM) in 1974.

Since the beginning of the 1980s, several extreme oxygen-depletion events with hydrogen sulfide development and killing of fish and macro-zoobenthos, as for example in the Belt Sea and the Western Baltic (Ehrhardt and Wenck 1984; Weigelt and Rumohr 1986), initiated a number of national governmental research projects in Denmark, Sweden and Germany.

At the end of the 1960s, and increasingly from the end of the 1970s, observations on changes of phytobenthic communities in the Baltic were reported and related to human impact on the Baltic ecosystem (Schwenke 1969; Rönnberg 1981; Kangas et al. 1982; Haathela 1984).

Institut für Meereskunde, Universität Kiel, Düsternbrooker Weg 20, 24105 Kiel, Germany

Ecological Studies, Vol. 123
Schramm/Nienhuis (eds) Marine Benthic Vegetation
© Springer-Verlag Berlin Heidelberg 1996

5.2   Physical and Chemical Characterization

The Baltic Sea including the transition zones to the North Sea (Belt Sea, Sound) comprises an area of 3.95×10^5 km^2 and has a volume of 2.12×10^4 km^3 (Fig. 5.1). The average depth is only 54 m, with a maximum depth of about 400 m north of the Island of Gothland. The entire Baltic can be described as a series of shallow basins, separated from each other by sills.

A salinity gradient can be observed, with decreasing salinities from polyhaline conditions in the Belt Sea (15–18 PSU) to oligohaline conditions in the northernmost Bothnian Bay with a mere 2–3 PSU (Fig. 5.2).

The salinity gradient is determined by the irregular inflow of salt-rich North Sea Water (32 PSU) and the land run-off of freshwater from the drainage basins around the Baltic. Usually, the freshwater discharge exceeds the seawater inflow, resulting in a continuous outflow of low-salinity surface water from the Baltic into the Kattegat.

As far north as the Aland Sea, the Baltic is stratified, with a permanent primary halocline between 50–70 m depth. During summer, a rather stable thermocline occurs at 10–15 m depth, below which a layer of cold "winter water" extends down to the permanent halocline.

The relative stability of these discontinuity layers determines the typical seasonal variation of the nutrient situation in the open Baltic. In spring, with increasing light intensities, plankton blooms develop, causing in the upper layer down to the thermocline a rapid decrease of nutrient concentrations close to detection limits in early summer. Usually, the breakdown of the spring bloom is followed by nutrient regeneration processes in the water column, often causing a second phytoplankton bloom in late summer. During autumn and winter, storms and thermal convection cause mixing of the nutrient-rich deep water from below the discontinuity layers with the surface layer, replenishing nutrients in the water column (von Bodungen 1986).

A different nutrient pattern has been observed in macrophytic communities, probably as a result of intensive nutrient regeneration through benthic heterotrophic activity with the possibility of rapid recycling within the phytobenthic communities. Compared to the surface water in offshore stations, in red algae or *Fucus* communities from Kiel Bight, for example, nutrient levels were significantly higher, and the spring-early summer minimum occurred several weeks later compared to the surface water in offshore stations (Fig. 5.3; Schramm et al. 1988). Kautski and Wallentinus (1980) showed that in red algal communities in the archipelago along the Swedish east coast (Askö area), nutrient regeneration

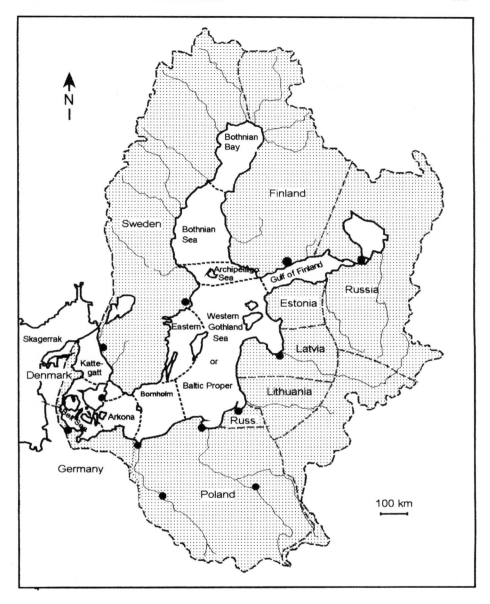

Fig. 5.1. The Baltic Sea with subareas and drainage basin

through associated *Mytilus edulis* exceeded considerably the nutrient demands of benthic algae. The authors estimated that the remaining nutrients can supply the pelagic system with about 6% of its nitrogen and 17% of its phosphorus demand. Estimates for the Baltic proper area

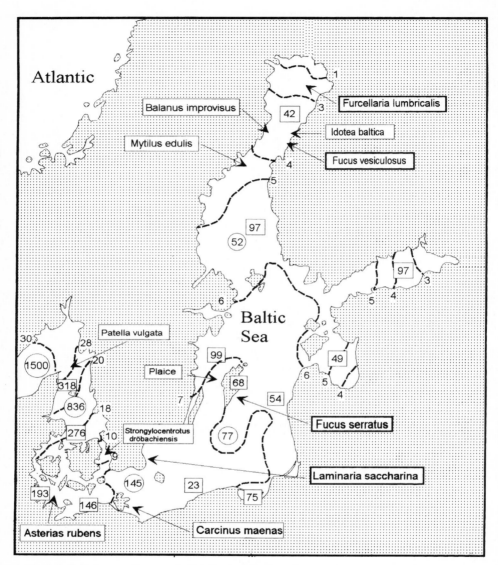

Fig. 5.2. Salinity gradient and distribution limits of some key species in the Baltic. Numbers of species of macrofauna (*circles*) and of macroalgae (*quadrats*)

predicted a recycling by mussels of about 250 000 t inorganic N, 77 000 t inorganic P and 97 000 t amino-N per year, which is several times higher than the input from terrestrial sources. In this context, Kautsky and Wallentinus speculate whether the absence of *Mytilus* in the Bothnian Bay and hence a reduction in nutrient regeneration can contribute to the

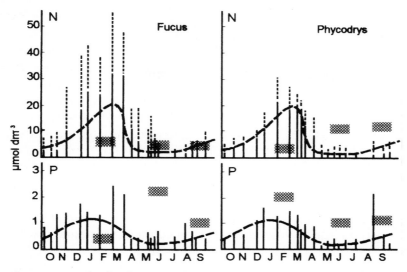

Fig. 5.3. Examples for the seasonal variations of seawater inorganic nitrogen and phosphate concentrations within a *Fucus vesiculosus* and a red algal community (*bars*), and offshore (*broken curve*). Kiel Bight (western Baltic). Inorganic nitrogen: *solid bars* $NO_3 + NO_2$-N; *broken bars* NH_4-(After Schramm et al. 1988)

fact that pelagic primary production in this area is five times lower than in the Baltic proper (Lassig et al. 1978).

The particular nutrient situation in phytobenthic communities may explain why many macrophytes in the Baltic florish even after the phytoplankton spring bloom, when nutrients are depleted in the water column, but light conditions are favourable. This fact is of great importance in judging possible effects of eutrophication on macrophytes or on primary producers in general. As nutrient availability in phytobenthic systems may differ considerably from pelagic sytems, even under "normal" conditions, i.e. when man-made eutrophication can be excluded or be assumed as negligible, marine macrophytes compared to phytoplankton probably react differently to additional supplies of nutrients.

5.3   Biota and Plant Communities

Salinity and substrate are the major factors that determine the biota, in particular the composition and distribution of the zoo- and phytobenthos in the Baltic (Schwenke 1969; Remane and Schlieper 1971; Jansson

1978; Wallentinus 1991). The marine flora of the Baltic in principle derives from the North Sea flora as already described by Reinke (1889). Most of the characteristic North Sea forms, however, penetrate only as far as the Kattegat. With decreasing salinity, the number of marine macroalgae decreases considerably. Of the more than 300 occurring macroalgal species along the Swedish and Danish coasts of the Kattegat, only 42 can be found in the innermost Baltic Sea (Bothnian Bay). As is typical for many brackish or estuarine systems, the number of marine species in the Baltic does not decrease continuously with the salinity gradient, but unsteadily. Of the main systematic macroalgal groups, the decline is more pronounced in the red and brown seaweeds compared to the greens. Of the total number of macroalgae the percentage of green algae even increases due to the higher proportion of euryhaline forms among the green algae and because of the incoming freshwater species (Wallentinus 1991; Nielsen et al. 1955; Fig. 5.4; Table 5.1).

A critical boundary for many fully marine organisms can be found in the Belt Sea with its unstable osmotic conditions due to salinity fluctuations as high as to ± 10 PSU. A further drop in the number of marine

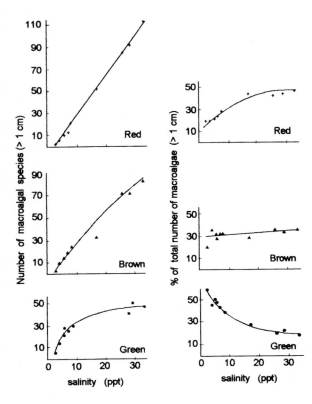

Fig. 5.4. Number and percentage of red, brown and green macroalgae (> 1 cm) along the average salinity gradient in the Baltic Sea. (After Wallentinus 1991)

Table 5.1. Number of macroalgal species along the salinity gradient in the Baltic. (Nielsen et al. 1995)

Location, area	Salinity (PSU)	Bangiophyceae	Fucophyceae	Chlorophyceae	Tribophyceae	Charophyceae	Total
Kattegatt west	30–18	126	106	70	12	5	318
east	28–10	121	92	69	7	4	293
Danish Belts	20–15	102	86	71	12	5	276
The Sound	10–8	103	73	65	8	6	255
Kiel Bight	17–10	63	66	53	5	6	193
Baltic south of the Sound	8	47	31	35	10	5	128
Arkona Sea south	8	45	45	43	6	7	146
Sweden south (Hanö Bay)	8	35	32	51	7	7	132
Bornholm Sea	8–7	27	28	18	3	3	79
Poland, Gulf of Gdansk	7–6	15	23	31	–	6	75
Poland, coastal waters	8–7	7	8	8	–	–	23
Baltic proper, west	7–6	18	21	16	–	7	62
Baltic proper, Gothland	8–7	22	22	17	–	7	68
Baltic proper, east	7–6	14	18	17	–	5	54
Gulf of Riga	5	11	14	17	–	7	49
Estonian Archipelago	6–5	12	19	23	–	7	61
Gulf of Finland	6–3	15	25	48	2	7	97
Baltic proper, northwest	7–5	20	32	37	1	9	99
Aland Sea	6–5	16	28	44	2	7	97
Aland, Finnish Archipelago	6–5	15	25	36	1	6	83
Bothnian Sea	6–4	16	26	42	3	10	97
Bothnian Bay	4–1	6	11	13	2	10	42
Sum		156	128	110	12	16	422

forms occurs in the Arcona Sea, the transition zone from the Western Baltic to the Baltic proper, the latter being characterized by rather stable salinities between 8 to 5 PSU, in contrast to the areas west of the Arcona Sea with salinities changing between 10 to 20 PSU. Finally, a third critical boundary can be observed in the Åland Sea, at the entrance to the innermost parts of the Baltic, the Bothnian and Finnish gulfs, where salinity decreases below 5 PSU down to 2–3 PSU.

Besides salinity, the substrate conditions in the Baltic determine to a great extent type and distribution of the macrophytobenthos. Great parts of the coasts of Denmark, Germany, Poland and eastern Baltic countries consist of sandy or muddy soft bottoms which are only suitable for growth of rooting plants. Hard substrates necessary for attachment of seaweeds occur in the form of solid rock mainly along the Swedish and Finnish coasts, or as glacial materials, particularly along sand cliffs, which are common along the coasts of Denmark, Germany and partly also of Poland.

The physiognomically important phytobenthic communities include the mostly filamentous annual forms in the lower level of the "eulittoral", followed further down by the *Fucus* belt and seagrass meadows, and the red algal communities down to the distribution limits of the sublittoral. Among these, the *Fucus* communities are most conspicuous. *Fucus vesiculosus*, the structurally most important seaweed in the Baltic, forms together with about 30 species of associated macrofauna and macrophytes the most diverse benthic subsystem, important as spawning and breeding grounds and shelter for many macrofauna and fish species (Segerstraale 1944).

In fully marine areas, *Fucus vesiculosus* and *Fucus serratus* are usually intertidal species, forming rather narrow belts as can still be observed on the west coast of Sweden. In the Baltic Sea, however, they are always submerged as a result of lack of tides, ice-scouring during winter and extended periods of low water levels in spring.

The vertical distribution in the Baltic usually ranged from 0 to 5–6 m depth. For some nearshore locations, maximum depth down to 8 m has been reported (Levring 1940; Hoffmann 1952; Waern 1952; Schwenke 1969; von Wachenfeldt 1975; Luther 1981; von Wachenfeldt et al. 1986); in offshore clear waters it may reach down to depths of 12 m (Black 1978; Kautsky et al. 1986).

While *Fucus serratus* penetrates as far as the Gothland Sea into the Baltic, *Fucus vesiculosus* together with *Chorda filum* or *Dictyosiphon foeniculaceus* extends community forming into the Gulf of Finland and the Bothnian Bay till the salinity borderline of the mesohaline (3 PSU). Below the *Fucus* belt, where substrate is suitable, red algal communities are common, and in the western Baltic, *Laminaria* beds can be found up to the Darss sill. In terms of extension and biomass, the red algal communities are certainly the most important phytobenthic systems in the Baltic, penetrating as far as the Åland Sea. The dominant species may occur in rather homogeneous populations, however, they mostly form mixed communities of red and brown algae.

Many of the stenohaline forms show size reduction, as for example *Delesseria sanguinea*, which together with *Phycodrys rubens* or *Polysiphonia stricta*, reach their northern boundary in the Baltic proper. Loose lying forms of *Phyllophora pseudoceranoides* or *Furcellaria lumbricalis* penetrate into the Bothnian Bay or the Gulf of Finland till the lower mesohaline borderline. In the innermost parts of the Baltic, including the Gdansk Bay and the German inshore "Bodden" waters, under oligohaline conditions (3–0.5 PSU) marine algae such as *Blidingia minima*, *Bangia atropurpurea* or species of *Enteromorpha* can be found. *Laminaria saccharina* and *L. digitata* together with *Desmarestia aculeata*, *Ascophyllum nodosum* and *Chondrus cripus* penetrate in size-reduced form not further into the Baltic than the Darss sill, the western border of the Arcona Sea. Like *Fucus*, the community forming *Laminaria* grow only submerged.

5.4   Eutrophication and Pollution Situation

This book is, on the first hand, concerned with the changes in phytobenthic communities as a result of eutrophication. Therefore, in this chapter, focus will be on changes in inorganic nutrient levels in the Baltic, although it cannot be excluded that other factors, either natural or manmade, may also have affected the phytobenthos (Wallentinus 1981). While little is known about natural medium- or long-term fluctuations in the structure and function of marine systems, as for example as a result of climatic or hydrographic changes, local effects of pollutants such as heavy metals, toxic organic compounds and other harmful substances, heat or dumping of solid materials have been observed.

Despite the fact that the Baltic Sea is one of the most intensively studied sea areas in the world, it is difficult to obtain a comprehensive up-to-data picture of the present pollution and eutrophication situation. The complex interaction of biological, chemical, physico- and geochemical processes in combination with the specific hydrographic conditions in the Baltic makes it difficult to quantify the actual eutrophication and pollution levels, or to prove the role of man-made contribution to eutrophication.

Unfortunately, the Helsinki Convention did not include the national coastal waters, and only a few Baltic countries have filled this gap through surveys of their own national territories. In addition, few long-term investigations exist to define "normal" unpolluted conditions.

Even less information is available about the natural nutrient conditions in nearshore systems, particularly within the phytobenthic systems, which may considerably differ from the pelagic offshore systems, as mentioned before (Kautsky and Wallentinus 1980; Schramm et al. 1988).

Despite lack of data on nutrient levels and still little insight into the mechanisms and pathways of nutrient flow in shallow water benthic systems, the fact that the relatively small water body of the Baltic Sea $(2.12 \times 10^4 \text{ km}^3)$ receives the land run-off from a drainage area of $1.72 \times 10^6 \text{ km}^2$ with a population of more than 70 million people postulates a disturbance of the natural nutrient budget. Every year huge amounts of additional nutrients and pollutants due to human activity enter the Baltic through the rivers, in the form of diffuse land run-off or dicharge and dumping, and all eventually pass through the coastal waters.

Various attempts have been made to estimate the land-based and air-borne load or input of nutrients and pollutants to the Baltic (Pawlak 1980; Larsson et al. 1985; HELCOM 1987, 1990). Based on the available data, the summarized and rounded estimates have been compiled in Table 5.2.

Another source of pollution and in particular eutrophication, although probably more important as locally confined point sources, are the increasing mariculture activities, mostly along the Scandinavian coasts. Rönnberg and co-workers (1990, 1992), for example, measured increased heavy metal and nutrient levels in *Fucus vesiculosus* in relation to the distance from fish farms in the Finnish Archipelago.

There are few long-term series of measurements available, which show the trend of eutrophication in the Baltic. The comparison of the nutrient data obtained in the framework of the various Baltic monitoring programmes from the early 1970s to 1988/1989, as summarized in the Second Periodic Assessment Report of HELCOM (1990), indicates in general increasing trends of P and N concentration, although to varying degrees, in the surface winter water in most of the Baltic regions.

The observed variations are probably coupled to periodic changes in the hydrographic conditions, typical for the Baltic. For example, in the surface mixed layer of the central Baltic (south-eastern Gothland Sea), the mean phosphate concentration increased from 0.27 to 0.6 μmol dm^{-3}

Table 5.2. Estimates of total annual input of nutrients and pollutants (in t/year) to the Baltic Sea. Rounded figures, adopted from various sources. (Pawlak 1980; Larsson et al. 1985; United Nations 1987; HELCOM 1987, 1990)

	Rivers	Municipal	Industrial	Atmospheric	N–fixation	Total
Total N	640 000	87 000	14 000	320 000	135 000	1 196 000
Total P	52 000	15 000	8 000	3 800	–	78 800
BOD	1110 000	230 000	370 000	–	–	1 710 000
Cadmium	46 300	3 200	9 300	–	–	58 900
Mercury	3 700	1 100	270	–	–	5 100
Copper				2 900	–	3 200
Lead	239	18	8	–	–	265
Chromium	127	–	78	–	–	205
Arsenic	72	4	101	–	–	177
Nickel	–	10	96	–	–	106
Oil	26 000	9 000	600	–	–	35 600

and inorganic nitrogen from 1.86 to 4.45 μmol dm^{-3} from 1969 to 1978. In the same period, as a result of intrusion of saltier North Sea water into the Baltic, an increase in salinity from 7.5 to 8.0 PSU occurred. In the following years until 1990, however, the nutrient values remained stable, although salinity dropped again to 7.5 PSU (Fig. 5.5; Nehring 1991; Nehring and Matthäus 1991). As there is no indication of a significant decrease in land- or air-borne inputs, Nehring suggests that this development may either indicate a steady state between inputs, biogeochemical sinks, recycling of nutrients and exchange through the inlets to the Baltic, or may be attributed to periodical 3–7 year cycles in atmospheric circulation. The generally positive trend of nutrient levels in the Baltic most likely results from the high accumulation rates in the deep water in the period before 1977. The phosphate content of the bottom water has increased considerably since the early 1950s (Nehring 1989; HELCOM 1990; Nehring and Matthäus 1991).

In nearshore areas, long-term series of nutrient levels are rare. A comparatively early series of nutrient measurements from a nearshore area is available for the station Boknis Eck in the western Kiel Bight from 1958–1975, followed by single data up till 1984 (Babenerd and Zeitzschel 1985; Babenerd 1986). While total P did not change during the period 1958–1975 (mean value 1.09 μmol dm^{-3}), in the following years till 1984 the mean value of 1.94 μmol dm^{-3} was significantly higher. Mean phosphate values did increase from 0.7 to 1.1 μmol dm^{-3} between 1958 and 1969, and remained stable. Dissolved inorganic nitrogen data from the same station, available from 1965–1985, showed little variation and no trend (von Bodungen 1986).

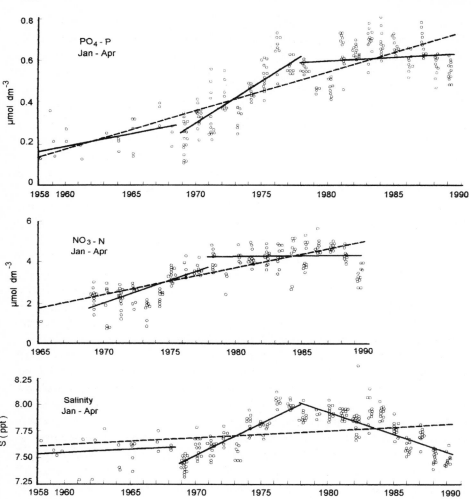

Fig. 5.5. Trends of nutrient concentration and salinity under winter conditions in the surface layer of the southeastern Gothland Sea from 1958–1990. (After Nehring 1991)

More comprehensive long-term studies of coastal eutrophication in the Baltic were undertaken in the waters of the Island of Funen county in the Belt Sea of Denmark (Funen County Council 1991). Over the 14-year study period form 1976–1990, no clear long-term trends in the winter levels of inorganic P and N were observed. During this period, nitrate-nitrogen values varied considerably, following closely the pattern of land run-off, which was to a great extent controlled by precipitation. Inorganic phosphorus concentrations, on the other hand, could not directly be related to land run-off. However, they followed the pattern of the annual salinity and wind velocity variations, suggesting variations in the degree of mixing between surface water and the phosphorus-rich bottom water.

More indicative than the observed trends in nutrient concentrations are the symptoms of eutrophication. During the past 30 years, in large areas of the Baltic a rapid decrease in oxygen in the deep water and prevailing anoxic conditions in the deep basins have been observed (Andersin and Sandler 1983). This phenomenon has been attributed to an increased inflow of saltier water into the deep basins in the Baltic since the beginning of this century, resulting in a higher stability of the halocline and thus decreased ventilation of the deep water. More recently, it appears more likely that the observed oxygen depletion can be interpreted as a sign of eutrophication, i.e. increasing production and thus sedimentation of organic matter results in increased oxygen consumption below the halocline (Larsson et al. 1985; HELCOM 1990; Nehring and Matthäus 1991).

Since the beginning of the 1980s, several extreme oxygen depletion events with hydrogen sulfide development and killing of fish and macro-zoobenthos have been reported for the Belt Sea and the Western Baltic (e.g. Ehrhardt and Wenck 1984; Weigelt and Rumohr 1986).

The increased nutrient levels in the photic layer of the Baltic should ultimately have caused an increased planktonic primary production. While the results of earlier investigations are contradictory, probably because the observation periods were not long enough and were partly overlapping with natural cyclic variations (Kayser et al. 1981), more recent studies seem to prove effects of eutrophication on plankton production in most areas of the Baltic. Phytoplankton biomass (chlorophyll a) as well as pelagic primary productivity increased in the western and southern Baltic and showed positive trends in the other parts, except for the Arkona Sea and the Gulf of Finland. Also, zooplankton biomass showed generally positive trends, which, however, were significant only in the Mecklenburg Bight (Schulz and Kaiser 1986; HELCOM 1990).

Toxic blooms, in particular red tides, have not occurred so far in the Baltic, although phytoplankton species otherwise known to produce toxins have eventually been observed, sometimes even mass development (Graneli et al. 1990). However, harmful blooms of dinoflagellates, chrysophyceans and prymnesiophyceans seem to become more frequent in the Kattegat-Belt-Arkona Sea region. In the inner Baltic, blooms were mostly caused by cyanobacteria. Several potentially toxic species of cyanobacteria, dinoflagellates, chrysophyceans and diatoms occur, which may increase the risk of development of noxious blooms.

Increased pelagic production as a result of eutrophication may have major effects on the bottom fauna. Increasing deposition of planktonic organic material, i.e. input of additional energy into the bottom systems, may result in growing numbers and biomass of the fauna. On the other hand, it will enhance oxygen-consuming processes, the cause for seasonal oxygen deficiency and the occurrence of hydrogen sulphide.

A decline in species number and biomass of the macrozoobenthos, or even extinction of the bottom fauna during the past 15 years, has been reported for all deeper areas of the Baltic, except for the Gulf of Finland, where the fauna of the deep areas deteriorated in the 1970s. A normal macrofauna community was, however, re-established in the period 1986–1989. In the northern Kattegat as well as in the Bothnian Bay, the abundance and biomass of the zoobenthos have increased, probably as a result of eutrophication. No trend was observed in the shallow areas of the Åland Sea or in the open Bothnian Sea during this period, however, macrofauna values are significantly higher compared to the 1920s (HELCOM 1990).

5.5   Changes in the Phytobenthos and Possible Causes

The effects of eutrophication on coastal systems, in particular on phytobenthic systems, have been well documented in a number of studies around the Baltic. The general picture, mostly derived from comparative studies along the nutrient gradient of receivers (e.g. Wallentinus 1979; Viitasalo 1984; Borum 1985; Funen County Council 1991), may be summarized as follows.

The primary effect of increasing nutrient levels on benthic vegetation is enhanced primary production, in particular of fast-growing epiphytic filamentous algae and green seaweeds (e.g. *Pilayella*, *Ceramium*, *Cladophora*, *Enteromorpha*, *Monostroma*).

The typical plant communities in the shallow waters of the Baltic, the *Fucus* belt, seagrass meadows (*Zostera*), and *Chara* stands, are reduced or disappear due to increased competition with fast-growing nutrient opportunists and epiphytic load. Increased plankton production in the water column causes greater turbidity and increased sedimentation of organic matter, which leads to decreasing light penetration. Due to the changing light climate, the boundaries of depth distribution, e.g. of *Fucus*, *Laminaria*, red seaweeds and seagrass move upward.

The most conspicuous change in phytobenthic communities is probably the decline of *Fucus* communities in most coastal areas of the Baltic. Reports that the bladder-wrack was declining or disappearing in many places along the Baltic coasts, particularly in Germany, Sweden and Finland, were first made by local fisherman in the early 1970s. Scientific investigations on the phenomenon were started in the mid-1970s in Finland (Kangas et al. 1982; Kangas 1983; Haathela and Letho 1982; Rönnberg 1984). Here, in the Finish waters, the decline of *Fucus vesiculosus* was most evident in moderately exposed localities in the central archipelago and in the semi-exposed outer archipelago areas. The changes in *Fucus* communities were in most cases connected with enhanced growth or mass development of filamentous algae. Since a direct influence of land-borne pollution could be excluded in these locations, Kangas (1983) and Haathela and co-workers (1984) suggested that increased nutrient levels in the 1970s originating from the open Baltic could be the primary cause of the observed alterations following the typical eutrophication pattern. The increase in nutrients may have resulted in an increase in the biomass of phytoplankton and periphyton, in suspended organic materials and sedimentation, which may have had adverse effects on the *Fucus* communities. The increased load of epiphytic filamentous and microscopic algae, as well as attached grazing animals, may have reduced the viability of *Fucus* by shading or even weighing it down. The increase in periphyton, on the other hand, offers more food for juvenile *Idotea*, which may result in more grazing on *Fucus* by the adults. Also, resettlement of *Fucus* may be hampered by algae, detritus and sediments on the bare rocks, so that the zygotes cannot attach permanently. In the early 1980s, in some of the affected Finnish waters a recovery of the *Fucus* populations was observed (Rönnberg et al. 1985).

A re-investigation in 1984 of locations in the archipelago of the Swedish east coast studied in the 1940s (Waern 1952) showed that the lower distribution of *Fucus vesiculosus* had moved upward by 2 m from 11.5 to 8.5 m, and the maximal development (bottom coverage) from 5–6 to 3–4 m, probably due to decreased light penetration (Fig. 5.6). Also, in the algal communities growing below *Fucus vesiculosus*, changes were observed: the formerly dominant *Sphacelaria arctica* had been replaced by red algae such as *Ceramium tenuicorne* and *Rhodomela confervoides* (N. Kautsky et al. 1986).

Similar observations were made in other parts of the archipelago along the Swedish coast. In the Askö area (northern Baltic proper), where extensive quantitative studies on the phytobenthic communities were performed in 1974–1975 (Jansson and Kautsky 1977), *Fucus* had disappeared in locations close to the mainland or declined particularly in

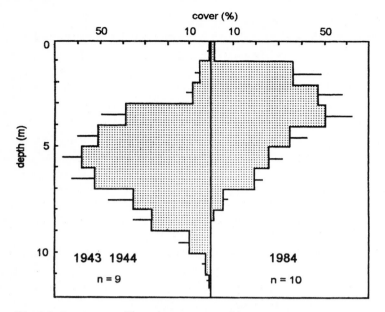

Fig. 5.6. Percentage of bottom area covered by *Fucus vesiculosus* at different depth intervals ($\pm$ SE of means) compared between 1943/1944 (Waern 1952) and 1984. (After Kautsky et al. 1986)

the intermediate, but also in the outer archipelago 10 years later in 1989 and 1990 (Fig. 5.7; H. Kautsky et al. 1992). Filamentous algae such as *Cermamium tenuicorne* had increased, although total plant biomass was reduced. The detritivorous snail *Hydrobia* sp. dominated the snail biomass, having replaced the formerly common grazing snail *Theodoxus fluviatilis*. Typical also is the considerable increase in the blue mussel *Mytilus edulis*.

While in most cases the changes of benthic vegetation were attributed to eutrophication as the primary cause, followed by a general deterioration of the environment, studies on the distribution of benthic macrophytes in various recipients of pulp mill waste discharges along the Swedish coast indicate that *Fucus vesiculosus* responded to toxic substances (e.g. chlorate) not only by absence or reduced occurrence, but also by morphological changes (Lindvall 1984; H. Kautsky et al. 1988, 1992). In one location (Mönsterås, Kalmar Sound), reduction of the chlorate content of the effluents resulted in partial recovery of *Fucus* in the receiving area.

Changes in phytobenthic communities, particularly a decline of *Fucus*, were also reported for other coastal areas of the Baltic. Macrophytic communities either changed in their structure, declined or moved towards the

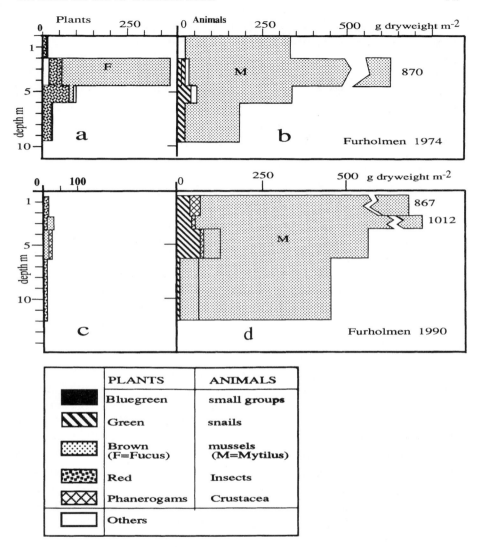

Fig. 5.7. Depth distribution and biomass of macrophytes (**a, c**) and animals (**b, d**) at Furholmen in the inner archipelago of the Askö area (northern Baltic proper) in 1974 and 1990. (After Kautsky et al. 1992)

more open parts of the bays along the southern Finnish and Estonian coasts in the Gulf of Finland (Lindgren 1975; Peussa and Ravanko 1975; Hällfors et al. 1984, 1987; Kukk 1985, 1995; Mäkinen et al. 1984; Trei et al. 1987). Along the Baltic coasts of Estonia, Latvia, Lithuania, and Russia changes in the littoral communities have been reported for the Gulf of Riga (Kukk and Martin 1992; Kukk 1993, 1995; Trei 1984).

A drastic decline of benthic vegetation was observed in some areas along the Polish coasts, especially in Gdansk Bay (Plinski and Tarasiuk 1992). In the Puck Lagoon (northwestern Gdansk Bay), the formerly rich vegetation of *Fucus vesiculosus* beds, seagrass meadows, and *Furcellaria lumbricalis*-dominated red algal communities, inhabited by abundant and diverse fauna, deteriorated rapidly from 1970. In 1987, *Fucus vesiculosus*, *Furcellaria lumbricalis* and *Coccothylus truncatus* (formerly *Phyllophora brodiaei*) had disappeared and the *Zostera* meadows had declined, whereas *Zanichellia palustris* and filamentous brown algae, such as *Pilayella littoralis*, spread out (Plinski and Florszyk 1984; Kruk-Dowgiallo 1991; Zmudzinski and Osowiecki 1991; Ciszewski et al. 1992; Fig. 5.8).

Distinct changes occurred also in the brackish, semi-enclosed inlets (Bodden) of the eastern German coasts. In the Greifswalder Bodden (Island of Rügen), fringing *Potamogeton* and *Zostera* meadows, and red algal communities, dominated by *Furcellaria lumbricalis* extending down to 8 m depth, covered approximately 50% of the area in the 1930s. In 1988, the red algae had nearly disappeared. The remaining communities of green algae (0–1.5 m), *Potamogeton pectinatus* (1–3 m), *Zostera marina* (2–4 m), and red algae (3–4 m) occupied only 3% of the bottom (Messner and von Oertzen 1991; Fig. 5.9).

A comprehensive, large-scale quantitative study on the phytobenthos in the Kiel Bay (western Baltic) employing SCUBA diving and underwater television was undertaken during 1985–1988 (Breuer and Schramm 1988; Schramm 1988; Vogt and Schramm 1991). The distribution and biomass of *Fucus* along the German coasts of Kiel Bay was investigated in 1987/1988 (Vogt and Schramm 1991). Comparison with earlier investigations in the 1950s and 1960s (Hoffmann 1952; Schwenke 1965) revealed a drastic decline in *Fucus* biomass from nearly 45 000 t wet weight down to only 2400 t within 35 years (Fig. 5.10). While *Fucus vesiculosus* as well as *Fucus serratus* were still frequent at depths below 2 m down to 10 m in the 1970s (Black 1978), no *Fucus* was found below 2 m depth during the investigation in 1987/1988.

A model including the major potential causes or their combination, and considering the particular conditions in Kiel Bight, is proposed to explain the observed decline (Fig. 5.11). Vogt and Schramm come to the conclusion that in Kiel Bay diseases, ice scoring (Rönnberg and Haathela 1988) or pollution seem to be of minor importance. Among the major factors is the loss of substrate suitable for settlement of *Fucus* through deposition of eroded cliff material, and to a certain extent as a result of commercial stone collection. The primary factor, however, seems to be eutrophication, enhancing growth of "opportunistic" annual seaweeds,

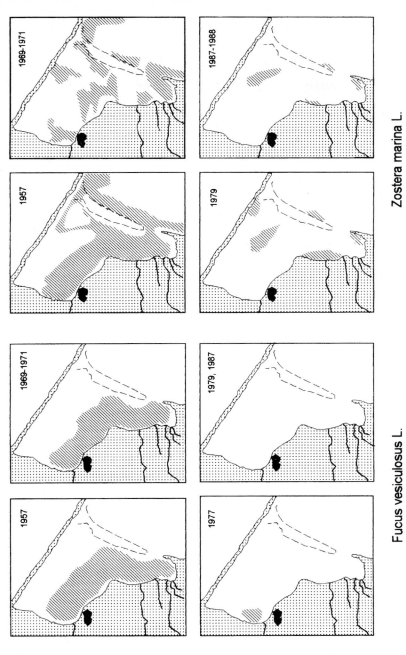

Zostera marina L.

Fucus vesiculosus L.

Fig. 5.8a

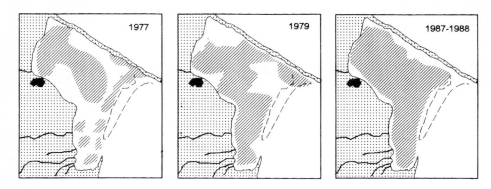

Pilayella littoralis Kjellm.

Fig. 5.8a,b. Changes in macrophytobenthic communities in Puk Lagoon (Gdansk Bay, Poland) from 1957 to 1988. (After Kruk-Dowgiallo 1991)

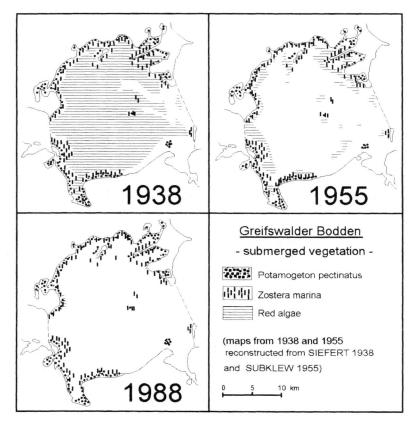

Fig. 5.9. Long-term changes in macrophytobenthic communities in the Greifswalder Bodden, southern Baltic. (After Messner and von Oertzen 1991)

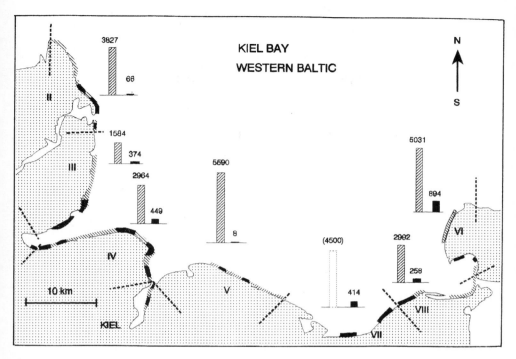

Fig. 5.10. Decline of *Fucus* spp. biomass (t wet weight) in Kiel Bight (western Baltic) from 1950/1952 (Hoffmann 1952; Schwenke 1964); hatched until 1987/1988 (solid). (After Vogt and Schramm 1991)

which are characterized by high nutrient uptake rates and saturation levels as well as by fast growth (Wallentinus 1984). Competition for nutrients seems to be of lesser importance, since nutrient levels in the phytobenthic communities are usually sufficient for saturated growth of *Fucus* throughout the year (Schramm et al. 1988). More probably, the impairment of the light climate through overgrowing and shading by the epiphytes is responsible for the decline of *Fucus*.

In this context, it is interesting to mention recent reports of the invasion of *Fucus evanescens* from Danish waters into Kiel Bay, which may be related to eutrophication processes (Schueller and Peters 1994). The epiphyte load in late spring was significantly higher on *F. vesiculosus* (20%) compared to *F. evanescens* (5%), which may be a competitive advantage for the latter in eutrophicated waters.

Apart from shading by epiphytes, there are some indications that as a result of increasing plankton growth and sedimentation (Babenerd 1986), the turbidity of the water in Kiel Bight has increased. While Hoffmann (1952) was able to carry out his survey down to 6–8 m depth using a

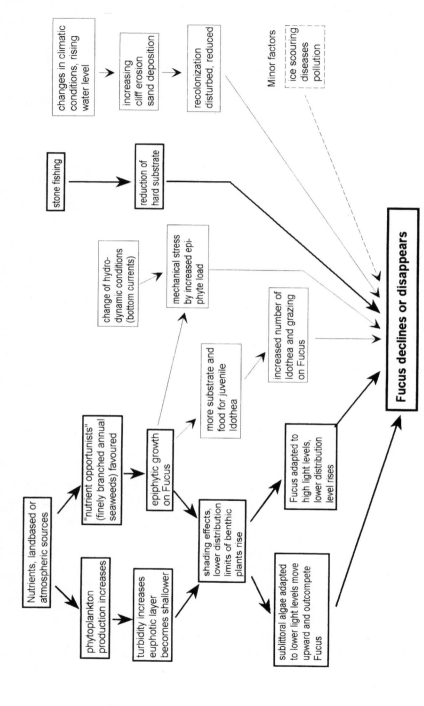

Fig. 5.11. Hypothetical model showing the major causes for the decline of *Fucus* spp. in Kiel Bight, western Baltic. (After Vogt and Schramm 1991)

simple water viewer, the visibility in 1987/1988 was usually reduced to 2-3 m depth.

Another indication of changes in light climate may be that since summer 1989 in the upper littoral of Kiel Fjord *Laminaria saccharina* can increasingly be observed, a seaweed which earlier was considered as a typical low-light adapted deep-water form in the Kiel Bay. Recent long-term registration of light conditions in Kiel Fjord, in connection with transplantation experiments and laboratory studies on light requirement of *Fucus vesiculosus*, revealed that annual light conditions at 2 m depth are not sufficient for a positive energy budget of the bladder-wrack (B. Schaffelke, pers. comm. 1994).

Apart from competition between plants, the impact of animals associated with *Fucus* may play a role. Settlement and mass development of young *Mytilus edulis* on *Fucus* plants, particularly when overgrown with epiphytes may have detrimental effects, as also observed for *Halydrys siliquosa* (Lundälv et al. 1986). The increased epiphytic filamentous algae favour mass development of *Idotea baltica*, the adults of which also graze on *Fucus* and may daily consume half of their body weight (Jansson 1974; Salemaa 1987). Grützmacher (1983) showed in in situ culture experiments that *Idotea* reduced the biomass of *Fucus* test plants by 30% within 2 months.

Parallel to the decline of *Fucus* and its decreasing depth distribution, significant changes were also observed for the sublittoral communities in Kiel Bight. A large-scale quantitative investigation employing SCUBA diving and underwater television during 1985-1986 (Breuer and Schramm 1988) was compared to a semi- quantitative survey carried out during 1962-1964 (Schwenke 1964, 1969). During this period, distinct changes in biomass, species composition and depth distribution occurred. The maximum depth distribution of the vegetation changed from 20 to 18 m. Biomass increased above the 12 m level with exception of the 6 m level (Fig. 5.12). Extended *Furcellaria lumbricalis* populations disappeared and have been replaced by *Coccotylus* (formerly *Phyllophora*) *truncata* and *Phycodrys rubens* which became the predominant species. These species together with filamentous red and brown algae penetrated to depths between 2-6 m, formerly inhabited by *Focus* and seagrass communities (W. Schramm, unpubl. data).

From the transition zone of the Baltic, i.e. the Belt Sea and the Sound, reports on changes in the benthic vegetation stem mainly from Swedish and Danish monitoring programmes, which were started in the mid-1970s and early 1980s. Along the coasts of southern Sweden (Skåne) and the Sound a strong decline of *Fucus vesiculosus* was observed from 1983 to 1985, probably as a result of ice-scouring during the hard winter

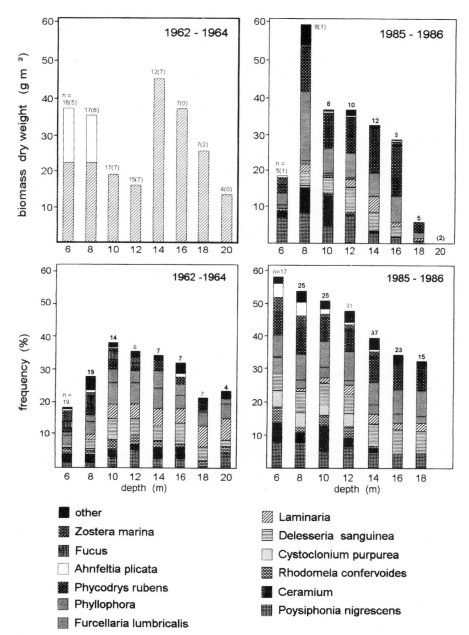

Fig. 5.12. Biomass and frequency of sublittoral seaweeds at different depths in Kiel Bight, western Baltic in 1962/1964 (Schwenke 1964) compared to 1985/1986. (After Breuer and Schramm 1988)

1984/1985. In the following years till 1991, the *Fucus* populations had nearly recovered (Carlsson and Gustavsson 1991).

A constant deterioration of the marine environment of the Danish islands as a result of eutrophication has been observed since the 1960s and early 1970s. For the waters covering the transition zone between the Baltic and the Kattegat, a comprehensive report covering the period 1976–1981 was published in 1991 (Funen County Council 1991). As this report has been quoted in detail in Chapter 7 (Part B.I.), we shall restrict ourselves to a summarized general picture. The extended seagrass beds declined to about 50% of the distribution area since the turn of the century. The lower distribution boundaries of both seagrass and seaweeds have moved upward by 4–5 m, and luxurious growth of filamentous algae is common.

Most affected are the shallow inlets, bays, coves or fjords, subject to heavily nutrient-loaded land run-off. In extreme cases such as the inner Odense Fjord, over $5000\,\mu g\,dm^{-3}$ inorganic nitrogen and over $300\,\mu g$ phosphate dm^{-3} have been measured. Mass development of sea lettuce (*Ulva lactuca*) up to $1020\,gm^{-2}$ dry weight resulted in extreme daily oxygen variation, with oversaturation in daytime, oxygen depletion at night. Release of H_2S from the sediments occasionally caused the killing of benethic fauna.

The biological and ecological mechanisms, or the ecophysiological effects of eutrophication of pollution on macrophytes, which ultimately lead to the observed changes in the structure of phytobenthic communities, have rarely been studied for the Baltic conditions.

Feldner (1976), for example, investigated the eutrophicating effects of nutrient components of domestic waste waters on Baltic seaweeds. Productivity and structure of *Zostera marina* communities in locations with different eutrophication levels were investigated in Kiel Bight (Feldner 1977).

Studies on the influence of different thallus morphologies on nutrient uptake suggest that biomass increase and sometimes mass development of finely branched, filamentous annual algae, which is symptomatic for eutrophication, are probably the result of competitive advantage of these forms over the slow growing annual seaweeds, such as *Fucus* or *Laminaria,* due to their capability to take up nutrients more rapidly and their fast growth (Wallentinus 1984). Also, the much higher uptake rates of ammonium compared to nitrate nitrogen are probably ecophysiologically relevant, because in eutrophicated water the NH_4-N concentration is usually higher, compared to unpolluted areas.

Schramm and co-workers (1988) have compared the nutrient requirements and productivity of *Fucus vesiculosus* and *Phycodrys rubens* (the

latter being a dominant species in red algal communities replacing
Fucus) in relation to the seasonal nutrient pattern in their respective
habitats in Kiel Bight.

In situ inorganic nitrogen concentrations in all locations investigated
were high enough to support nutrient-saturated growth of *Fucus*
throughout the year, whereas *Phycodrys* appeared to be growth-limited
during the summer months (cf. Fig. 5.3). In situ phosphate concentrations were mostly below growth saturating levels, except for *Fucus* in
winter.

However, as seaweeds are capable of utilizing their internal nutrient
reserves, the tissue nutrient levels are probably more relevant for optimal
nutrient supply than the external nutrient concentrations. Figure 5.13
shows that internal nutrient levels are always almost in the range of
growth-saturated levels, except for N in *Phycodrys* during summer, and P
during winter. In view of the possible response of the two seaweeds to
eutrophication, i.e. an increase in nutrients, we may conclude that under
the present conditions additional nutrients would increase productivity
probably only in *Phycodrys*. Altogether it appears that eutrophication, i.e.
additional nutrient supply, would be more advantageous to *Phycodrys*
than to *Fucus*.

Long before structural changes in phytobenthic communities can be
observed, indicating disturbances as a result of eutrophication or pollution, a functional response will most likely occur. The measurement of

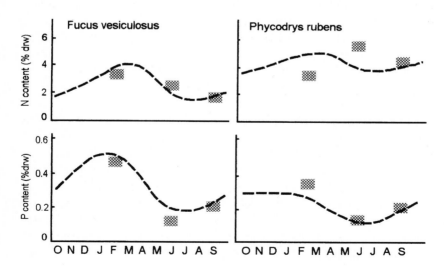

Fig. 5.13. Seasonal variations of total nitrogen and phosphorus content in the tissue of
Fucus vesiculosus and *Phycodrys rubens* (*broken curves*) in relation to tissue nutrient
contents at which growth is saturated (*shaded areas*). (After Schramm et al. 1988)

disturbances of community metabolism is an approach, which has been used to study the effects of increased nutrient levels on the functional efficiency of seaweed or seagrass communities in enclosure systems (Schramm and Booth 1981).

Similarly, the effects of pollutants (antifouling paint containing tributyltin TBT; chlorate) on gas and nutrient exchange of various *Fucus* communities along the Swedish east and west coast were studied employing in situ flow-through enclosure systems (Lindblad et al. 1986, 1988, 1989; H. Kautsky et al. 1992). Both chlorate and TBT had distinct negative effects on photosynthesis, respiration or nutrient exchange (Fig. 5.14). These effects were characterized by using a perturbation index (PI) that normalizes the differences in biomass and physiological status of the tested communities (Linblad et al. 1986). As a measure of the summary effect on various parameters such as gross production (GP), respiration (R) etc., the absolute disturbance index (ADI) was introduced, which allows comparison of communities from different areas (Lindblad et al. 1988). For example, Kautsky and co-workers (1992) showed that a *Fucus* community under more marine conditions on the west coast was less disturbed by TBT compared to the Baltic community, whereas in the case of chlorate treatment, the absolute disturbation index was slightly higher in the *Fucus* community from southern Sweden (Kämpinge), despite the slightly higher salinity (Fig. 5.15). The authors relate these differences mainly to additional osmotic stress for the marine seaweed *Fucus*, which may be even higher in the rougher osmotic environment of the southern Baltic with its changing salinities (8–10 PSU) compared to the brackish conditions in the northern Baltic (5–6.5 PSU).

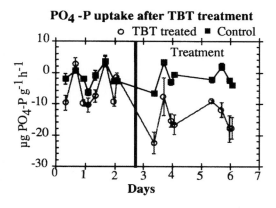

Fig. 5.14. Uptake or release of PO_4-P by *Fucus vesiculosus* communities at Askö (northern Baltic proper) treated with tributyltin (*TBT*). (Kautsky et al. 1992, modified from Lindblad et al. 1989)

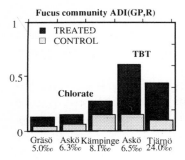

Fig. 5.15. The Absolute Disturbance Index (*ADI*) on the basis of changes in community gross production (*GP*) and respiration (*R*) of *Fucus vesiculosus* communities from the Swedish Kattegat and Baltic coasts, treated with tributyltin (*TBT*) or *chlorate* compared to untreated controls. (Kautsky et al. 1992)

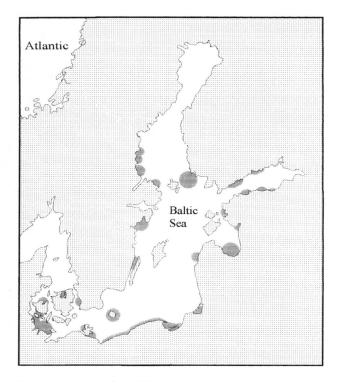

Fig. 5.16. Areas in the Baltic where changes in benthic macrovegetation have been reported

5.6 Summary and Conclusions

The specific geographic, geomorphological and hydrographic features of the Baltic Sea favour eutrophication processes, including increasing plankton productivity, deposition and accumulation of organic matter, followed by oxygen depletion and hydrogen sulphide development. The phytobenthic communities are either directly affected by increased nutrient levels, which favour fast growing seasonal nutrient opportunists in competition with comparatively slowly growing, perennial macrophytes, or indirectly by increased turbidity and inpairment of the light climate. The present situation shows (Fig. 5.16) that eutrophication-related changes in the marine vegetation and the mass development of algae occur particularly in sheltered, enclosed water bodies, being the recipients of nutrient-loaded land runoff, such as lagoons, bights, coves, fjords, or archipelagos. Also, the estuaries of rivers and streams, which carry the wastes of the bordering population, industries, or agriculture, are the actual and potential sites of changes in the benthic vegetation. From the existing observations as well as from the resutls of experimental work it may concluded that in the Baltic the primary and secondary effects of eutrophication, rather than of other factors, is the main cause for the widespread changes in benthic vegetation. In some areas with high levels of pollution due to the proximity of certain industries or high population densities, as for example along the Swedish or Polish coasts, direct effects of pollutants toxic to marine macrophytes may be the primary cause of observed changes in the vegetation.

References

Andersin AB, Sandler H (1983) Occurrence of hydrogen sulphide and low oxygen concentrations in the Baltic deep basins. Proc 16th CBO Symp, Kiel 1988, 1: 102–111

Babenerd B (1986) Long-term observations of some hydrographical, chemical and planktological variables in Kiel Bay, 1957–1975. ICES CM. 1986/L 19: Biol Oceanogr Committee 1–8

Babenerd B, Zeitzschel B (1985) Trends für eintragsrelevante Faktoren und für die Nährsalzkonzentrationen im Wasser der Kieler Bucht. Ein Beitrag zur Erfassung der Eutrophierung der Nord- und Ostsee. Ber Inst Meereskd, Univ Kiel 148: 1–45

Black H (1978) Vegetationsdynamische Untersuchungen an epilithischen Algengemeinschaften im sublitoral der westlichen Ostsee unter Berücksichtigung der produktionsbiologischen Bestandsabschätzung. Rep Sonderforsch Ber 95, Univ Kiel 44: 1–144

Borum J (1985) Development of epiphytic communities on eelgrass (Zostera marina) along a nutrient gradient in a Danish estuary. Mar Biol 87: 211–218

Breuer G, Schramm W (1988) Changes in macroalgal vegetation in Kiel Bight (Western Baltic) during the past 20 years. Kiel Meeresforsch Sonderh 6: 241–255

Carlsson L, Gustavson BS (1991) Förändringar i förekomsten Fucus vesiculosus längs Skånes Kuster under 1983–1991. Växtsamhällen i Östersjön, Bertil Lindvall Symp, Kalmar 1991, Univ Kalmar (summary in Swedish)

Ciszewski P, Kruk-Dowgiallo L, Zmudzinski L (1992) Deterioration of the Puck Bay and biotechnical approaches to its reclamation. Proc 12th Baltic Mar Biol Symp 1991, Helsingör, pp 43–46

Ehrhardt M, Wenck A (1985) Wind pattern and hydrogen sulfide in shallow waters of the Western Baltic Sea, a course and effect relationship? Meeresforschung 30: 101–110

Feldner J (1977) Ökologische und produktionsbiologische Untersuchungen am Seegrass *Zostera marina* L. in der Kieler Bucht (Westliche Ostsee). Rep Sonderforsch Ber 95, Univ Kiel 30: 1–170

Feldner R (1976) Untersuchungen über die eutrophierende Wirkung einiger Nährstoffkom-ponenten häuslichr Abwässer auf die Benthosalgen der Kieler Bucht (Westliche Ostsee). Diss Univ Kiel, Kiel, 134 pp

Funen County Council (1991) Eutrophication of coastal waters. Coastal water quality management in the County of Funen, Denmark, 1976–1990. Legislation, nutrient loads and eutrophication effects–causes and trends. Funen County Council, Department of Technology and Environment, Odense, 280 pp

Graneli E, Sundström B, Edler L (eds) (1990) Toxic marine phytoplankton. Elsevier, Amsterdam, 554 pp

Grützmacher M (1983) Produktionsbiologische Untersuchungen an *Fucus*-Beständen der Kieler Bucht (Westliche Ostsee). Diss Univ Kiel, Kiel, 108 pp

Haathela I (1984) A hypothesis of the decline of the bladder wrack (*Fucus vesiculosus* L.) in SW Finland in 1975–1981. Limnologica 15 (2): 345–350

Haathela I, Letho J (1982) The occurrence of bladder-wrack (*Fucus vesiculosus*) in 1975–1980 in the Seili area, Archipelago Sea. Memo Soc Fauna Flora Fenn 58: 1–5

Hällfors G, Kangas P, Niemi A (1984) Recent changes in the phytal at the south coast of Finland. Ophelia (Suppl) 3: 51–59

Hällfors G, Viitasalo I, Niemi A (1987) Macrophyte vegetation and trophic status of the Gulf of Finland: a review of Finnish investigations. Meri, Helsinki 13: 111–158

HELCOM (1987) Baltic Marine Environment Protection Commission. First Baltic Sea pollution load compilation. Balt Sea Environ Proc 20, Helsinki, 56 pp

HELCOM (1990) Baltic Marine Environment Protection Commission. Second periodic assessment of the state of the marine environment of the Baltic Sea, 1984–1988; background document. Balt Sea Environ Proc 35 B Helsinki, 432 pp

Hoffmann C (1952) Über das Vorkommen und die Menge industriell verwertbarer Algen der Ostküste Schleswig-Holsteins. Kiel Meeresforsch 9: 5–14

Jansson AM (1974) Community structure, modelling and simulation of the *Cladophora* ecosystem in the Baltic Sea. Contr Askö Lab Univ Stockholm 5: 1–130

Jansson BO (1978) The Baltic – a system analysis of a semi-enclosed sea. In: Charnock H, Deacon G (eds) Advances of oceanography. Plenum Press, New York, pp 131–183

Jansson AM, Kautsky N (1977) Quantitative survey of hard bottom communities in a Baltic archipelago. In: Keegan F, O'Ceidigh P, Boadan PJS (eds) Biology of benthic organisms. Proc 11th Eur Symp Mar Biol, Pergamon Press, New York, pp 359–366

Kangas P (1983) Recent changes in the littoral algal communities of the Baltic Sea with particular reference to the decline of the bladder wrack, *Fucus vesiculosus* L. Helsinki Commission, Scientific Technical Committee STC/INF/8, Helsinki, pp 1–13

Kangas P, Autio H, Hällfors G, Luther H, Niemi A, Salemaa H (1982) A general model of the decline of *Fucus vesiculosus* at Tvärminne, south coast of Finland in 1977–1981. Acta Bot Fenn 118: 1–27

Kautsky H (1991) Influence of eutrophication on the distribution of phytobenthic plant and animal communities. Int Rev Gesamten Hydrobiol 76: 423–432

Kautsky H (1992) The impact of pulp mill effluents on phytobenthic communities in the Baltic Sea. Ambio 21 (4): 308–313

Kautsky H, Kautsky U, Nellbring S (1988) Distribution of flora and fauna in an area receiving pulp mill effluents in the Baltic sea. Ophelia 28: 139–155

Kautsky H, Kautsky U, Kautsky N, Kautsky U, Lindblad C (1992) Studies on the *Fucus vesiculosus* in the Baltic Sea. Acta Phytogeogr Suec 78: 33–48

Kautsky N, Wallentinus I (1980) Nutrient release from a Baltic *Mytilus*-red algal community and its role in benthic and pelagic productivity. Ophelia (Suppl) 1: 7–30

Kautsky N, Kautsky H, Kautsky U, Waern M (1986) Decreased depth penetration of *Fucus vesiculosus* (L.) since the 1940's indicates eutrophication of the Baltic Sea. Mar Ecol Prog Ser 28: 1–8

Kayser W, Renk H, Schulz S (1981) Die Primärproduktion der Ostsee. Geod Geoh Veröff R IV 33: 27–52

Kruk-Dowgiallo L (1991) Long-term changes in the structure of underwater meadows of the Puck Lagoon. Acta Ichthyol Piscator (Suppl) 22: 77–84

Kukk H (1985) The influence of anthropogenous factors on the composition and distribution of bottom vegetation in the Gulf of Finland. Hydrobiol Res, Tallinn 15: 123–126

Kukk H (1993) Floristic composition of the phytobenthos and its long-term changes in the Riga Gulf, the Baltic Sea. Proc Estonian Acad Sci, Ecology 3 (2): 85–91

Kukk H (1995) Phytobenthos. In: Ojaveer E (ed) Ecosystem of the Gulf of Riga. Estonian Academy Publishers, Tallinn, pp 131–138

Kukk H, Martin G (1992) Long-term dynamics of the phytobenthos in Pärnu Bay, the Baltic Sea. Proc Estonian Acad Sci, Ecology 2/3: 110–118

Larsson U, Elmgren R, Wulff F (1985) Eutrophication and the Baltic Sea: causes and consequences. Ambio 14(1): 9–14

Lassig J, Leppänen JM, Niemi A, Tamelander G (1978) Phytoplankton primary production in the Gulf of Bothnia in 1972–1975 as compared with other parts of the Baltic Sea. Finn Mar Res 244: 101–115

Levring T (1940) Studien über die Algenvegetation von Blekinge. Diss Univ Lund, Lund, 178 pp

Lindblad C, Kautsky N, Kautsky U (1986) An in situ method for bioassay studies of functional response of littoral communities to pollutants. Ophelia (Suppl) 4: 159–165

Lindblad C, Kautsky N, Kautsky U (1988) An in situ system for evaluating effects of toxicants to the metabolism of littoral communities. In: Cairn J, Pratt J (eds) Functional testing of aquatic biota from estimating hazards of chemicals. American Society for Testing and Materials, Philadelphia, pp 97–105

Lindblad C, Kautsky U, André C, Kautsky N, Tedengren M (1989) Functional response of *Fucus vesiculosus* communities to tributyltin measured in an in situ continuous flow through system. Hydrobiologia 188/189: 277–283

Lindgren L (1975) Algal zonation on rocky shores outside Helsinki as a basis for pollution monitoring. Merentutkimuslaitoksen Julk/Havforskning Inst Skrift 239: 344–347

Lindvall B (1984) The condition of a *Fucus*-community in a polluted archipelago area on the east coast of Sweden. Ophelia 3: 147–150

Lundälv T, Larsson CS, Axelson L (1986) Long-term trends in algal-dominated subtidal communities on the Swedish west coast – a transitional system? Hydrobiologia 14: 81–95

Mäkinen A, Haathela I, Ilvessalo H, Lehto J, Rönnberg O (1984) Changes in the littoral rocky shore vegetation in the Seili area, SW archipelago of Finland. Ophelia (Supply) 3: 157–166

Messner U, von Oertzen JA (1991) Long-term changes in the vertical distribution of macrophytobenthic communities in the Greifswalder Bodden. Acta Ichthyol Piscator (Suppl) 21: 135–143

Nehring D (1989) Phosphate and nitrate trends and the ratio oxygen consumption to phosphate accumulation in central Baltic deep waters with alternating oxic and anoxic conditions. Beitr Meereskd 59: 47–58

Nehring D (1991) Recent nutrient trends in the western and central Baltic Sea, 1958–1989. Acta Ichthyol Piscatoria (Suppl) 21: 153–162

Nehring D, Matthäus W (199) Current trends in hydrographic and chemical parameters and eutrophication in the Baltic Sea. Int Rev Gesamten Hydrobiol 76: 297–316

Nielsen R, Kristiansen A, Mathiesen L, Mathiesen H (eds) (1995) Distributional index of the benthic macroalgae of the Baltic Sea area. Acta Bot Fenn 155: 1–51

Pawlak J (1980) Land-based inputs of some major pollutants to the Baltic Sea. Ambio 9 (3–4): 163–167

Peussa M, Ravanko O (1975) Benthic macroalgae indicating changes in the Turku area. Merentutkimuslaitoksen Julk/Havforskning Inst Skrift 239: 339–343

Plinski M, Florszyk I (1984) Changes in the phytobenthos resulting from the eutrophication of the Puck Bay. Limnologica (Berl) 15: 325–327

Plinski M, Tarasiuk J (1992) Changes in composition and distribution of benthic algae on the Polish coast of the Baltic Sea 1986–1991. Oceanologia 33: 183–190

Reinke J (1889) Algenflora der westlichen Ostsee deutschen Antheils VI. Ber Komm Unters Dtsch Meere in Kiel. Parey, Berlin, 101 pp

Remane A, Schlieper C (1971) Biology of brackish water, 2nd edn. Binnengewässer 25: 1–372

Rönnberg O (1981) Features in the present distribution of Fucus vesiculosus (L.) in the Archipelago Sea. Rep Dept Biol, Univ Turku, Turku 2: 11–14

Rönnberg O (1984) Recent changes in the distribution of Fucus vesiculosus (L.) around the Aland Islands (N Baltic). Ophelia (Suppl) 3: 189–193

Rönnberg O, Letho J, Haathela I (1985) Recent changes in the occurrence of Fucus vesiculosus (L.) in the Archipelago Sea, SW Finland. Ann Bot Fenn 22: 231–244

Rönnberg O, Haathela I (1988) Does anchor ice contribute to the decline of Fucus in the Baltic? Mar Pol Bull 19: 388–389

Rönnberg O, Aadjers K, Ruokolathi C, Bondestam M (1990) Fucus vesiculosus as an indicator of heavy metal availability in a fish farm recipient in the northern Baltic Sea. Mar Poll Bull 21: 388–392

Rönnberg O, Aadjers K, Ruokolathi C, Bondestam M (1992) Effect of fish farming on growth, epiphytes and nutrient content of Fucus vesiculosus L. in the Aland archipelago, northern Baltic Sea. Aquat Bot 42: 109–120

Salemaa H (1987) Herbivory and microhabitat preferences of Idotea spp. (Isopoda) in the northern Baltic Sea. Hereditas 88: 165–182

Segerstraale S (1944) Weitere Studien über die Tierwelt der Fucus-Vegetation an der Südküste Finnlands. Soc Sci Fenn Comment Biol 9(4): 1–28

Schramm W (1988) Untersuchungen zur Rolle benthischer Primärproduzenten im organischen und anorganischen Stoffhaushalt der Kieler Bucht. UBA Ber Wasser 10204215/127, 34 pp

Schramm W, Booth W (1981) Mass bloom of the alga Cladophora prolifera in Bermudas: productivity and phosphorus accumulation. Bot Mar 24: 419–426

Schramm W, Abele D, Breuer G (1988) Nitrogen and phosphorus nutrition of two community forming seaweeds (Fucus vesiculosus, Phycodrys rubens) from the western Baltic (Kiel Bight) in the light of eutrophication processes. Kiel Meeresforsch Sonderh 6: 221–241

Schueller GH, Peters AF (1994) Arrival of Fucus evanescens (Phaeophyceae) in Kiel Bight (western Baltic). Bot Mar 37: 471–477

Schulz S, Kaiser W (1986) Increasing trends in plankton variables in the Baltic Sea – a further sign of eutrophication. Ophelia (Suppl) 4: 249–257

Schwenke H (1964) Vegetation und Vegetationsbedingungen in der westlichen Ostsee (Kieler Bucht). Kiel Meeresforsch 20: 157–168

Schwenke H (1965) Beiträge zur angewandten marinen Vegetationskunde der westlichen Ostsee (Kieler Bucht). Kiel Meeresforsch 21: 144–152

Schwenke H (1969) Meeresbotanische Untersuchungen in der westlichen Ostsee als Beitrag zu einer marinen Vegetationskunde. Int Rev Gesamtem Hydrobiol 54 (1): 35–94

Trei T (1984) Long-term changes in the bottom flora of Haapsalu Bay. Limnologica 15 (2): 351–352

Trei T, Kukk H, Kukk E (1987) Phytobenthos as an indicator of the degree of pollution in the Gulf of Finnland and neighbouring sea areas. Meri, Helsinki 13: 63–110

United Nations (1987) Environment statistics in Europe and North America. Part 2. Statistical monograph of the Baltic Sea environment. United Nations, New York, 87 pp

Viitasalo I (1984) Changes in the littoral vegetation of a brackish-water bay near Helsinki, Finland, following conversion of the sewage outlet system. Ophelia (Supl) 3: 253–258

Vogt H, Schramm W (1991) Conspicuous decline of *Fucus* in Kiel Bay (western Baltic): what are the causes? Mar Ecol Prog Ser 69: 189–194

von Bodungen B (1986) Annual cycles of nutrients in a shallow inshore area, Kiel Bight – variability and trends. Ophelia 26: 91–107

von Wachenfeldt T (1975) Marine benthic algae and the environment in the Öresund I–III. Diss Univ Lund, Lund, 328 pp

von Wachenfeldt T, Waldemarsson S. Kangas P (1986) Changes in the littoral communities along the Baltic coasts. Helsinki Comm, Baltic Sea Environm Proc 19: 394–403

Waern M (1952) Rocky shore algae in the Öregrund archipelago. Acta Phytogeogr Suec 30: 1–298

Wallentinus I (1979) Environmental influences on benthic macrovegetation in the Trosa-Askö area, northern Baltic proper. II. The ecology of macroalgae and submersed phanerogams. Contr Askö Lab, Univ Stockholm, Stockholm 25: 1–210

Wallentinus I (1981) Phytobenthos. In: Melvasalo T, Pawlak J, Grasshoff K, Thorell L, Tsiban A (eds) Assessment of the effects of pollution on the natural resources of the marine Baltic Sea, 1980. Baltic sea Environ Proc 5 B: 322–342

Wallentinus I (1984) Comparison of nutrient uptake rates for Baltic macroalgae with different thallus morphologies. Mar Biol 80: 215–225

Wallentinus I (1991) The Baltic Sea gradient. In: Mathieson AC, Nienhuis PH (eds) Intertidal and littoral ecosystems. Goodall DW (ed) The ecosystems of the world, vol 24. Elsevier, Amsterdam, pp 83–108

Weigelt M, Rumohr H (1986) Effects of wide-range oxygen depletion on benthic fauna and demersal fish in Kiel Bay, 1981–1983. Ber Inst Meerskd Kiel 138: 1–136

Zmudzinski L, Osowiecki A (1991) Long-term changes in the macrofauna of the Puck lagoon. Acta Ichthyol Piscator (Suppl) 21: 259–264

6 The Northern Atlantic Coasts (The Swedish West Coast, Norway and Iceland)

I. WALLENTINUS

6.1 Introduction

The macroalgal communities are highly significant elements in rocky shore areas, especially on coasts with many bays or archipelagos such as the Swedish Skagerrak coast and the Norwegian fjords.

Shift in the composition of these communities caused by eutrophication is especially important for the marine ecosystem, since several functions are altered:

1. The fixing of solar energy and nutrients, which is changed from going into the comparatively large biomasses of mainly long-lived perennial seaweeds, to being mainly bound to ephemeral algae. The latter fluctuate considerably over the year and thus by their breakdown give pulsed releases of both carbon and nutrients back to the marine ecosystem, which causes reduced stability of the system.

2. The loss of an architecturally important structure comprised especially of the large perennials (mainly brown algae). Instead the communities become totally dominated by smaller species (filamentous and sheet-like algae), reducing the system from a three-dimensional to a more or less two-dimensional structure. This in turn has a high impact on macrofauna and fish in the community and also affects commercially important species, which use the large plants for foraging and shelter. However, the filamentous algae are of great importance for the micro- and meiofauna as well as for smaller macrofaunal species, since they provide a huge increase in the available surface area.

Dept. of Marine Botany and Göteborg Marine Research Centre, University of Göteborg, Carl Skottsbergs gata 22, 41319 Göteborg, Sweden

Ecological Studies, Vol. 123
Schramm/Nienhuis (eds) Marine Benthic Vegetation
© Springer-Verlag Berlin Heidelberg 1996

3. The often mosaic pattern of different macroalgal species in areas with only a small degree of anthropogenic impact, which creates several niches for the fauna, will change into a more homogeneous structure and since many species show a trend towards moving upwards, the deeper areas become poor in or devoid of macroalgae.

4. The great diversity of macroscopic organisms is lost which may also have an impact on the biodiversity of the system at the genetic level since the gene pools might differ between the populations on the various Atlantic shores due to isolation and/or sterility.

5. Eutrophication might also enhance the establishment of introduced foreign species, because other competitive species may already have been lost as a result of the anthropogenic impact. This may have favoured for example the establishment of the Japanese brown alga *Sargassum muticum* in some areas. In other cases, the introduced species might itself be favoured by nutrient enrichment as observed for example in *Fucus evanescens* in several areas (e.g. the Oslo fjord, the Swedish west coast, Denmark).

Most of these functional and structural changes caused by eutrophication are in principle the same worldwide, although the dominant species may differ. The phenomenon of large amounts of drifting algae, mainly filamentous or sheet-like species, which can be a hindrance to both fishery and recreation, is another feature recognized in most eutrophicated areas. Nutrient enrichment favouring the opportunistic macroalgae may also affect other subsystems. Seagrasses may be affected by increased amounts of epiphytes on the plants or by sedimenting material covering the leaves. The microphytobenthic communities in shallow sediments may become covered by accumulation of drifting macroalgae (e.g. Sundbäck et al. 1990) with resultant damage to their structure and function or sediment areas may be converted to communities dominated by macroalgae (e.g. Nilsson et al. 1991).

The life span of macroalgae makes them suitable for detecting integrated effects caused by eutrophication, and like all sessile organisms they reflect changes in the area from which they are sampled. Such responses are either due to direct effects, such as increase or decrease in abundance of species by increased nutrient loads and increased incorporation of nutrients in the algal tissues, or to indirect effects, such as decreased depth distribution of the algae. These changes in depth may be caused by the greater turbidity in eutrophicated waters, but also by increased epiphytism as well as changed substrate conditions as sedimenting material accumulates on rocky bottoms.

Interpreting the recorded changes in the composition of the seaweed communities, however, is not always without problems. The nutrients available are not only a result of direct discharges of sewage, or waste-waters from industries and fish farms, land runoff or atmospheric deposition. Nutrients can also be brought by deep-water influxes (e.g. Rydberg et al. 1990). Silt carried by rivers or resuspended through storms or by dredging also has a similar detrimental effect on the communities by covering the plants or rocks. Furthermore, climatic changes may have comparable effects, especially if severe winters make the inner bays freeze and drifting ice in early spring may then erode the rocks. Combined effects of discharges of nutrients together with detrimental substances such as TBT and pulp mill effluents (see e.g. Kautsky et al. 1992 and references therein), or acid metal wastes (see e.g. Bokn 1990) may also mask the effects of nutrient load, which may not be obvious until the discharges of the other substances have been stopped. Extremely high loads of ammonium can have negative effects on algal abundance. For example, in the Baltic Sea the enhancement effect of a nitrogen load from a factory on the green alga *Enteromorpha* was not seen until the amounts of ammonium discharged into the sea had been reduced (Kautsky 1982). On the other hand, nutrient inputs in areas with low ionic strength may favour growth of marine species. For example, the innermost record in the Bothnian Bay (the Baltic Sea) of *Fucus vesiculosus* is from a location close to a sewage outlet (Pekkari 1973).

Changes in the macroalgal communities caused by eutrophication often include disappearance or decrease of fucoids (for references see below), as well as increasing amounts of the opportunistic filamentous or sheet-like and tubular species or groups, such as the green algae *Blidingia minima*, *Cladophora* spp., *Percursaria percursa*, *Rhizoclonium* spp., *Enteromorpha* spp., *Ulva* spp. and *Ulvaria fusca*; the brown algae *Ectocarpus* spp. and *Pilayella litoralis*; and the red algae *Ceramium* spp., *Polysiphonia* spp. and *Porphyra purpurea*. These algae with high surface area to volume ratios are characterized by high production and nutrient uptake rates compared to the much lower rates of the late successional species such as fucoids (cf. Wallentinus 1984 and references therein).

Structural changes in the macroalgal communities are conspicuous and can easily be observed by the general public. Thus, there will always be a high demand for information on changes in these communities. However, in many countries monitoring programmes for the macroalgal communities are lacking or much reduced in comparison to programmes on phytoplankton or the macrofauna on the deeper bottoms. Also, when programmes have been carried out, the results are very often not published in international journals, which hampers comparative analyses.

6.2 The Swedish West Coast

The Swedish west coast (including the shores of the Kattegat and the
Skagerrak; cf. Fig. 6.1) represents a transitional area between the fully
marine Atlantic Ocean and the brackish Baltic Sea. The influences of
tides are moderate. The tidal variations of water level are a maximum of
ca 0.3 m on the northern Swedish west coast, while the water levels may
change about 1 m or more during the year (cf. Johanneson 1989). Much
of the northern part of the Swedish west coast is characterized by archipelago areas, while the Swedish Kattegat coast is mainly open, to a large
extent consisting of sandy beaches which harbour communities of sea-

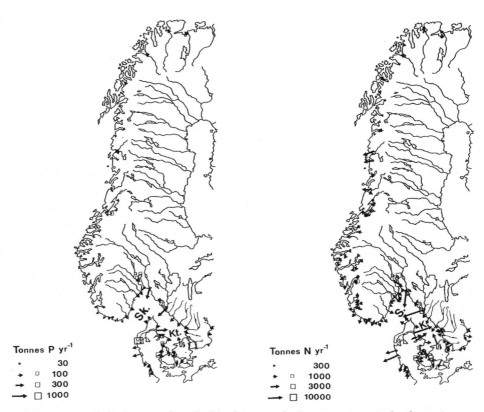

Fig. 6.1. Annual discharges of total phosphorus and nitrogen, respectively, from rivers
and fjords (*arrows*), and industrial and/or urban discharges (*open squares*) along the
Norwegian and Danish coasts, and the Swedish west coast. Sk and S denote the Skagerrak; Kt and K the Kattegat. (After Bernes 1993)

grasses (mainly *Zostera marina*, but species of the genera *Ruppia* and *Zannichellia* are also common).

Although consisting of fewer species, the algal zonation in the littoral and sublittoral belts is roughly the same as on the northern Atlantic shores (e.g. Söderström 1965; Wærn 1965), except for the laminarians which are closely restricted to the sublittoral zone as is *Fucus serratus*.

Many species also are reduced in size, a phenomenon more accentuated the further one approaches the Baltic (e.g. Wærn 1965; Wallentinus 1991). In comparison to the fully marine Atlantic shores, the marine flora thus is somewhat impoverished, even in areas not affected by pollution. Several factors are responsible. The reduced salinity, caused by the outflow of the Baltic water and in some regions also by the many rivers, is one of the main reasons for this reduction in number of species. But also higher water temperatures are responsible for the lack of some species (see e.g. Sundene 1962). Prominent species missing are brown algae such as *Pelvetia canaliculata*, *Fucus ceranoides* (cf. Bäck et al. 1992), *Himanthalia elongata*, *Saccorhiza polyschides* (see Norton 1977) and *Alaria esculenta* (cf. Sundene 1962), and the red alga *Mastocarpus stellata*, while others occur in the area only rarely (see e.g. species listed by Wærn 1958, 1961; Karlsson et al. 1992a).

For the Swedish west coast, monitoring studies which include the seaweed communities have been carried out only on a pilot scale since 1989 (Karlsson et al. 1992b). Before, only temporary research projects were performed. Within this programme changes in macroalgal vegetation have been analyzed statistically in time and space. However, unfortunately, we still have too poor a knowledge of the quantitative response of the macroalgal communities to a particular load of nutrients. This is especially true for the effects of altered balance between nitrogen and phosphorus as well as for the role of trace elements.

Introduced species have changed the macroalgal communities in this area during the last century, too (e.g. Karlsson et al. 1992a; Wallentinus, in press). During the last decade, the Japanese brown alga *Sargassum muticum* was first found attached in the northern part of the province of Bohuslän in 1987. Now in the early 1990s, it grows along many of the shores of the Skagerrak, from the Norwegian border southwards to the Göteborg archipelago, the northern Kattegat (Karlsson et al. 1992a). They also reported that it has locally started to be a nuisance to recreation and fishery, but so far it has not outcompeted native algal species.

Other events also influence the composition of the algal vegetation. The toxic bloom of the prymnesiophyte *Chrysochromulina polylepis* in the late spring of 1988 had an impact on several macroalgae, mainly red and some green algae, in both the Skagerrak and the Kattegat area (e.g.

Rosenberg et al. 1988; Lindahl and Rosenberg 1989). However, most of the species had recovered within 1 year (Pedersén et al. 1990; J. Karlsson, pers. comm.). The reason for the bloom will probably never be fully clarified (Granéli et al. 1993) and climate, nutrients, trace elements and biotic factors may have contributed. The macroalgal communities were also affected indirectly by the large amounts of mussels settling on the rocks in mainly exposed areas after the bloom, as well as on the seaweeds. These extensive mussel belts started to disappear to a large extent during the winter of 1990/1991.

6.2.1 The Swedish Kattegat Coast

Considerable changes in the macroalgal communities caused mainly by eutrophication have occurred in some areas of the southern Kattegat over the last decades, the most thoroughly studied being those in Laholm Bay (Wennberg 1987, 1992; Baden et al. 1990; Rosenberg et al. 1990). Other areas along the Kattegat coast have been affected, too, although less drastically in most areas, except for shallow bays and harbours. During the last century, there has been an estimated increase in nitrogen and phosphorus loads in the Kattegat by six and ten times, respectively. The nitrogen increase has mainly occurred during the last 25–30 years, while the phosphorus load during that period showed a slight decrease (e.g. Rosenberg et al. 1990). An annual load of about 81 000 tonnes of total nitrogen and 3100 t of total phosphorus has been estimated to enter the Kattegat (cf. Fig. 6.1), including 26 000 t of nitrogen from the atmosphere (data reported to the Helsinki Commission as quoted in Anonymous 1991). Other sources in the same publication report values about twice as high.

Laholm Bay. This bay in the southern Kattegat has been much studied from several aspects, and the nutrient loads from land, atmospheric deposition and exchanges with the deep water and the Baltic Sea are well documented (Rosenberg et al. 1990; Rydberg et al. 1990 and references therein). The nitrogen load from land to Laholm Bay amounts to an average of about 5000 t total nitrogen (ca. 50% inorganic nitrogen) per year (Joelsson and Stilbe 1993), which to a very high extent depends on land runoff from agriculture (ca. 40%) but also to an increasing extent from leakage from forested areas (ca. 40%), while sewage has been estimated to contribute with around 15%. On top of that comes the atmospheric deposition, which amounts to roughly 10–15% of the total load from land. In addition, around 20% of the nitrogen is estimated to originate from deep water entrainment and up to 30% from the Baltic Sea (Rydberg et al. 1990). For phosphorus, the deep water contributed between about 30–50% and the Baltic Sea around 35–65% of the inorganic phosphorus brought to the area. Sweden has internationally committed itself to reduce the nitrogen and phosphorus loads before 1995 by 50% compared to 1985. However, in 1992 still no decrease in total nitrogen load to the bay had been measured (Joelsson and Stilbe 1993), although in 1992 a highly efficient nitrogen reduction system in the sewage treatment plants in the area had been implemented according to the law.

The most thorough qualitive and semiquantitative studies of macroalgae are those by Wennberg (1987, 1992) who has followed the changes in the algal vegetation in the Båstad area, southern Laholm Bay since the 1950s. In the mid and late 1970s, a sharp decline could be seen in the abundance of fucoids, which also became heavily overgrown by epiphytes, and in some areas they totally disappeared, culminating in 1984 (Fig. 6.2). This was coupled with an increase in opportunistic algae on the jetties, and occurred in the same period as the load of nitrogen to the area about tripled. The green algae also generally have a dark green colour (I. Wallentinus, own observations), characteristic of areas with high nitrogen load. Large amounts of drifting, opportunistic algae are commonly washed ashore, often building up wracks, which have to be removed from the beaches by tractors. However, the macroalgal drift is also coupled to a large extent to the wind pattern, and the same batch of seaweeds can sometimes be washed in and out from the beaches several times (Holm 1987). In 1992 some fucoids, mainly *Fucus evanescens*, but also some specimens of *F. serratus* and *F. vesiculosus* reappeared on the jetties in the Båstad area, although these occurred much more sparsely than in the 1960s, and still no real fucoid belts could be found in 1993 (T. Wennberg, pers. comm.).

The Skälderviken. Changes similar to those in Laholm Bay have also occurred in the neighbouring Skälderviken bay, which is highly influenced by nutrients from agricultural areas because of the discharges of the rivers Rönneå and Vegaån. The nitrogen load of the bay is only slightly less than that of Laholm Bay (Rydberg et al. 1990). The banks of seaweeds accumulating on the northern shores of the Skälderviken, previously mainly consisting of fucoids, have been replaced by a dominance of filamentous algae. In some areas *Fucus evanescens* can be seen in great abundance (I. Wallentinus, own observations).

The Central Kattegat. Expeditions along the coast of Kattegat were undertaken in 1988 and 1989, after the *Chrysochromulina* bloom

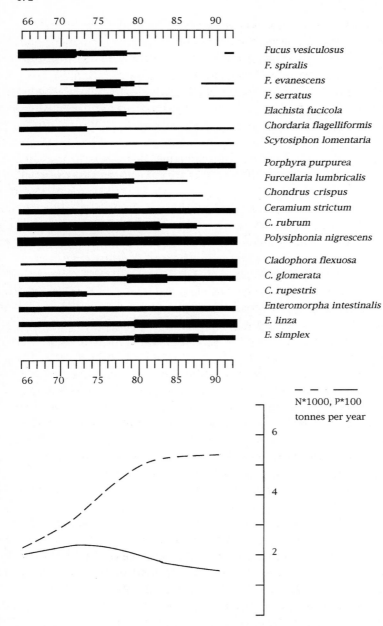

Fig. 6.2. *Above* observed changes in the occurrence of a number of macroalgal species at the piers, Båstad, Laholm Bay, during the period 1965–1992 (after Wennberg 1987; T. Wennberg, pers. comm.); *below* average annual loads of nitrogen and phosphorus to the bay during 1965–1990. (After Rosenberg et al. 1990)

(Pedersén et al. 1990; M. Pedersén, pers. comm.). In the first year negative effects especially on red algae were observed above the halocline (ca. 15 m depth) due to the bloom. However, the algae had recovered in the second year. Looking at longterm changes, no decrease in depth distributions in offshore areas was found compared to earlier investigations, but the vegetation was quite meager compared to the highly diverse and richly developed communities of the 1950s and 1960s. In some areas, the common red alga *Furcellaria lumbricalis* had decreased drastically (Pedersén et al. 1990). In the shallow areas in the mid-Kattegat ("Middlegrund" and "Fladen") there was still a rich vegetation of macroalgae down to a depth of more than 25 m in the late 1980s. However, in one of the areas industrial removal of sand and stones had been carried out, creating large craters with decaying drift seaweeds (Pedersén et al. 1990; Söderlund and Björk 1990).

A decline in the abundance of the red alga *Corallina officinalis* was observed in the mid-1980s (Karlsson 1986) in an area outside the town of Varberg, which might indicate changes due to eutrophication. Most other species, however, appeared to have quite normal abundances including *Furcellaria lumbricalis*, which has been reported as declining in some other areas.

The Northern Kattegat. The large rivers Göta Älv and the Nordre Älv bring nutrient-rich effluents to the northernmost Kattegat coast, which together with land runoff and discharges from sewage treatment plants have a high impact on the archipelago on the Swedish side (cf. Fig. 6.1).

The rivers are the main contributors of nitrogen to the area, bringing about 10 000 – 20 000 t of total nitrogen annually (mainly in the form of nitrate), of which about 35% is organic (Selmer and Rydberg 1992). About one fourth of the water passes through the southern branch, the river Göta Älv, into the estuary where another 2000 t of nitrogen, mainly ammonium, is added from the sewage treatment plant. On the other hand, the amount of phosphorus brought by the rivers is low, and with the high removal efficiency of phosphorus in the sewage treatment only small amounts are added, making the area even phosphorus limited. For many decades the nutrient enrichment has had a large impact on the algal vegetation in the estuaries and part of the archipelago area to the north, contributing to a dominance of green algae and cyanobacteria in the inner archipelago area and to a great abundance of *Fucus evanescens* in the outer archipelago (e.g. Lindgren 1965; Kuylenstierna 1989–1990; I. Wallentinus, own observations). Selmer and Rydberg (1992), however, found that there was a large sink for nitrogen in the outer archipelago.

They suggested that this could be due to denitrification in the sediments or internal storages of nitrogen in the algae, while low nitrogen uptake and denitrification rates were measured in the water phase. Unfortunately, there are no studies of the biochemical response of macroalgae in this area, but in the estuary dark green opportunistic algae are often seen as a borderline on the rocks, indicating excess of nitrogen.

Several of the localities in Danish waters sampled by Professor Rosenvinge about 70–100 years ago have been revisited. One of the stations is situated on a stone reef in the northern Kattegat, between Sweden and Denmark, where Nielsen (1991) found a high diversity of macroalgae and only a very slight impact of eutrophication, as indicated by an abundance of filter-feeding bryozoans. Several red algae had luxurious growth, which might indicate plenty of available nutrients, but no decrease in light transmission. However, the perennial *Furcellaria lumbricalis* was absent, as also observed in other areas. Although no specimen of *Corallina* was found in the survey, she commented that it did occur in the area at the time of the survey.

6.2.2   The Swedish Skagerrak Coast

So far, there are few clear effects of eutrophication in the offshore northern areas (cf. Fig. 6.1) where non-crustose algae could still be found down to a depth of at least 32 m in the 1980s (Karlsson et al. 1992a). Only little change was observed during the short period of the pilot study (Karlsson et al. 1992b), except for a significant decrease of *Corallina officinalis* in shallow depths. *Halidrys siliquosa*, which was reported to have decreased during the late 1970s and then increased (Lundälv et al. 1986) did not show any significant changes.

On the other hand, there are several indications of changes in the macroalgal communities in bays and inner archipelago areas along this coast (Michaneck 1967; Rex 1976; Lundälv et al. 1986; Svane and Gröndahl 1989; Isaksson and Pihl 1992; Pihl et al. 1996; H. Kautsky, pers. comm.; C. Larsson, pers. comm.). Michaneck (1967) was one of the first to emphasize the upward shift in the distribution of macroalgae as a result of eutrophication, based on revisits to Gislén's diving sites of about 40 years earlier. The other studies, some of these also based on revisited sites, have shown the same trend in several areas as well as a general increase in opportunistic species at the expense of the late successional ones or large amounts of drifting ephemeral macroalgae.

6.3   The Border Area Between Sweden and Norway

6.3.1   The Hvaler Area

A joint Swedish-Norwegian study programme was carried out from 1988 to 1990 from Hvaler down to the Koster area in northern Bohuslän, with the aim of clarifying which factors determine the quality of the water in the outer part of the area. The area is influenced by the large Norwegian river Glomma (cf. Fig. 6.1), by industrial discharges as well as by water from the Baltic Current. This programme (with measurements of nutrients, phytoplankton, particles and soft-bottom fauna) did not include macroalgae. In the early 1980s, Norwegian studies including macroalgae were carried out in the area close to the river outlet (Bokn 1984). Bokn observed that several macroalgae did not occur in the inner area, despite high nutrient concentrations. This was especially noteworthy for opportunistic green algae such as species of the genera *Enteromorpha*, *Cladophora* and *Ulva*, whereas cyanobacteria were common. Also *Fucus serratus* and *Ascophyllum nodosum* had disappeared from the area. The innermost area was then heavily polluted by nutrients, heavy metals and organic pollutants, while the outer area was only slightly or moderately polluted (Skei 1984). Also large amounts of silt, particles and humic substances were brought by the river, influencing the macroalgae negatively. Later experimental studies in mesocosms and transplantation experiments in the field (Bokn 1990) showed that discharged acid waste from titanium dioxide production was highly detrimental to the macroalgae in the estuary.

6.3.2   The Idefjorden Area

This strongly stratified threshold fjord (thresholds of 8–10 m in the outer area and a maximum depth around 30–40 m in the fjord) constitutes the border between Sweden and Norway. The upper layer of the fjord is brackish due to freshwater outflow from the river Tista (for salinity values see Lein et al. 1974), whereas the bottom water is anoxic. Since the 1950s, the fjord has also been heavily polluted by discharges from a paper and pulp mill (dissolved and particulate organic material, toxic substances) as well as from sewage from the town of Halden and land runoff from agricultural areas, all of which had strongly reduced the marine flora and fauna. The macroalgae of the fjord were studied in the early 1970s (Lein et al. 1974) and in the early 1980s (Rueness et al., unpubl. data). In the late 1970s, sewage treatment was introduced in Halden, and from 1978 the pulp mill effluents started to decrease and stopped in 1991, when the plant was closed down. With decreased load of organic material and toxic substances the nutrients from land runoff may enhance algal growth in the area in the future. The localities for macroalgal studies from the 1980s were revisited in 1992 (Rueness et al., unpubl. data). The distribution of several algae had moved inwards with the improved water conditions (Lein et al. 1974; Rueness et al., unpubl. data; L. Afzelius, pers. comm.). Thus in the early 1990s green algae such as *Cladophora* sp., *Enteromorpha* sp. and *Gayralia oxysperma* could be found very close to the river mouth, while in the 1970s the first two seaweeds only occurred outside the bridge of Svinesund. In the 1970s, *Fucus vesiculosus* was seen only in the outermost part of the bay, but in the early 1980s had moved close to the bridge and 10 years later small specimens occurred even inside the bridge. Also *F. serratus* had moved slightly inwards and *F. evanescens* could be found in the outer area. Cyanobacteria, which were the dominating autotrophs in the inner area in the 1970s nevertheless were prominent, indicating that high concentrations of organic pollution still occurred.

6.4   The Norwegian South and Southwest Coasts

The macroalgal composition along the southern coast of Norway does not differ significantly from that of the Swedish Skagerrak coast (cf. Rueness et al. 1990). Several macroalgae have their eastern/southern limits on the SW coast of Norway due to e.g. too low winter or too high summer temperatures as well as a result of narrow tides. Examples are the brown algae *Alaria esculenta*, *Dictyota dichotoma*, *Fucus distichus* ssp. *anceps*, *Himanthalia elongata*, *Pelvetia canaliculata*, and the red algae *Ceramium shuttleworthianum*, *Mastocarpus stellata*, *Porphyra miniata*, and *Vertebrata* (= *Polysiphonia*) *lanosa* (Rueness et al. 1990).

In 1990, a 10-year monitoring programme of hard-bottom communities was started by NIVA (Norsk institutt for vannforskning). The programme includes 27 stations from the Swedish border in the east to north of Bergen, studied by transects, sampled squares, and stereo photos. In some areas additional analyses of the epibionts and the length of the stipes and lamina of *Laminaria hyperborea* are made (Pedersen et al. 1991; cf. also Moy and Walday 1992). Most of the stations are located in the outer coastal zone, including four areas for intensive studies. No impact of eutrophication was found in these outer areas at the start of the monitoring programme (Pedersen et al. 1991).

After the toxic algal bloom of the prymnesiophyte *Chrysochromulina polylepis* in the late spring of 1988, a large number of localities along the southern and southwestern coasts were investigated for effects of the bloom on the hard-bottom communities (e.g. Berge et al. 1988; Pedersen et al. 1989). The main effects were about the same as on the Swedish west coast. In 1990, almost normal community composition was reported for those areas with an increase in the number of macroalgal species occurring compared to the period directly after the bloom (Pedersen et al. 1991).

Changes in the macroalgal communities have also been caused by introduced species (e.g. Rueness 1989; Wallentinus in press), the most prominent being that of the Japanese brown alga *Sargassum muticum*. It was first reported as drift plants in southern Norway in 1984, and in 1988 the first attached plants were found (Rueness 1989 and references therein). During 1989 it grew attached at several new localities along the Norwegian coast of Skagerrak, from west of the mouth of the Oslofjord in the east, to as far west as Mandal, occurring partly in large quantities and reaching sizes of 1–2 m (Rueness 1990). Rueness stated that it mainly occupies areas where other species of the Fucales are less well developed, and that it can be a nuisance in marinas and recreation areas. He also predicted its further dispersal along the Norwegian coast towards north and east, including into the inner Oslofjord, where so far only drift plants had been found. In 1993, it was reported growing at the island of Herdla outside Bergen on the west coast (T. Lein and J. Rueness, pers. comm.).

Experimental studies of the impact of fish farming (discharges of nutrients and warm-water effluents) have been carried out showing that both the metabolism of the algae and the community structure were influenced by the heated water and that the effects were accentuated (synergistic or antagonistic effects) when combined with nutrients (Pedersen et al. 1992).

6.4.1 The Oslo Fjord

Along the Norwegian south coast, the most well-known area with respect to changes in macroalgal communities is that of the Oslo Fjord. Especially the inner part of this fjord has been extensively studied (for references see Bokn et al. 1992). They referred to reports on macroalgae in this area dating back 100 years from which it can be concluded that *Fucus evanescens* did not occur in the area until the end of last century when it was introduced. This species constituted an important part of the fucoid belts in the inner fjord in the late 1980s (Bokn et al. 1992 and references therein). There is no large river entering the area, so the effects are mainly due to impact from direct anthropogenic discharges (of which a large part has been sewage) and atmospheric deposition. During the 1960s and 1970s there was a considerable decline in the fucoids and an increase in ephemeral green algal species (of the genera *Enteromorpha, Blidingia, Cladophora* and *Ulva*) mainly due to pollution from sewage (Rueness 1973; Bokn and Lein 1978; Klavestad 1978). Revisits in the late 1980s (Bokn et al. 1992) showed an increase in fucoid abundance in some part of the inner Oslo fjord, especially for *Fucus spiralis*. On the other hand, a decrease had occurred for *Fucus evanescens*, a species which earlier had been favoured by high nutrient concentrations. Bokn et al. (1992) also quoted a simultaneous increase in Secchi depths and decreases in pelagic chlorophyll *a* concentrations and in phosphorus load to the area, where previously the loads of phosphorus and nitrogen had increased 13 and 6 times, respectively, between 1910 and the 1970s.

In the outer Oslo fjord area, however, Rueness and Fredrikssen (1991) found that the macroalgal communities were more affected in the late 1980s than 40 years earlier, which they attributed to higher loads of nutrients and organic matter. They noticed an almost total dominance of the filter-feeding mussel *Mytilus edulis* at depths down to 6–10 m, and that several species had disappeared or become rare in the area during the last 40 years such as e.g. the belt forming *Chordaria flagelliformis, Halidrys siliquosa, Laminaria digitata* and *Corallina officinalis*. It is not clear whether the dominance of the mussels was due to the effects of the *Chrysochromulina* bloom in 1988 (cf. the Swedish west coast). In addition, several species had decreased in depth distribution substantially, although the observed decrease in Secchi depths was not of the same magnitude as the upward shift of the vegetation. The decreased depth distribution is a common feature in other eutrophicated areas also and thus might indicate an impact of eutrophication, seen also in the changes of the soft-bottom fauna in the area. Further offshore macroalgae are less affected (Karlsson 1995).

6.4.2   Other Areas Along the South and Southwest Coasts

Effects of eutrophication in several inner fjord areas have been reported during the last decades (cf. Fig. 6.1). Examples are the Sandefjordsfjorden and Mefjorden (Iversen 1981), the Grenland fjords (Bokn 1979) and the Kristiansfjorden (Green et al. 1985) on the south coast as well as the Gandsfjorden, Riskafjorden and Byfjorden in the Stavanger area (Bokn and Molvær 1988). In the report after the *Chrysochromulina* bloom (Pedersen et al. 1989), clear effects of sewage discharges in deep water (> 20 m) at a station in the outer area of Trysfjorden (Søgne, east of Kristiansand) were mentioned, however, it was not reported in which way the communities were influenced.

Iversen (1981) in his study tried to apply the use of a species pollution index (cf. also Wallentinus 1979), by which he could identify affected sites in the inner fjords, although he also found that the effects of wave exposure might influence the results. However, the amount of sewage being discharged directly into these fjords at the time of the study in the late 1970s as well as the industrial discharges and land runoff certainly suggests a substantial impact on the macroalgal communities. In Bokn's study (1979), where the inner fjord area had been receiving discharges from among others an industry producing artificial fertilizers, the differences in percentage composition of the main macroalgal groups were used to determine the impact. A similar approach has been suggested as part of the future Norwegian monitoring programme for rocky substrates together with parameters such as diversity and dominance indices, as well as ratios between ephemeral and perennial organisms and depth distribution data (Moy and Walday 1992).

6.5   The Norwegian West and North Coasts

Several macroalgae have their northern or southern distribution limits along the Norwegian west coast, where in the northernmost part some arctic species are common (e.g. Jaasund 1965; Lein and Küfner 1990; Rueness et al. 1990). There are few published reports on the impact of nutrient discharges on macroalgal communities along the coastline north of Bergen (cf. Fig. 6.1), whereas effects of fish wastes have been studied in some areas (e.g. Oug et al. 1991 and references therein).

The nutrient enrichment effects in the early 1980s in the Glomfjord, slightly north of the Arctic circle, were described by Molvær and Knutzen (1987). The fjord received discharges from an industry producing ammonia and artificial fertilizers, while the freshwater inflow was nearly constant due to water level regulation in a power plant. The average annual discharges per day were calculated as 0.73 t of P and 5.2 t of N, nearly all in inorganic form. The fjord is stratified into a slightly brackish surface layer of around 1–2 m deep, an intermediate layer down to the threshold at about 100 m, and deep water below. High nutrient concentrations could be measured in the surface layer as far as 10 km from the discharge, being highest on the northern side of the fjord, and even 25 km from the discharge there were indications of somewhat increased nutrient levels. Reduced oxygen levels were measured in the deep water in late summer.

Molvær and Knutzen found an almost total dominance of green algal associations (species of *Ulothrix*, *Enteromorpha* and *Urospora*) in the littoral zone along 4–5 km on the northern side. Fucoids were absent, whereas they occurred in the fjord inside the point of discharge. In addition, the filter-feeding *Semibalanus balanoides* and *Mytilus edulis* were present in large amounts, probably being favoured by high plankton concentrations and partly competing for space with the algae. Nitrogen and phosphorus content in the fucoid tissues were also increased at stations where the nutrient concentrations were high and showed a linear correlation with the concentrations in the water. By the mid-1980s, the discharges had been reduced by 54 and 26% for nitrogen and phosphorus, respectively, resulting in lower chlorophyll *a* concentrations in the water (Molvær and Knutzen 1987).

6.6   Iceland

The composition of macroalgal communities along the Icelandic coasts have been described by Munda in several papers, and she has emphasized the differences along the coasts, where several Atlantic species disappear and are replaced by arctic and subarctic species depending on the influence of different water masses as well as climate (e.g. Munda 1992 and references therein). However, there are few published studies referring to effects of nutrients on macroalgae along the Icelandic coasts. Gunnarsson and Pórisson (1976) described the impact of sewage discharge on the cover and distribution of littoral algae near Reykjavik. They found fewer species and less coverage close to the outlet with reduction of species such as *Pelvetia canaliculata*, *Ascophyllum nodosum*, *Vertebrata* (= *Polysiphonia*) *lanosa*, *Sphacelaria radicans* and *Elachista fucicola*. On the other hand, species of *Ulothrix* and *Prasiola* were found exclusively near the discharge, and moreover, *Fucus distichus* grew only near that area.

The risk of eutrophication from discharges, including aquaculture, in Icelandic coastal waters was pointed out during a symposium on Iceland in 1986 (Jónsson 1987), although no examples proving effects on macro-algae were given.

References

Anonymous (1991) Hur mår Sverige? En rapport om miljösituationen. Miljödepartementet, Stockholm, 279 pp

Bäck S, Collins JC, Russell G (1991) Recruitment of the Baltic flora: The *Fucus ceranoides* enigma. Bot Mar 35: 53–59

Baden SP, Loo L-O, Pihl L, Rosenberg R (1990) Effects of eutrophication on benthic communities including fish: Swedish west coast. Ambio 19: 113–122

Berge JA, Green N, Rygg B, Skulberg O (1988) Invasjon av planktonalgen *Chrysochromulina polylepis* langs Sør-Norge i mai-juni 1988. Akkutte virkninger på organismesamfunn langs kysten. Del. A Sammendragsrapport. STF Overvåkningsrapport 328a/88,. Rapp 88115, NIVA, Oslo (mimeogr)

Bernes C (1993) Nordens miljö-tillsånd, utveckling och hot. Monitor 13: 1–211

Bokn T (1979) Use of benthic macroalgal classes as indicators of eutrophication in estuarine and marine waters. In: Hytteborn H (ed) The use of ecological variables in environmental monitoring. Swed Environ Protect Agency Rep 1151: 138–146

Bokn T (1984) Basisundersøkelse i Hvalerområdet og Singlefjorden. Gruntvannsorganismer 1980–1982. Overvåkningsrapport 135/84. Rapp 0–8000303, NIVA, Oslo (mimeogr)

Bokn T (1990) Effects of acid wastes from titanium dioxide production on biomass and species richness of benthic algae. Hydrobiologia 204/205: 197–203

Bokn T, Lein TE (1978) Long-term changes in fucoid association of the inner Oslofjord, Norway. Norw J Bot 25: 9–14

Bokn T, Molvær J (1988) Overvåkning av Gansfjorden, Riskafjorden og Byfjorden, Stavanger 1987. Rapp 0–87003, NIVA, Oslo (mimeogr)

Bokn TL, Murray SN, Moy FE, Magnusson JB (1992) Changes in fucoid distribution and abundances in the inner Oslofjord, Norway: 1974–80 versus 1988–90. Acta Phytogeogr Suec 78: 117–124

Granéli E, Paasche E, Maestrini SY (1993) Three years after the *Chrysochromulina polylepis* bloom in Scandinavian waters in 1988: Some conclusions of recent research and monitoring. In: Smayda TJ, Shimizu Y (eds) Toxic phytoplankton blooms in the sea. Elsevier, Amsterdam, pp 23–32

Green NW, Knutzen J, Åsen PA (1985) Basisundersøkelse av Kristiansandfjorden. Delrapport 3. Gruntvannssamfunn 1982–1983. STF Overvåkningsrapport 189/85. Rapp 0–8000354, NIVA, Oslo (mimeogr)

Gunnarsson K, Pórisson K (1976) The effect of sewage on the distribution and cover of littoral algae near Reykjavik. Acta Bot Isl 4: 58–66

Holm A-L (1987) Kvantifiering av drivande makroalger längs Laholmsbuktens södra stränder – en pilotstudie samt diskussion om eutrofieringens orsaker och effekter. MSc Thesis, Dep Mar Bot, Univ Göteborg, 58 pp (mimeogr)

Isaksson I, Pihl L (1992) Structural changes in benthic macrovegetation and associated epibenthic faunal communities. Neth J Sea Res 30: 1–7

Iversen PE (1981) Benthosalgevegetasjonen i Sandefjordsfjorden og Mefjorden, Søndre Vestfold. Del I-II. MSc Thesis, Dep Mar Bot, Univ Oslo, 157 + 173 pp (mimeogr)

Jaasund E (1965) Aspects of the marine algal vegetation of North Norway. Bot Gothoburg 4: 1–174

Joelsson A, Stilbe L (1993) Fortsatt hög kvävebelastning. In: Carlberg A, Lindblom R, Pettersson K, Rosenberg R (eds) Havsmiljön 1992. Rapport om miljötillståndet i Kattegatt, Skagerrak och Öresund. Kontaktgrupp Hav, Göteborg, p 14

Johanneson K (1989) The bare zone of the Swedish rocky shores: why is it there? Oikos 54: 77–86

Jónsson GS (1987) Kilder till marine eutrofiering i Island. Nordforsk Miljövårdsserien 1987(1): 147–151

Karlsson J (1986) Marina makroalger i Varbergs kommun. MSc Thesis, Dep Mar Bot, Univ Göteborg, 100 pp (mimeogr)

Karlsson J (1995) Inventering av marina makroalger i Østfold 1994: Området Heia-Torbjørnskjær. Tjärnö Mar Biol Lab, Univ Göteborg, 36 pp (mimeogr)

Karlsson J, Kuylenstierna M, Åberg P (1992a) Contributions to the seaweed flora of Sweden: New or otherwise interesting records from the west coast. Acta Phytogeogr Suec 78: 49–63

Karlsson J, Nilsson P, Wallentinus I (1992b) Monitoring of the phytal system on the Swedish west coast – A pilot study. Dep Mar Bot, Univ Göteborg, 37 pp (mimeogr)

Kautsky H, Kautsky L, Kautsky N, Kautsky U, Lindblad C (1992) Studies on the *Fucus vesiculosus* community in the Baltic Sea. Acta Phytogeogr Suec 78: 33–48

Kautsky L (1982) Primary production and uptake kinetics of ammonium and phosphate by *Enteromorpha compressa* in an ammonium sulphate industry outlet area. Aquat Bot 12: 23–40

Klavestad N (1978) The marine algae of the polluted inner part of the Oslofjord. A survey carried out 1962–1966. Bot Mar 21: 71–97

Kuylenstierna M (1989–90) Benthic algal vegetation in the Nordre Älv estuary (Swedish west coast) vols 1–2. PhD Thesis, Dep Mar Bot, Univ Göteborg, 244 + 162 pp

Lein TE, Küfner R (1990) Quantitative investigations of littoral communities dominated by knobbed wrack (*Ascophyllum nodosum*) in southwestern and northern Norway. Blyttia 48: 45–51 (in Norwegian with English summary and legends)

Lein TE, Rueness J, Wiikø (1974) Algological observations in the Iddefjord and the adjacent fjord areas, SE Norway. Blyttia 32: 155–168 (in Norwegian with English summary and legends)

Lindahl O, Rosenberg R (eds) (1989) The *Chrysochromulina polylepis* algal bloom along the Swedish west coast 1988. Physico-chemical, biological and impact studies. Swed Environ Protect Agency Rep 3602: 1–71 (in Swedish with English summary)

Lindgren P-E (1965) Coastal algae off Göteborg in relation to gradients in salinity and pollution. Acta Phytogeogr Suec 50: 92–96

Lundälv T, Larsson CS, Axelsson L (1986) Long-term trends in algal-dominated rocky subtidal communities on the Swedish west coast – a transitional system? Hydrobiologia 142: 81–95

Michaneck G (1967) Quantitative sampling of benthic organisms by diving on the Swedish west coast. Helgol Wiss Meeresunters 32:403–424

Molvær J, Knutzen J (1987) Eutrofiering i Glomfjord, Norge. Nordforsk Miljövårdsserien 1987(1): 157–168

Moy F, Walday M (1992) Marine vannkvalitetskriterier – hardbunn. Vurdering av utvalgte indeksers egnehet som grunnlag for fastsettinng av vannkvalitet. Høringsutkast. Rapp 0–8612602, NIVA, Oslo, 64 pp (mimeogr)

Munda IM (1992) The gradient of the benthic algal vegetation along the eastern Icelandic coast. Acta Phytogeogr Suec 78: 131–141

Nielsen R (1991) Vegetation of Tønneberg Banke, a stone reef in the northern Kattegat, Denmark. Int J Mar Biol Oceanogr 17 (Suppl 1): 199–211

Nilsson P, Jönsson B, Lindström Swanberg I, Sundbäck K (1991) Response of a marine shallow-water sediment system to an increased load of inorganic nutrients. Mar Ecol Prog Ser 71: 275–290

Norton TA (1977) Experiments on the factors influencing the geographical distribution of *Saccorhiza polyschides* and *Saccorhiza dermatodea*. New Phytol 78: 625–635

Oug E, Lein TE, Küfner R, Petersen I-B (1991) Environmental effects of a herring mass mortality in Northern Norway. Impacts on and recovery of rocky-shore and soft-bottom biotas. Sarsia 76: 195–207

Pedersen A, Wikander PB, Oug E, Green N (1989) Invasjon av planktonalgen *Chrysochromulina polylepis*. Virkninger på organismesamfunn langs kysten. NIVA's undersøkelser i november 1988. SFT Overvåkningsrapp 355/89, NIVA, Oslo (mimeogr)

Pedersen A, Green N, Walday M, Moy F (1991) Langtidsövervåkning av trofiutviklingen i kystvannet langs Sør-Norge. Haardbunnsunderskelsene 14.mai-9. juni 1990. Årsrapp 1990. SFT Overvåkningsrapp 447/91, NIVA, Oslo (mimeogr)

Pedersen A, Moy F, Bakke T, Walday M (1992) The effects of elevated temperatures and discharges from landbased fishfarms on a sublittoral mesocosm. Abstr 14th Int Seaweed Symp, Brittany, France, August 1992, p 113

Pedersén M, Björk M, Larsson C, Söderlund S (1990) Ett marint ekosystem i obalans – dramatiska förändringar av hårdbottnarnas växtsamhällen. Fauna Flora 85: 202–211

Pekkari S (1973) Effects of sewage water on benthic vegetation. Nutrients and their influence on the algae in the Stockholm archipelago during 1970. Oikos 6 (Suppl 15): 185–188

Pihl L, Magnusson G, Isaksson I, Wallentinus I (1996) Distribution and growth dynamics of ephemeral macroalgae in shallow bays on the Swedish west coast. Neth J Sea Res 35

Rex B (1976) Benthic vegetation in Byfjorden 1970–73. In: Söderström J (ed) The By fjord: marine botanical investigations. Swed Environ Protect Agency Rep 684: 67–153 (in Swedish with English summary)

Rosenberg R, Lindahl O, Blanck H (1988) Silent spring in the sea. Ambio 17: 289–290

Rosenberg R, Elmgren R, Fleischer S, Jonsson P, Persson G, Dahlin H (1990) Marine eutrophication case studies in Sweden. Ambio 19: 102–110

Rueness (1973) Pollution effects on littoral algal communities in the inner Oslofjord, with special reference to *Ascophyllum nodosum*. Helgol Wiss Meeresunters 24: 446–454

Rueness J (1989) *Sargassum muticum* and other introduced Japanese macroalgae: Biological pollution of European coasts. Mar Poll Bull 20: 173–176

Rueness J (1990) Spredning av japansk drivtang (*Sargassum muticum*) langs Skagerrak-kysten. Blyttia 48: 19

Rueness J, Fredrikssen S (1991) An assessment of possible pollution effects on the benthic algae of the outer Oslofjord, Norway. Int J Mar Biol Oceanogr 17(Suppl 1): 223–235

Rueness J, Jacobsen T, Åsen PA (1990) Seaweed distribution along the Norwegian coast with special reference to *Ceramium shuttleworthianum* (Rhodophyta). Blyttia 48: 21–26 (in Norwegian with English Abstr and legends)

Rydberg L, Edler L, Floderus S, Granéli W (1990) Interaction between supply of nutrients, primary production, sedimentation and oxygen consumption in SE Kattegat. Ambio 19: 134–141

Selmer J-S, Rydberg L (1992) Effects of nutrient discharge by river water and waste water on the nitrogen dynamics in the archipelago of Göteborg; Sweden. Mar Ecol Prog Ser 92: 119–133

Skei J (1984) Basisundersøkelse i Hvalerområdet og Singlefjorden, 1980-1983. Konklusjonsrapport. Overvåkningsrapp 171/84. Rapp 0–8000303, NIVA, Oslo (mimeogr)

Söderlund S, Björk M (1990) Rädda våra undervattensskogar. Sveriges Natur 1990 (6):
16-19
Söderström J (1965) Vertical zonation of littoral algae in Bohuslän. Acta Phytogeogr Suec
50: 85-91
Sundbäck K, Jönsson B, Nilsson P, Lindström I (1990) The impact of accumulating
drifting macroalgae on a shallow-water sediment system: an experimental study. Mar
Ecol Prog Ser 58: 261-274
Sundene O (1962) The implication of transplant and culture experiments on growth and
distribution of *Alaria esculenta*. Nytt Mag Bot 9: 155-174
Svane I, Gröndahl F (1989) Epibiosis of Gullmarsfjorden: An underwater stereophoto-
graphical transect analysis in comparison with the investigations of Gislén in 1926-29.
Ophelia 28: 95-110
Wærn M (1958) Phycological investigations of the Swedish west coast. I. Introduction
and study of the Gåsö shell-bottom. Svensk Bot Tidskr 52: 319-342
Wærn M (1961) Tillägg till Sveriges rödalgsflora (preliminärt meddelande). 1. Om
Bertholdia neapolitana, *Antithamnion tenuissimum* och *Polysiphonia nigra* i Bohuslän.
2. Om *Dumontia*-krustan och andra rödalgskrustor i Bohuslän. Svensk Bot Tidskr 55:
234-236 (in Swedish with English summary)
Wærn M (1965) A vista on the marine vegetation. Acta Phytogeogr Suec 50: 15-28
Wallentinus I (1979) Environmental influences on benthic macrovegetation in the Trosa-
Askö area, northern Baltic proper. II. The ecology of macroalgae and submersed
phanerogams. Contrib Askö Lab Univ Stockholm 25: 1-210
Wallentinus I (1984) Comparisons of nutrient uptake rates for Baltic macroalgae with
different thallus morphologies. Mar Biol 80: 215-225
Wallentinus I (1991) The Baltic Sea gradient. In: Mathieson AC, Nienhus PH (eds)
Ecosystems of the world 24. Intertidal and littoral ecosystems. Elsevier, Amsterdam, pp
83-108
Wallentinus I Introductions and transfers of plants. In: Munro A (ed) Status 1990 of
introductions and transfers in ICES member countries. Coop Res Rep (in press)
Wennberg T (1987) Long-term changes in the composition and distribution of the mac-
roalgal vegetation in the southern part of Laholm Bay, south-west Sweden, during the
last thirty years. Swed Environ Protect Agency Rep 3290: 1-47 (in Swedish with
English summary)
Wennberg T (1992) Colonization and succession of macroalgae on a breakwater in
Laholm Bay, a eutrophicated brackish water area (SW Sweden). Acta Phytogeogr Suec
78: 65-77

Part B.I
The North Sea and the Atlantic Coasts
of Western and Southern Europe

7 The North Sea Coasts of Denmark, Germany and The Netherlands

P. H. Nienhuis

7.1 Introduction

Eutrophication is defined here as the process of increasing concentration and load of inorganic nutrients, inducing changes in the aquatic communities. In general terms the large European rivers provide the main source for nutrient enrichments in coastal waters. The main rivers loading the surface waters along the eastern North Sea coasts are, from north to south, the Elbe, Weser, Ems, Rhine, Meuse and Scheldt (Fig. 7.1). Roughly 50% of the total N load comes from the rivers Rhine and Meuse together, strongly connected to the total discharge of fresh water, which is larger in wet years (1981, 1987,1988) and smaller in dry years (1985, 1989, 1990). The river Elbe gives the second most important fresh-water flow into the North Sea (30%) and the river Ems and Weser together are responsible for 20% of the N load. The river Scheldt adds only an insignificant amount (less than 5%) to the overall loading. Although there is a recent trend of decreasing loads of N on the North Sea, a significant decrease over the period 1980–1990 cannot be calculated (Fig. 7.2), De Vries et al. 1993). A comparable explanation accounts for the total P load (Fig. 7.2), where the Rhine-Meuse discharge is responsible for roughly 60%. A decreasing trend in P-loading over the period 1980–1990 can be observed, but after corrections for discharges and connected salinities, the decline is not significant (De Vries et al. 1993).

An important hydrographic characteristic of the southern North Sea is the northerly directed residual current along the coast of the Netherlands, Germany and Denmark. Water from the river Rhine for example is transported along the coast in a 50- to 70 km- wide zone and

Netherlands Institute of Ecology, Centre for Limnology, Rijksstraatweg 6, 3631 AC Nieuwersluis, The Netherlands

Ecological Studies, Vol. 123
Schramm/Nienhuis (eds) Marine Benthic Vegetation
© Springer-Verlag Berlin Heidelberg 1996

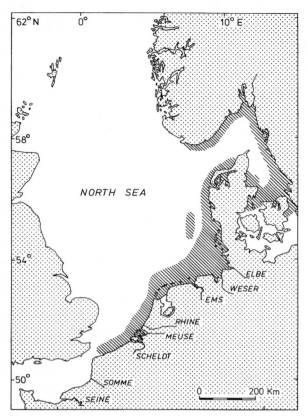

Fig. 7.1. Areas (*hatched*) identified as eutrophication problem areas in the convention waters as covered by the Paris Commission. The map was prepared by the Working Group on Nutrients. (Paris Commission 1992)

after more than 200 days 1 to 2% of this discharge water arrives at the stretch of sea north of Denmark (Fig. 7.3): the Skagerrak functions ultimately as a sink for nutrients and particles from the Dutch and the German rivers (Fig. 7.4)

The nitrogen loads of the Rhine and Meuse have increased two- to fourfold, over the period 1950–1985, whereas the phosphorus load has increased five- to sevenfold. As a result of the reduced pollution of the rivers, nitrogen and phosphorus loads decreased somewhat after 1980. In the past, concentrations of nutrients increased even more markedly than the load: a fivefold increase for N and a tenfold increase for P, with recently a slight decrease. The increased concentrations of nutrients in the Rhine-Meuse water resulted in a three- to fivefold increase in N and P concentrations in Dutch coastal waters (van der Veer et al. 1988). Roughly the same trend is apparent for the river Elbe.

The interest in eutrophication of the Baltic Sea environment has existed since the 1970s. Anoxia and changes in pelagic and benthic

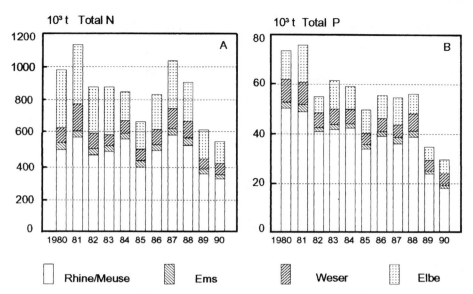

Fig. 7.2. A. Total N and **B.** total P loadings of the rivers Rhine-Meuse, Ems, Weser and Elbe onto the North Sea. (De Vries et al. 1993)

communities, as effects of increasing eutrophication are a major concern for scientists and policymakers in the Baltic States. Two recent and significant special issues of scientific journals bear witness to this: Kieler Meeresforschungen Sonderheft 6, 1988 pages 54–172 and Ambio 19(3), 1990 pages 1–176. The main interest is focused on the pelagic system: nutrient dynamics, phytoplankton primary production and effects on macrozoobenthos. Comparatively little information is available on the relation between eutrophication and submerged macrophytes. Accumulation of macrophytes appears to be a local problem, connected to the nearshore loadings of inorganic nutrients and to the specific hydrographical conditions of enclosed coastal areas, fjords, bays, harbours, etc. For example the Baltic is a non-tidal water mass with a long residence time, comprising many sheltered lagoonal habitats and hence has many actual and potential sites for mass accumulation of macroalgae. The North Sea coasts of Denmark, Germany and The Netherlands comprise tidal high-energy beaches in the south and extensive tidal flats and shallow foreshore areas in the north. Eutrophication-related macrophyte accumulations occur increasingly, and these have only recently been regarded as an environmental problem, far smaller, however, than blooms of toxic and non-toxic nuisance phytoplankton algae in the open North Sea (Paris Commission 1992).

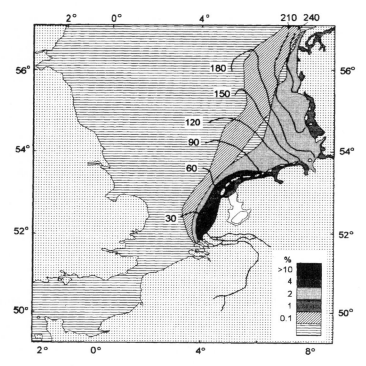

Fig. 7.3. Flow of the Rhine-Meuse water along the coasts of the Netherlands, Germany and Denmark; SW wind 4.5 m s^{-1}, water mass fractions in %, age distribution indicated as *solid lines* in days (derived from De Ruyter et al. 1987). At NW winds a maximum fraction of 10% of Rhine-Meuse water enters the Oosterschelde estuary. (Nienhuis 1993)

In this Chapter the status of macrophyte-related eutrophication will be recorded along the coasts of Denmark, Germany, The Netherlands and Belgium. The phenomena along the North Sea coast will be stressed and Danish examples from the outlets of the Baltic area will be cited. The North Sea extends its hydrographic impact to the Skagerrak-Kattegat-Belt Sea area, where deep, saline water enters the Baltic and low saline surface water flows in the opposite direction. The Baltic proper is beyond the scope of this chapter. (See Chap. 5, this Vol.)

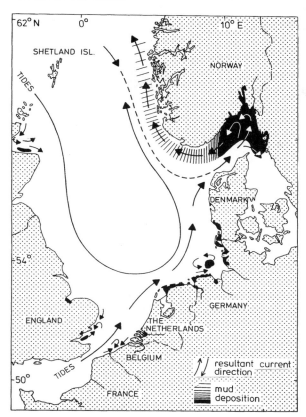

Fig. 7.4. Distribution of suspended matter and resultant transport directions, location of recent mud deposits (*black*) and older deposits. (Eisma 1987)

7.2   Eutrophication Along the North Sea Coasts

7.2.1   Denmark and the Outer Baltic

Very little information is available on the relation between eutrophication and macrophytes along the Danish North Sea coastline. I am not aware of any comprehensive study in the open literature. Obviously eutrophication is not recognized as a problem in the Danish Wadden Sea. In sharp contrast with the dearth of knowledge from the tidal coasts of Denmark, stands the considerable amount of information collected in the Kattegat-Belt-W. Baltic area between Sweden and Denmark and in the Danish fjords.

Increased inputs of nutrients to the outer Baltic coastal areas over the last decades caused severe eutrophication of these waters. In combination with a relatively low water exchange in these vertically stratified and almost non-tidal waters, local and regional effects of increased macroalgal biomass, and decreased oxygen concentrations in bottom water leading to mortalities of benthic animals and decreased fish catches have at times been observed. The effects were first noted in the Baltic proper, but are now obvious also in Swedish and Danish coastal areas in the Kattegat and the Belt Sea (Rosenberg 1985).

The coast of Denmark is characterized by a number of fjords (Fig. 7.5), complicated brackish-water systems with gradients in salinity and eutrophication stress. The inner parts of the fjords are less saline and contain higher inorganic nutrient concentrations than the outer parts. The composition of the littoral and sublittoral vegetation and the areas they occupy in these fjords have changed significantly since the turn of the century. The areas covered by eelgrass (*Zostera marina*) have been reduced by half because the depth to which sunlight penetrates has been reduced (Funen County Council 1991).

Many studies have been devoted to eelgrass distribution and ecology along the coast of Denmark, giving us a clear historical picture of the wax and wane of this macrophyte in coastal waters. The study of eelgrass (*Zostera marina* L.) has a long tradition in Denmark. Petersen (1891) pointed out the importance of the vast meadows covering the bottom in

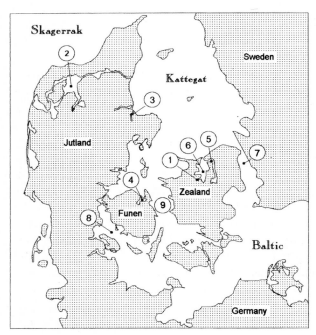

Fig. 7.5. Map of Denmark with coastal sites, mainly fjords, mentioned in the text. *1* Holbak Fjord; *2* Limfjord; *3* Randersfjord; *4* Odense fjord; *5* Roskilde fjord; *6* Isefjord and Vellerup Vig; *7* Oresund; *8* Little Belt; *9* Great Belt

shallow Danish waters when he published his results of Holback Fjords. The importance of *Zostera* was further stressed in papers by Ostenfeld (1908), Petersen and Jensen (1911), and Petersen (1913, 1915 and 1918). These authors considered *Zostera* to be the most important primary producer in Danish waters and the basis for life, especially for the coastal fisheries. However, the wasting disease which struck the Danish *Zostera* populations in 1932 and 1933 showed that this was not the case, since fishing continued at the normal level, and consequently Danish research on eelgrass more or less stopped before the second World War. *Zostera* populations re-established at several localities in the 1940s and the 1950s, but at a much lower level than before the wasting disease (Rasmussen 1977).

Rasmussen (1973) described the fluctuations of eelgrass abundance in Isefjord: in the period 1900-1932 eelgrass was very common. The wasting disease in 1932–1933 reduced the vegetation to almost zero. There was a slow recovery from 1940 onwards, but it was only in the period 1960 – 1967 that the *Zostera* vegetation recovered to pre-1930 levels. After 1967 again a serious deterioration of the eelgrass communities was observed.

It is difficult to disentangle natural fluctuations of *Zostera* populations and changes caused by eutrophication. Eelgrass populations undoubtedly are exposed to large fluctuations, of which the causes are hardly known. According to Rasmussen (1973), high water temperatures may have weakened the eelgrass population in the 1930s and hence may have contributed directly or indirectly to their destruction. Brackish-water populations (less than $12-15‰$ S) showed a resistance against the wasting disease, and it may be assumed that low salinities are favourable to generative reproduction. Seeds of *Zostera marina* germinate best in water with low salinity. On the other hand, nutrients flow from the land into the sea, thus brackish waters are more exposed to eutrophication than saline waters.

There is accumulating evidence that the decrease of eelgrass populations after 1960 – 1965 is caused by eutrophication. Eutrophication of Danish coastal waters and consequent deterioration of eelgrass vegetation, gave a new stimulus to the investigations of this macrophyte. Sand-Jensen (1975) described biomass, net production and growth dynamics in Vellerup Vig., Isefjord. Pinnerup (1980) and Bak (1979) studied growth and production of eelgrass at different localities in the Limfjorden. Wium-Andersen and Borum (1980) presented data for biomass and primary production in an eelgrass (*Zostera marina* L.) population in the Øresund, Denmark. It is shown that eelgrass grows throughout the year. Growth is directly related to the incident light intensity and nearly stops when light intensity is reduced under snow-covered ice. These results are not in agreement with the often cited statements of Setchell (1929) that growth of *Zostera* is primarily controlled by temperature and stops above 20°C and below 10°C. According to Wium-Andersen and Borum (1980) Setchell's statement should be rejected as speculative and without any modern experimental support.

The decrease of submerged macrophytes between 1950 and 1975 and the increase in biomass of epiphytes has been described for Randersfjord by Mathiesen and Mathiesen (1976). Borum (1985) followed the effect of nutrient enrichment on epiphyte development on eelgrass (*Zostera marina* L.) at four localities along a nutrient gradient in Roskilde Fjord, Denmark between March and December 1982. In the most nutrient-poor area, epiphyte biomass followed a distinct bimodal seasonal pattern with maxima in spring and early fall. Low nutrient availability and a high rate of eelgrass leaf renewal kept epiphyte biomass at a low level throughout the summer period. Unlike phytoplankton, the epiphytic community was not stimulated by nutrient enrichment during spring, however, from May through August, the biomass of both components increased exponentially with increasing concentrations of total N in the water. Along the nutrient gradient, phytoplankton biomass increased 5- to 10- fold, while epiphyte biomass increased 50- to 100- fold. Thus differences in nutrient conditions among study sites were more clearly reflected by epiphytes than phytoplankton.

The nitrogen supply to the Sound and the Kattegat has increased fourfold from 1930 to 1980 and doubled in the period 1950–1980. A review of the data shows that this doubling took place mainly after 1970 and that P supply was constant or decreased slightly during this latter period (Anderson and Rydberg 1988). An increasing number of Danish publications refer to macroalgal blooms as the consequence of continuing eutrophication (Steffensen 1976; Frederiksen 1987; Josefson 1990; Rosenberg et al. 1990). Geertz-Hansen et al. (1993) described the seasonal regulation of growth rates of free-floating *Ulva lactuca* along a nutrient gradient in Roskilde Fjord. Grazing by invertebrates was negligible in the inner part of the estuary and allowed biomass accumulations of the green macroalga, whereas grazing pressure in the outer, more saline part matched the growth rate during summer and exceeded it by more than twofold during autumn. Reduced grazing control is apparently an important and often overlooked factor for biomass accumulation of free-floating macroalgae under eutrophic conditions.

Strong and persistent eutrophication leads to anoxia. According to Jørgensen (1980) the bottom water in local areas of Limfjorden

(Denmark) frequently becomes anoxic during warm summer periods. The water may be stagnant for one to several weeks due to stratification in temperature and salinity. The anoxia stimulates the anaerobic metabolism in the sediment and H_2S accumulates. Benthic animals react to lack of oxygen by moving out of the mud and may survive lying on the mud surface. Mussel beds increase the benthic respiration per m^2 tenfold and thereby enhance oxygen depletion of the bottom water. Their high metabolic rate may thus regulate the size of the mussel beds to the limit at which animals in the center die off. Frequent anoxia lead to mass mortality of the bottom fauna in Limfjorden. Due to the seasonal oxygen depletion, the composition of the benthic community is in some areas regulated by alternating sequences of extinction and recolonization.

In 1991 an important and comprehensive report appeared (Funen County Council 1991) on the eutrophication of the Danish coastal waters between the Kattegat and the western Baltic Sea. The area of the County of Funen is 3538 km^2 or approximately 8% of the total area of Denmark. The population on January 1989 was 458000 equal to about 9% of the population of Denmark. Sewage from homes and industry accounts for approximately 18% of the nitrogen load and 78% of the phosphorus load. The non-point runoff (primarily from agriculture) is responsible for 82% of the nitrogen load and 22% of the phosphorus load.

Measurements of the nitrogen concentration in the coastal waters during the years 1976 to 1989 show that nitrogen concentration follows the variations in runoff from the land, not just on an annual basis, but also on a monthly and even weekly basis. Because of the variability in the nitrogen discharge, the 15-year period, 1976 to 1990, is too short to conclude whether there has been a general trend in the increase in nitrogen concentrations in the sea. The instantaneous nitrogen reservoir in farm soils, however, is so large that even a small peak in precipitation can trigger a massive nitrogen discharge to the coastal waters. These peaks result in an increase in nitrogen concentrations and a burst of algal production followed by oxygen depletion in the coastal waters.

The time sequential development in the phosphorus concentrations in the open coastal waters does not follow the variations in runoff from the land. Concentrations of phosphorus in the coastal waters were actually higher after the period of oxygen depletion in 1981, even though the phosphorus runoff after 1981 was less than in the years before. This may be the result of internal loading, the release of phosphorus from marine sediments under deoxygenated conditions. Winter concentrations of phosphorus at the surface of the coastal waters are to a large degree controlled by the rate of mixing and therefore by wind conditions. During the years 1976 to 1989, the wind conditions in the winter months varied a great deal, but showed no clear trend, showing that the increased level of phosphorus concentration during this period cannot be accounted for by the changes in the hydrographic conditions.

The composition of plant species found on the sea bottom around Funen and the areas they occupy have changed significantly since the turn of the century. The eelgrass-covered areas in coastal waters have been reduced by 50% owing to the reduced penetration of sunlight. The depth limit for eelgrass in open waters has been reduced from 9-10 m in 1890 to 5-6 m today and in fjords from 6-7 to 2-3 m. For seaweeds the depth limit has been reduced from 30-35 m to 10-12 m during the same period. Filamentous algae and sea lettuce, *Ulva* species, increase, particularly in enclosed waters with high nutrient loads. In many areas perennial macrophytes have disappeared completely.

Large areas of the coastal waters of Funen, where the depth exceeds 15 m, are often affected by oxygen depletion in the bottom waters. Every year some coastal water areas experience oxygen depletion during August to October; in 1988 and 1989 this occurred from June to October. From 1976 to 1989, oxygen concentrations in the bottom waters of the accumulation basins around Funen and in the deep channels of the Great Belt have shown a downward trend. These changes in the oxygen content of the bottom waters are the result of a combination of increased eutrophication in the coastal waters and "unfortunate" variations from year to year in the meteorological and hydrographic conditions (Funen County Council 1991).

In all deep basins in the Little Belt, the benthic animal communities have been reduced, and in one case completely eliminated between the 1930s and today. During the last couple of years, the catches of the most important commercial fish species in the Belt Sea have declined significantly. The reduced catches reflect a significant decline in the stocks of plaice and cod, which is possibly the result of higher local organic loads.

In many nearshore areas around Funen, the natural vegetation is reduced to a few, pollution-tolerant species, typically filamentous algae and sea lettuce. Around the islands of the Southern Funen Archipelago the vegetation in large areas is dominated by floating filamentous algae while sea lettuce (*Ulva* spp.) completely covers the inner part of Odense Fjord (Seden Strand). Assuming a doubling time of 3 days, a net primary production of more than 10000 tons dry weight *Ulva* spp. can be calculated for an area of 2×2 km^2 in Odense Fjord (Frederiksen 1987). In the inner areas of Helnaas Bay filamentous algae are common. During the warm summer of 1989, the growth of filamentous algae together with anoxia and blooms of sulphur bacteria increased exponentially in the waters of the Southern Funen Archipelago, areas of extensive natural and recreative value.

According to the Funen County Council (1991) Funen is representative of the rest of the Danish catchment area to the Kattegat with respect to human population, fertilizer consumption and average nitrogen runoff. Conclusions based on these investigations can therefore be applied nationally, also because Funen is situated between the Belt Seas, where the water of the Baltic and the North Sea meet. Independent investigations in the Limfjord, Laholm Bay (on the Swedish west coast) and the waters of the Southern Funen Archipelago indicate that the marine environment was in reasonable ecological balance in the mid-1950s. Deterioration started in the 1960s and has continued until now. Reduction of the nutrient loading to the aquatic environment by at least 50% for nitrogen and 80% for phosphorus as enacted by the Danish Parliament in the spring of 1987 in its Aquatic Environment Plan is the main goal for water managers now. This will result in a predicted runoff of nitrogen and phosphorus from agricultural lands equal to the runoff in the mid 1950s (Funen County Council 1991).

7.2.2   Germany

The North Sea coastal areas of the German territory were regarded as eutrophication problem areas by the Paris Commission (1992). The coastal sea suffers from elevated nutrient concentrations during winter, caused by the northerly directed currents from the Channel, loaded with fresh water from the river Rhine, and by the rivers Elbe, Weser and Ems. Anoxic conditions in the subtidal sediments of the German Bight have been measured since 1989 and hypoxia (< 2 mg 0_2 1^{-1}) regularly occurred from 1981 onwards, together with temporary die–off of the macrobenthic fauna (Gerlach 1990; Rachor 1990).

Knowledge of the nearshore effects of eutrophication is incomplete and scanty. The Germans have a long tradition of descriptive ecological research in the Wadden Sea (Nienburg 1925, 1927; Nienburg and Kolumbe 1931; Linke 1939) but unbiased quantitative documentation on the historic changes in the macrophyte communities is rare. One of the best described areas is Königshafen, an intertidal area or of 2×2 km on the island of Sylt (Fig. 7.6) in the northern part of the German Wadden Sea, described by Reise and co-workers.

In 1924, seagrasses covered a large area. They showed a restricted distribution in 1934 and 1974, and were fairly common again in 1988 (Fig. 7.7). Wohlenberg (1935) reported a disease-related decline of

Fig. 7.6. Map of the Wadden Sea bordering Denmark, Germany and The Netherlands

Zostera marina L. in a subtidal bed east of Königshafen during 1933–34. This bed vanished entirely and never recovered. In 1974 *Z. noltii* Hornem. occurred in small beds with distinct boundaries and *Z. marina* was a rare associate. Since 1979 *Z. noltii* occurred more scattered and over a wider range in the tidal zone. At the same time *Z. marina* reappeared and gained dominance in 1988.

Changes observed in Königshafen partly correspond to changes observed elsewhere and thus deserve more than local interest. The dramatic decline of *Zostera marina* in the 1930s occurred throughout the North Atlantic and the prevailing interpretation of that phenomenon is that of an epidemic disease (Short et al. 1988). In Königshafen, as in the Dutch Wadden Sea further south, subtidal beds never recovered (Van den Hoek et al. 1979).

Apparently, *Zostera noltii* was not affected by the wasting disease. However, in the 1970s it covered a smaller area than in previous decades. The same was observed in the Dutch Wadden Sea (Den Hartog and Polderman 1975) and in Jadebusen, and intertidal area 150 km south of

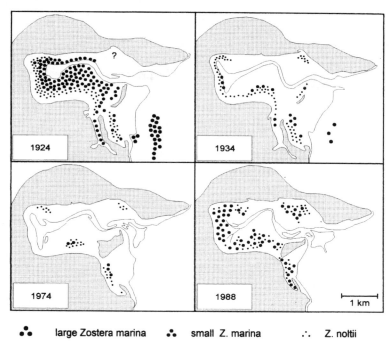

● large Zostera marina   ● small Z. marina   ∴ Z. noltii

Fig. 7.7. Distribution of seagrasses in Königshafen *1924* from map in Nienburg (1927); Zostera augustifolia and Z. stenophylla are assumed here to be small growth forms of Z. marina; *1934* from partial maps and text notes in Wohlenberg (1935, 1937), complemented by interpretation of aerial photographs from August 1936; *1974, 1988* our own grid mapping and aerial photographs. (Reise et al. 1989)

Königshafen (Michaelis 1987). Michaelis also observed a shift in dominance from *Z. noltii* to *Z. marina*.

The pattern of seagrass changes is complex and no conclusive interpretation can be offered. Some of the more recent declines in seagrass have been linked to pollution, i.e. to decreased transparency in the water column and to massive epiphytal growth as a response to nutrient enrichment. A coincidence of warm summers and mild winters with declines in *Z. marina* was pointed out by Rasmussen (1973). Den Hartog (1970) suggested that frost, grazing, and sedimentation cause pluriannual cycles in intertidal seagrass beds. None of these suggestions alone offers a consistent explanation for the observed long-term changes in Königshafen. Continuous multifactorial studies appear to be necessary. The composition and frequency of macroalgae in dredge samples from the North Frisian Wadden Sea for the period 1932 to 1940 was recorded by Hagmeier (1941). In 1932, he described most oyster beds as having a rich stock of red algae, while brown and green algae occurred sporadically. In contrast, in dredge hauls of the 1980s red algae were rare and drifting brown and green algae were relatively frequent (Fig. 7.8). The difference in the relative frequencies of red algae is highly significant. The depth range of Hagmeier's sampling sites (mostly oyster beds) with red algae was 0.4 to 7.8 m below mean low tide level. In the lower intertidal of Königshafen, red algae were common in the 1940s (Kornmann 1952) and also in the 1980s (Reise et al. 1989).

According to Reise and coworkers (1989) two opposing trends occurred in Königshafen: green algae exhibited massive growth and brown algae showed a moderate increase, while red algae became rare in the subtidal zone. Both trends may be attributed to coastal eutrophication. In contrast to algal development in the intertidal zone, red algae diminished in the subtidal zone. Since red algae remained common in the lower intertidal zone, a reduction in transparency provides a plausible explanation.

Similar to seagrass meadows, beds of *Mytilus edulis* belong to the dominant benthic communities in the Wadden Sea in terms of biomass production and nutrient cycling (Asmus 1987; Dame and Dankers 1988). Changes in the distribution and coverage of mussel beds could have great effects on the entire ecosystem. Comparing the 1920s with the

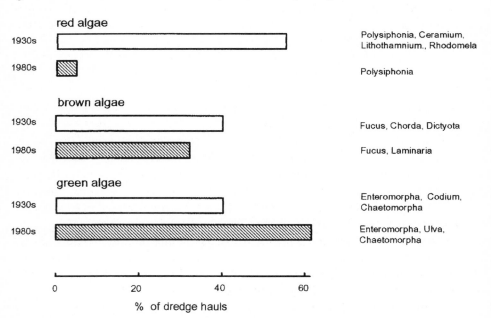

Fig. 7.8. Frequency (%) of macroalgae in subtidal channels of the North Frisian Wadden Sea, dredged in 1932 to 1940 (n = 27 samples) and in 1987–1988 (n = 78 samples) during summer months (June to August). (Reise et al. 1989)

1980s, Riesen and Reise (1982) and Reise and Schubert (1987) showed an increase of mussel beds in the subtidal zone down to a depth of 20 m and, moreover, an increase for the intertidal zone was observed by Reise and coworkers (1989). In former times mussel beds occurred mainly around the low-water line but repeatedly invaded oyster beds living in the upper subtidal zone (Hagmeier and Kandler 1927). Further south in the Wadden Sea, in the lower intertidal of Jadebusen, Michaelis (1987) found a 2.5-fold increase in coverage of mussel beds going from the 1930s to the 1970s. An increase in phytoplankton over the last decades is known from the western Wadden Sea (Cadée 1986) and Helgoland Bight (Radach and Berg 1986) and mussels may have benefitted from this improved food supply.

Few beds of *Mytilus edulis* were present at Königshafen in the period 1924–1934 compared to the 1980s (Fig. 7.9). After the severe winter 1962–1963 new mussel beds spread along the low water line in eastern Königshafen (Ziegelmeier 1977).These became partly destroyed by ice scouring in March 1969 and February 1979 but they always recovered quickly. Since 1986, these intertidal mussel beds have been subjected to heavy exploitation by the mussel fishery. Mussels are dredged from a ship during high tide, leaving only a few small clumps of mussels on the bare sediment. *Fucus vesiculosus, f. mytili* occurs in association with mussels and concomitant with the mussel density, this alga has become more abundant (Fig. 7.9.)

Nienburg (1927) described a bloom *Enteromorpha* spp. at Königshafen in spring and early summer, restricted to the upper intertidal zone (Fig. 7.9.). In August, these algae disappeared. At the same time garlands of *Chaeotomorpha linum* (O.F. Müller) occurred in muddy depressions of southeastern Königshafen. With respect to green algae, the situation has changed dramatically since 1979 (Reise 1983a,b). Explosive growth of *Enteromorpha* spp. in June/July led to the formation of thick algal mats. They were anchored in the sand and covered a wide range of the tidal zone until late August or September when storms removed them. According to Reise (1983a,b, 1985) small snails (*Hydrobia*) and lugworms (*Arenicola marina*) together play a significant role in the development of green algal mats in the Wadden Sea. *Enteromorpha* spores germinate on sand grains and on the shells of *Hydrobia*. The growing thalli of *Enteromorpha* become attached in the feeding burrows of *Arenicola* and this anchoring prevents the algae from being drifted away by tidal currents and wave action.

Nutrient concentrations in the North Frisian Wadden Sea are considerably lower than further south along the German coast (Martens 1989). Nevertheless, several of the long-term changes observed in the

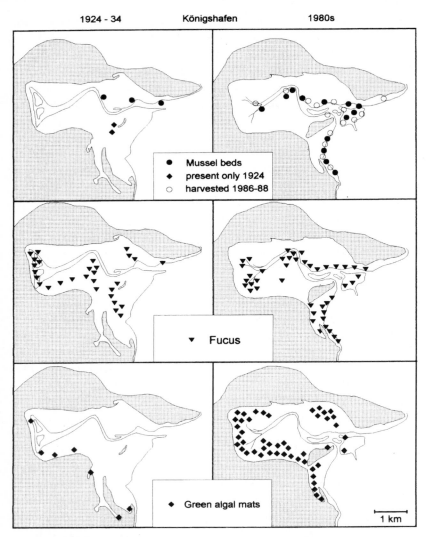

Fig. 7.9. Distribution of mussel beds, *Fucus vesiculosus* and green algal mats (spp. of *Enteromopha*, *Chaetomorpha* and *Ulva*) in Könighafen. *1924–34* From map of Nienburg (1927) and partial maps and text notes in Wohlenberg (1937); 1980 s our own mapping and aerial photographs. (Reise et al. 1989)

Wadden Sea around the island of Sylt are consistent with the assumption that increased nutrient concentrations in the coastal waters have caused an enrichment effect: (1) massive growth of green algal mats in sheltered intertidal zones; (2) loss of red algae from the deeper bottoms; (3) expansion of mussel beds in the intertidal zone and over the entire depth range of the subtidal zone; (4) increased abundance of infaunal species in the sandy bottoms of the intertidal and subtidal zone.

Positive responses by the benthos to enhanced eutrophication have also been observed in other regions of the Wadden Sea (Michaelis 1978, 1986; Beukema and Cadée 1986; Madsen and Jensen 1987). On the other hand, deleterious effects attributed to hypoxia are known from sites close to effluents of organic wastes (Essink 1984; Essink and Beukema 1986).

In the North Frisian Wadden Sea, the prevailing physical conditions with strong tidal currents and wave action normally prevent hypoxia. In the vicinity of mussel beds, however, resuspended biodeposits may cause oxygen deficiencies in the interstitial water. Unfortunately, in the deep tidal channels where the sessile epifauna has declined, oxygen concentrations have not been measured (Reise et al. 1989).

Not all of the observed long-term changes in the benthos around Sylt can be attributed to eutrophication. Erosion contributed to the loss of narrow-zoned communities in the upper intertidal, and possibly also to losses in the epifauna of the deep channels. Fisheries activities almost certainly affected oyster- and musselbeds and *Sabellaria* reefs. The historical comparisons presented cannot be conclusive with regard to the causes of the observed long-term changes (Reise et al. 1989).

The Wadden Sea area between the rivers Ems, Weser and Elbe has also been quite intensively studied from the 1930s onwards (Linke 1939; Michaelis et al. 1992). Much information is not published in international journals, and hence is rather inaccessible.

Until 1980, macroalgae represented a constant but inconspicuous element in the intertidal Wadden Sea ecosystem (Michaelis 1969, 1970, 1978, 1987; Hauser and Michaelis 1975; Koeman 1975). In some of the older reports a tidal-flat blooming is mentioned, caused by an explosive spring growth of green algae. Early symptoms of increasing macroalgal abundance were already seen by Michaelis (1978), but it is only from 1989 onwards that green algae increasingly reach high densities in many localities.

Aerial surveys from 1990 onwards revealed that each year during summer roughly 15% of the intertidal area between the rivers Ems and Weser are covered by green algae. The maximum biomass varies between 20 to 110 g dry weight m^{-2}. The dominating green algae are *Ulva scandinavica, U. lactuca, Enteromorpha prolifera* subsp. *prolifera* and subsp. *radiata* and *E. intestinalis*. The seasonal dynamics of the green algal vegetation changes from year to year, with regard to the start of the growing season and the duration of the vegetative period (May–October; Kolbe et al. 1995). Data from the German Wadden Sea suggests that in "macroalgae years" the concentrations of nitrate in the watercolumn are significantly lower than in normal years, implying excess N uptake by macroalgae during specific summers (Kolbe et al. 1995). Until 1980, the tidal flats of the German Wadden Sea were regarded as not endangered by oxygen deficiency. During benthos surveys around 1984, however, attention was drawn to small black spots occurring in sandy sediments without any visible traces of decaying plant or animal material. Within these spots the oxidized surface layer has disappeared, exposing the black reduced sediment layer. Surveys from 1987 onwards revealed that the number and size of the black spots increased dramatically on the tidal flats between Ems and Elbe (Michaelis et al. 1992). The presence of these black spots, which were recently also found in the Dutch Wadden Sea, is connected to eutrophication. An increased loading of the sediments with organic matter can easily lead to a flow of anaerobic seepage water from sandy, elevated spots in intertidal areas to the sediment surface (Philippart and Brinkman 1992).

The situation in the German Wadden Sea can be summarized as follows:

1. There is a decrease of seagrass distribution and abundance during and after the "wasting disease" in the 1930s, but to a far lesser extent than in the Dutch Wadden Sea. Some littoral populations of *Zostera marina* re-established themselves, but all sublittoral populations disappeared, owing to increased turbidity and eutrophication.
2. There is an increase of green macroalgal mats (mainly *Enteromorpha prolifera*) from the late 1970s onwards. From 1989 onwards, the increase becomes alarming, especially between the outlets of the rivers Ems, Weser and Elbe. Quantitative data to substantiate this 'increase', however, is scarce, but a monitoring system has been started recently (Kolbe et al. 1995).
3. There are changes in the infauna composition and abundance. An increased abundance of polychaetes in intertidal and subtidal sandy bottoms is manifest.
4. Königshafen is the well-studied northernmost locality of the German Wadden Sea. Seagrasses decreased in the 1930s but were rather common again in the late 1980s. The sublittoral *Zostera marina* populations disappeared, and a dramatic increase in green algal mats has recently been observed.

7.2.3 The Netherlands

Microphytobenthos (mainly benthic diatoms) and phytoplankton are the main primary producers of the Wadden Sea (Fig. 7.6), providing a net primary production of about $100\,g\,Cm^{-2}$ year $^{-1}$, in the Dutch Wadden sea. The primary production per average m^2 by littoral seagrass (*Zostera marina, Z. noltii*) and by macroscopic algae (mainly *Ulva* spp. and *Enteromorpha* spp.) is probably negligibly low, but actual biomass and production data for these macrophytes are too limited for sound generalizations. Before the outbreak of the "wasting disease" of the seagrass in 1932, the vast sublittoral *Zostera* beds (15 000 ha) in the Dutch Western Wadden Sea contributed considerably to primary production in this part of the Wadden Sea, probably about $20\text{--}40\,g\,Cm^{-2}$ year $^{-1}$ (Van den Hoek et al. 1979).

In the records of the nineteenth century, the total area occupied by *Zostera* in the Dutch western Wadden Sea and the Zuiderzee was estimated at 6500 ha, and by Van Goor (1921) at 15 000 ha (Den Hartog and Polderman 1975). Position and size of the sublittoral *Zostera marina* beds underwent long-term fluctuations in the period 1869–1930. The mapping of 1930 shows roughly a doubling of the area occupied by *Zostera marina*, compared with 1969. *Z. marina* also occurred in the former Zuiderzee, where it even grew in water with salinities lower than 6‰. *Z. marina* penetrated into eulittoral sand and mudflats as a form with narrow leaves, where it was accompanied by *Zostera noltii*. The latter species penetrated into the Zuiderzee as far as the 10‰ isohaline (Den Hartog 1970). *Z. noltii* at present occurs on eulittoral flats in the Wadden Sea from about mean high water neap downwards and *Z. marina* from about mean low water neap downwards. Where *Z. marina* grows in higher eulittoral positions it tends to occupy shallow depressions in the flats where water is retained during ebb; in such places *Z. noltii* occupies small elevations. Eulittoral *Zostera*-stands are found on nearshore flats and on tidal watersheds. The historic contribution (up to 1932) of sublittoral *Z. marina* to the production of organic matter in the Wadden Sea can be only very roughly approximated, because neither exact biomass nor primary production measurements from that period are available (Van den Hoek et al. 1979).

Upto 1930, there was quite an important local seagrass industry along the southwestern coasts of the Dutch Wadden Sea. Washed up seagrass was collected, and the beds which were sufficiently shallow during ebb, were harvested with scythes. The seagrass was used as packing material and for stuffing cushions and mattresses. In the Dutch Eastern Wadden Sea, and in the German Wadden Sea such an industry did not exist, apparently because extensive sublittoral *Zostera* meadows were lacking. However, in the German Northern Wadden Sea an important sublittoral *Z. marina* stand has been described for the coast of the isle of Sylt (Nienburg 1927; Wohlenberg 1935, 1937). It is likely that relatively high water transparency and the available vast shallow sublittoral flats promoted the development of the sublittoral *Zostera* meadows in the Dutch Western Wadden Sea and the German Northern Wadden Sea. The intervening stretch of Wadden Sea lacked sublittoral *Z. marina* beds probably as a result of its more turbid water, caused by the inflowing rivers, and by the larger tidal amplitudes resulting in stronger tidal currents.

In 1932 the "wasting disease" of *Z. marina*, which started in North America, reached the Netherlands (Den Hartog and Polderman 1975) and the German Northern Wadden Sea (Wohlenberg 1935, 1937), and it exterminated almost completely the sublittoral *Z. marina* beds. The eulittoral *Zostera noltii* was not affected. Only scattered eulittoral *Z. marina* stands and populations in brackish water ditches and ponds on the Wadden islands survived. From about 1940 onwards the decimated *Z. marina* populations all along the North Atlantic gradually increased in size and regained their lost grounds. Only in the Wadden Sea did this species never succeed in re-establishing its sublittoral beds. At present, only intertidal *Z. marina* stands occur. As *Z. marina* is incapable of surviving frost, its eulittoral populations are mostly annual and start their growth from seed each year (Van den Hoek et al. 1979).

Hydrographic changes caused by the closure of the Zuiderzee in 1932 are thought to have prevented the re-establishment of sublittoral *Z. marina* in the Dutch Western Wadden Sea. It is not clear which hydrographic changes were responsible, but erosion of the silty substratum favourable for the growth of *Zostera marina*, currents too strong for the establishment of seeds, and a rise of the average flood level by 20 cm are thought to be the main factors (Den Hartog 1970). It is, however, difficult to envisage that these three factors could have completely prevented the re-establishment of new favourable sublittoral habitats for *Z. marina*. One other possible cause could be a considerable increase in the turbidity of the seawater diminishing the light intensity below low water level to values lower than *Zostera's* light compensation value (Giesen et al. 1990a, b).

After the outbreak of the "wasting disease" in 1932 the significance of *Zostera* as a primary producer went down to almost nothing (Table 7.1). The presently remaining stands have only a local interest in the Dutch western Wadden Sea. In the German Wadden Sea they are much more important. The permanent disappearance of the vast sublittoral *Z. marina* beds in the Dutch western Wadden Sea is an interesting example of a large-scale effect on an estuarine ecosystem by human interference, in this case the closure of the Zuiderzee. A comparable interference, closure of Grevelingen estuary in the southwestern Netherlands, has had

Table 7.1. Rough approximations of biomass and production of *Zostera* in the Dutch Western Wadden Sea before and after the outbreak of the "wasting disease" in 1932

	1920–1932	1950–1960	1972–1973
Surface area of *Zostera* in ha	15 000	400–450(a)	160 (b)
Biomass in g C m^{-2}	85–350	10	4.5
Total biomass in tons C for Dutch Western Wadden Sea	12 750– 52 500	40–45	7
Production as g C m^{-2} year^{-1} in *Zostera* stands	100–500	14	6
Total production in tons C year^{-1} for Dutch Western Wadden Sea	15 000– 75 000	56–70	10
Production as g C m^{-2} year^{-1} computed for an average m^2 of the Dutch Western Wadden Sea (160 000 ha)	10–50	0.04	0.006

Only the recent eulittoral *Zostera* stands on the Balgzand and along the southern coast of Terschelling are quantitatively important: (a) mainly stands on Balgzand (100–150 ha) and along the southern coast of Terschelling (300 ha); (b) mainly stands on Balgzand (10 ha) and along the southern coast of Terschelling (150 ha) (Van den Hoek et al. 1979).

the reverse effect: it caused the establishment of vast sublittoral seagrass beds where these were originally lacking (Nienhuis 1983).

In recent years the eulittoral vegetations of *Z. marina* and *Z. noltii* in the Dutch western Wadden Sea have also been gradually declining (Table 7.1). Both the size of the stands and their density are gradually declining; those on the Balgzand near Den Helder (Fig. 7.6) have virtually disappeared. Den Hartog and Polderman (1975) speculated about pollution of the Wadden Sea as a complex cause of this decline. Water and silt from the river Rhine could indirectly introduce toxic materials that are perhaps noxious to the seagrasses, such as heavy metals, pesticides, PCBs and detergents. However, if this should be true, one would expect a comparable effect in the vicinity of the large German river mouths, where littoral *Zostera* stands still occur (Van den Hoek et al. 1979). An increase in the amount of floating algae (*Ulva*, *Enteromorpha*, *Chaetomorpha*) caused by the increasing nutrient concentrations might also act as a local cause of the death of *Zostera* because these algae sometimes form thick deposits on *Zostera* at low tide (Den Hartog and Polderman 1975).

The vanishing littoral seagrass stands have an important function, together with *Ulva-* and *Enteromorpha* stands, of food for brent geese (*Branta bernicla*) and wigeons (*Anas penelope*) during the winter half-year. The increasing grazing pressure by the growing populations of these birds in winter might also be responsible for the recent gradual decline of the eulittoral stands. These waterfowl also feed on the subterranean parts of *Zostera* and they are thus capable of eradicating *Zostera* stands (Van den Hoek et al. 1979).

In the intertidal as well as the subtidal zone of the Wadden sea multicellular algae play a subordinate role. The sand and mudflats are mostly devoid of any obvious, macroscopic algal growth. Concentrations of macroscopic algae tend to occur where the tidal currents have low velocities and wave action is limited. Intertidal seagrass vegetations as well as macroscopic algae congregate in near-shore zones and on the tidal watersheds. The algae grow attached either to *Zostera* leaves and rhizomes or to scattered shells and shellfragments (Michealis 1969, 1970).

The macroalgal vegetation consists mainly of *Enteromorpha* and *Ulva* species starting their development in April/May, and having explosive growth in June/July, and declining in August/September. Small, scattered plants and full-grown sediment-covered plants ensure survival during winter, showing vegetative proliferation during springtime after being uncovered. Full-grown plants, whether or not growing attached to shells, are repeatedly removed by the incoming and receding tidal streams which causes continuously changing vegetation patterns (Van den Hoek et al. 1979).

Recent data show that the biomass of *Ulva* species may vary with a factor 15 within the course of a year. Development of *Ulva* in the Wadden Sea seems to depend largely on the course of the summer. Bright and calm summers are favourable for its development. In 1988 and 1989, both years having a bright summer, *Ulva* was notably present. Mussel beds are an important substrate for *Ulva* species. In June, many of the plants become afloat, mainly due to the activities of the mussel farmers starting to harvest mussels with dredge nets. Currents determine whether *Ulva* can be found on the flats or in the tidal channels. In the Dutch Wadden Sea there are indications for an increase in the mean annual biomass of *Ulva* over the last 30 years (De Jonge et al. 1993).

Recent information on seagrasses is summarized by De Jonge and De Jong (1992) Philippart et al. (1992) and Philippart (1995). Seagrasses have become rare in the Dutch Wadden Sea. The sublittoral populations never recovered after the 1930s but also the intertidal populations are very thin nowadays (only several hundreds of hectares over an area of hundreds of km^2). The recent destruction of small and patchy habitats by shellfish-fisheries activities, is mainly held responsible for the almost total extinction of both *Zostera marina* and *Z. noltii*. The habitat map published by Dijkema (1991) allows a comparison between the Danish, German and Dutch Wadden Sea: nowadays seagrasses are rare in the

Dutch Wadden Sea and in the estuaries of the rivers Elbe, Weser and Ems, and rather common in the German Wadden Sea.

The Dutch Wadden Sea did not show a recent shift from the seagrass-dominance to green algae-dominance (compare Königshafen). However, local accumulations of macroalgae such as *Enteromorpha* and *Ulva* can be frequently observed (Beukema 1992). No attempt to quantify the amount of green algae and to compare historic data with recent data have been undertaken. Obviously, local accumulations of green algae are not considered as a management problem in the Dutch Wadden Sea.

The stretch of 200 km of sandy beaches along the west coast of the Netherlands does not suffer from undesired accumulations of green algae. The beaches are exposed to the open sea, and sand is transported continuously by wind and waves. Amenity beaches are mechanically cleaned during summer of debris and other floating objects. Seaweeds washed ashore do not cause a serious problem.

Because of the northerly-directed residual current along the coast of The Netherlands (Fig. 7.3) less than 5% of the original Rhine – Meuse water reaches the estuaries and brackish lagoons in the southwest Netherlands. Oosterschelde, Grevelingen, and Veerse Meer are therefore mainly loaded with nutrients from diffuse sources, such as agricultural runoff, treated waste water, and drainage canals. The saline water bodies in the southwest Netherlands are spatially separated (compartmentalization) by dikes and sluices constructed during the "Deltaworks", a civil engineering scheme that was in operation between 1957 and 1987, and aimed to restrict the sea-born flow of the main estuaries for safety reasons. Consequently, each of these (former) estuaries has its own eutrophication history and its own specific water regime. These characteristics mean that each unit has to be managed according to its particular conditions (Nienhuis 1992; Fig. 7.10).

Some properties of present-day Dutch estuaries vary considerably (Table 7.2). The residence time of the water masses in non-tidal Grevelingen and Veerse Meer is long compared to the same characteristic in tidal estuaries. The net freshwater load directly derived from the Rhine and Meuse rivers on Oosterschelde, Veerse Meer, and Grevelingen is very small: 1-2% of the discharge of the rivers. The Wadden Sea receives approximately 400 m^3 s^{-1} of water from the Rhine; Westerschelde and Ems–Dollard each receive 100 to 150 m^3 s^{-1} of fresh water from their tributaries. Veerse Meer experiences almost permanent stratification, whereas in Grevelingen and Ems–Dollard only a few deep channels are stratified during summer. The Westerschelde (marine section), Oosterschelde, and Wadden Sea estuaries are completely mixed

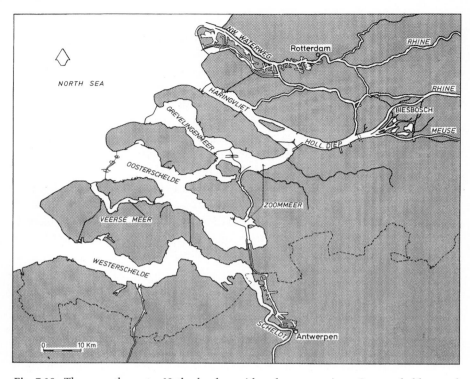

Fig. 7.10. The south-west Netherlands with the estuaries Oosterschelde and Westerschelde and the brackish lagoons Grevelingenmeer and Veerse Meer. The Biesbosch, Hollands Diep, Haringvliet and Nieuwe Waterweg form the lower reaches of the Rhine and Meuse. The Westerschelde estuary is connected to the river Scheldt. The Delta works started in 1960 and were finished in 1987. The Volkerakdam was closed in 1969 between Hollands Diep-Haringvliet and Zoommeer, and blocked the main southern exit of Rhine water

tidal systems. Westerschelde and Ems–Dollard are extremely turbid, Grevelingen has very clear water, and Oosterschelde, Veerse Meer, and Wadden Sea are in between (Nienhuis 1992, 1993).

There is no specific factor that limits growth of algae in Dutch estuaries. In the turbid Westerschelde the availability of light limits phytoplankton dynamics. In part of the Western Wadden Sea, phosphate may be limiting (Veldhuis 1987). Nitrogen may be the limiting factor for phytoplankton growth in Oosterschelde and Grevelingen (de Vries et al. 1988a; Westeijn et al. 1990), and occasionally also in Veerse Meer. The range of nutrient concentrations in Dutch estuaries differs greatly. Values above $1\,\mathrm{mg\,l}^{-1}$ for N, P and Si seldom occur in Oosterschelde and Grevelingen, and nutrient concentrations frequently approach zero during blooms of phytoplankton. Veerse Meer has higher maximum

Table 7.2. Properties of the Dutch estuaries (Nienhuis 1992) load = freshwater load

	Westerschelde	Oosterschelde	Grevelingen	Veerse Meer	Ems-Dollard	Wadden Sea
Area (km^2)	300	350	108	22	460	1200
Residence time (days)	30–90	5–40	180–360	±180	14–70	8–15
Load (m^3 s^{-1})	100	20	5	3	150	400
Tides	+	+	–	–	+	+
Stratification	–	–	±	+	±	–
Extinction coefficient (m^{-1})	0.5–7	0.4–1.5	0.2–0.5	0.3–1.4	1–7	0.5–3
Chlorinity ‰	0–17	15–17	14–16	8–12	0–17	10–17

values for N and Si, but depletion occurs during the growing season (de Vries et al. 1988b). Westerschelde and Ems–Dollard have the highest nutrient concentrations, never approaching zero during spring and summer and reaching high values during winter ($8\,mg\,l^{-1}N$, $4\,mg\,l^{-1}P$, and $9\,mg\,l^{-1}$ Si, Nienhuis 1992).

The total nitrogen loadings to Dutch coastal waters vary by two orders of magnitude (4–$235\,g\,N\,m^{-2}$ year^{-1}). Mean N concentrations show a much smaller range of 0.5–$4.6\,mg\,l^{-1}$. Obviously nitrogen loads of 40–$200\,g\,N\,m^{-2}$ year^{-1} do not give rise to extremely high chlorophyll concentrations during summer in the turbid Dutch estuaries and in the coastal waters. In such places light availability rather than nutrient supply may limit net primary production. In the clear waters of Veerse Meer, a load of $34\,g\,N\,m^{-2}$ year^{-1} results in high chlorophyll concentrations ($100\,mg\,chl\ a\ m^{-3}$ or even higher; according to de Vries et al. 1988b). Production of phytoplankton biomass in Veerse Meer, and consequently deposition of particulate organic carbon on the bottom sediments, was large enough during 1980–1983 to increase the area of anaerobic sediment from 4 to 25% of the bottom surface. Veerse Meer Lagoon is vulnerable to eutrophication because of the long residence time of the water mass, the low extinction coefficient, and the almost permanent stratification (Nienhuis 1992).

In Fig. 7.11 a tentative carbon budget of the main categories of primary producers in all Dutch estuaries is given. It has to be realized that the "pies" depict average, annual, integrated data, useful for reasons of comparison, but do not reflect temporal and spatial variability. Phytoplankton are the dominant primary producers in Dutch estuaries, contributing 45–71% to the overall annual budget of organic material. Notwithstanding the large differences in nutrient loadings of the separate waters, primary production of phytoplankton only shows a 2.4-fold difference between the lightlimited, turbid Ems-Dollard ($100\,g\,C\,m^{-2}$ year^{-1} and Westerschelde ($125\,g\,C\,m^{-2}$ year^{-1}), and the clear, presumably not nutrient-limited Veerse Meer ($240\,g\,C\,m^{-2}$ year^{-1}). Production levels in Wadden Sea, Oosterschelde, and Grevelingen are intermediate. Production of microphytobenthos (benthic diatoms, green algae, etc.) is roughly 30–70% of the production of phytoplankton.

High turbidity and exposure to waves and tides prevent the potential sediment habitats in Westerschelde and Ems-Dollard from being invaded by macrophytes. Oosterschelde and Western Wadden Sea have only local growth of macrophytes on sediment substrates in sheltered regions (*Zostera marina*, *Z. noltii*, *Enteromorpha prolifera*, *Ulva* species). In the brackish lagoons Grevelingen and Veerse Meer, macroalgae (mainly green algae) and seagrasses contribute significantly to the carbon budget. In Grevelingen the rooted seagrass *Zostera marina* covered 20% of the surface area of the lagoon and had a net production of $150\,g\,C\,m^{-2}$ $year^{-1}$. Eelgrass contributes only 14% to the annual carbon budget, while phytoplankton provides 60% (Fig. 7.11). Recently the population of *Zostera marina* in Grevelingen drastically decreased: from an area of 4000 ha in 1986 to 300 ha in 1993. Water quality did not change during that period. The water is still very clear and rather poor in nutrients. Spatial competition between seagrasses and macroalgae is out of the question. Macroalgae play a subordinate role in Grevelingen lagoon (cf. Fig. 7.12). There is much speculation on the causes of the recent seagrass decline in the Grevelingen lagoon. Obviously, the decrease cannot be attributed to eutrophication phenomena. One of the stronger hypotheses connects nitrogen limitation in the environment to the retention of the generative reproduction in *Zostera marina* (Van Lent and Verschuure 1994) and hence to the decline of the entire population.

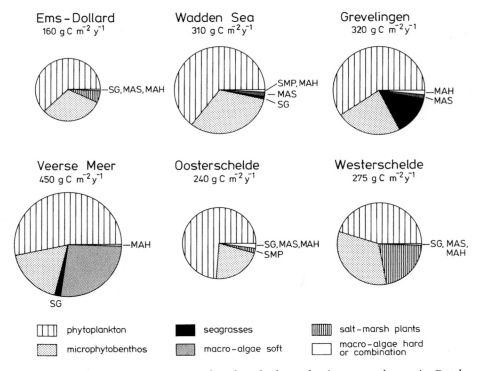

Fig. 7.11. Preliminary average annual carbon budget of primary producers in Dutch estuaries. *SG* Seagrass; *MAS* macroalgae on soft substrates; *MAH* macroalgae on hard substrates; *SMP* salt-marsh plants. (Nienhuis 1992)

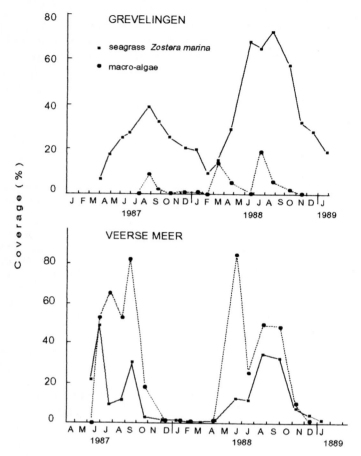

Fig. 7.12. Percentage coverage of *Zostera marina* and macroalgae in vertical projection on the sediment in permanent sample plots in Veerse Meer lagoon (*below*) and Grevelingen lagoon (*above*). Size of sample plots is $15 \times 15 \, m^2$, water depth at both localities is 1.25 m. (Nienhuis 1989)

In eutrophic waters, such as Veerse Meer, macroalgae became more prominent. Here, seagrasses cover only 3% of the surface area of the lagoon while macroalgae – mainly *Ulva* – cover 20%. The lagoon is dominated by *Ulva* species during summer, producing roughly $500 \, g \, C \, m^{-2} \, year^{-1}$ at a maximum in shallow areas. The averaged contribution to the annual carbon budget of Veerse Meer is $120 \, g \, C \, m^{-2} \, year^{-1}$, which is 27% of the lagoon's budget (Fig. 7.11). The high nitrogen load of Veerse Meer not only resuls in a high producttion of phytoplankton, but also in mass growth of *Ulva* in shallow areas.

Figure 7.12 shows the increase and decrease in biomass of macrophytes, expressed as percentage coverage, in permanent sample plots in the Grevelingen and in the Veerse Meer. In the Grevelingen the annual cycle of changes in biomass of seagrass is not disturbed by macroalgae, which have only a low presence. In the Veerse Meer the seagrass plot becomes completely dominated by quickly growing *Ulva*, suppressing the growth of *Zostera marina*.

7.2.4 Belgium

The Belgian North Sea coast south of the Netherlands (Fig. 7.1) consists of 75 km of sandy beaches with local jetties and harbour constructions as the only hard substrate. No published information is available on the distribution of seagrasses or green algae along the Belgian coast (E. Coppejans, pers. comm., State University, Gent). From my own experience I know that stone piers and jetties are covered with green algae, showing the 'normal' picture of a semi-exposed sandy beach. I have no knowledge of recent exceptional green algal growth of floating algal accumulations on beaches.

Some information is available on the Spuikom of Oostende, an enclosed eutrophic brackish-water basin connected with the North Sea by sluices. Mass developments of *Ulva scandinavica* dominate the basin during summer, accompanied by *Enteromorpha linza* (Coppejans and Gillis 1983).

7.3   Conclusions

1. Accumulations of green macroalgae appear to be a local management problem, connected to concentrated loadings of inorganic nutrients and to the specific hydrographical conditions of enclosed coastal areas, such as fjords, lagoons, bays and harbours.
2. Eutrophication related macroalgae accumulations increasingly occur along the coasts of Denmark, Germany and The Netherlands. A conservative attitude towards differences between historic and recent data sets and data interpretations, however, is needed for the North Sea coasts.
3. The Baltic habitats of Denmark comprise non-tidal water masses, sheltered lagoons with long residence time, and hence actual and potential sites for eutrophication related mass occurrence of

macroalgae. The eulittoral and sublittoral vegetation of macrophytes changed significantly after World War II, owing to increased eutrophication. *Zostera marina* communities disappeared or reduced substantially and were replaced by green algal accumulations (*Ulva* spp.) in shallow habitats.

4. In the German Wadden Sea a decrease in seagrass distribution and abundance was observed during and after the wasting disease in 1931–1933, but to a lesser extent than in the Dutch Wadden Sea. Some eulittoral populations of *Zostera marina* re-established themselves, but all sublittoral populations disappeared, owing to increased turbidity and eutrophication.

5. There was an increase in green macroalgal mats (mainly *Enteromorpha prolifera*) from 1978 onwards. After 1989, the increase becomes alarming, especially between the outlets of the river Ems, Weser and Elbe. Quantitative evidence to substantiate this algal "increase", however, is scarce.

6. In the Dutch Wadden Sea sublittoral seagrasses (*Zostera marina*) have become extinct after the "wasting disease" in 1932–1933. In the 1970s eulittoral *Z. marina* and *Z. noltii* stands also gradually declined. Increased eutrophication, turbidity and bottom trawling by fisherman are mentioned as responsible factors.

7. Communities in the Dutch Wadden Sea did not show a shift from seagrass-dominance to green-algal dominance (cf. German Wadden Sea). Local accumulations of macroalgae (*Enteromorpha* and *Ulva*) frequently occur and are seemingly increasing but these vegetations are not considered as a management problem.

8. The 300 km of exposed sandy beaches along the west coast of the Netherlands and Belgium do not suffer from undesired accumulations of green algae.

9. The estuaries and lagoons in the SW Netherlands show a diversified picture: Grevelingen lagoon is a stagnant mesotrophic watermass, where *Zostera marina* is the (potential) dominant and green macroalgae are virtually absent. Veerse Meer lagoon is a stagnant eutrophic lake, where *Ulva* spp. dominates and creates nuisance to water managers. Oosterschelde is a mesotrophic clear tidal estuary where green algae are subordinate. Westerschelde is a turbid, eutrophic tidal estuary, where turbidity is presumably limiting macroalgae mass development.

Acknowledgments. I thank Dr. H. Michaelis (Nordeney), Mr. D. de Jong (Middelburg) and Mr. H. Elgershuizen (Leeuwarden); they provided me with unpublished data from German and Dutch Wadden Sea and from the estuaries in the SW Netherlands. Publication no.2132 Netherlands Institute of Ecology, Centre for Limnology, Nieuwersluis, The Netherlands.

References

Anderson L, Rydberg L (1988) Trends in nutrient and oxygen conditions within the Kattegat: effects of local nutrient supply. Estuar Coast Shelf Sci 26: 559–579

Asmus H (1987) Secondary production of an intertidal mussel bed community related to its storage and turnover compartments. Mar Ecol Prog Ser 39: 251–266

Bak HP (1979) Alegras, *Zostera marina* L.i Limfjorden 1978. MS Thesis, Bot Inst, Arhus Univ, 124 pp (in Danish)

Beukema JJ (1992) Long-term and recent changes in the benthic macrofauna living on tidal flats in the western part of the Wadden Sea. Neth Inst Sea Res Publ Ser 20: 135–141

Beukema JJ, Cadée GC (1986) Zoobenthos responses to eutrophication of the Dutch Wadden Sea. Ophelia 26: 55–64

Borum J (1985) Development of epiphytic communities on eelgrass (*Zostera marina*) along a nutrient gradient in a Danish estuary. Mar Biol 87: 211–218

Borum J, Wium–Andersen S (1980) Biomass and production of epiphytes on eelgrass (*Zostera marina* L.) in the Øresund, Denmark. Ophelia (Suppl) 1: 57–64

Cadée GC (1986) Increased phytoplankton primary production in the Marsdiep area (Western Dutch Wadden Sea). Neth J Sea Res 20: 285–290

Coppejans E, Gillis J (1983) Quelques Clorophyceae et Phaeophyceae marines nouvelles pour la flore Belge provenant du Bassin de Chasse d'Oostende. Biol Jaarb Dodonaea 51: 55–66

Dame RF, Dankers N (1988) Uptake and release of materials by a Wadden Sea mussel bed. J Exp Mar Biol Ecol 118: 207–216

De Jonge VN, De Jong DJ (1992) Role of tide, light and fisheries in the decline of *Zostera marina* L. in the Dutch Wadden Sea. NIOZ Publ Ser. 20: 161–176

De Jonge VN, Essink K, Boddeke R (1993) The Dutch Wadden Sea: a changed ecosystem. Hydrobiologia 265: 45–71

Den Hartog C (1970) The seagrasses of the world. Verh K Ned Akad Wet (Afd Natuurk 2 R) 59: 1–275

Den Hartog C, Polderman PJG (1975) Changes in the seagrass populations of the Dutch Wadden Sea. Aquat Bot 1: 141–147

De Ruyter WPM, Postma L, de Kok JM (1987) Transport Atlas of the Southern North Sea. Rijkswaterstaat Tidal Waters Division, Delft Hydraulics, Den Haag, pp1–34

De Vries I, van Raaphorst W, Dankers N (1988a) Extra voedingsstoffen in zee: gevolgen, voordelen, nadelen. Landschap 5: 270–285

De Vries I, Hopstaken F, Goossens H, de Vries M, de Vries H, Heringa J (1988b) GREWAQ: an ecological model for Lake Grevelingen. Rep T–0215–03. Rijkswaterstaat DWG, Delft Hydraulics, pp1–159, 1–83

De Vries I, Los H, Jansen R, Cramer S, vd Tol M (1993) Risico-analyse eutrofiëring Noordzee. Rijkswaterstaat-Waterloopkundig Laboratorium. Rep DGW 93.029: 1–89

Dijkema KS (1991) Towards a habitat map of the Netherlands, German and Danish Wadden Sea. Ocean Shoreline Manage 16: 1–21

Eisma D (1987) The North Sea: an overview. Philos Trans R Soc Lond B 316: 461–485

Essink K (1984) The discharge of organic waste into the Wadden Sea: local effects. Publ Ser Neth Inst Sea Res 10: 165–178

Essink K, Beukema JJ (1986) Long–term changes in intertidal flat macrozoobenthos as an indicator of stress by organic pollution. Hydrobiologia 142: 209–215

Frederiksen OT (1987) The fight against eutrophication in the inlet of Odense Fjord by reaping of sea lettuce (*Ulva lactuca*). Wat Sci Tech 19: 81–87

Funen County Council (1991) Eutrophication of coastal waters. Coastal Water Quality Management in the County of Funen, Denmark, 1976–1990, 288pp

Geertz-Hansen O, Sand-Jensen K, Hansen DF, Christiansen A (1993) Growth and grazing control of abundance of the marine macroalga *Ulva lactuca* L. in a eutrophic Danish estuary. Aquat Bot 46;101–109

Gerlach SA (1990) Nitrogen, phosphorus, plankton and oxygen deficiency in the German Bight and in Kiel Bay. Kiel Meeresforsch Sonderh 7: 1–340

Giesen WBJT, van Katwijk MM, den Hartog C (1990a) Temperature, salinity, insolation and wasting disease of eelgrass (*Zostera marina* L.) in the Dutch Wadden Sea in the 1930s. Neth J Sea Res 25: 395–404

Giesen WBJT, van Katwijk MM, den Hartog C (1990b) Eelgrass condition and turbidity in the Dutch Wadden Sea. Aquat Bot 37: 71–85

Hagmeier A (1941) Die intensive Nutzung des nordfriesischen Wattenmeeres durch Austern- und Muschelkultur. Z Fisch 39: 105–165

Hagmeier A, Kandler R (1927) Neue Untersuchungen im nordfriesischen Wattenmeer und auf den fiskalischen Austernbanken. Wiss Meeresunters Abt Helgol 16: 1–90

Hauser B, Michaelis H (1975) Die Makrofauna der Watten, Strände, Riffe und Wracks um den Hohen Kechtsand in der Wesermündung. Jahresber 1974, Forschungsstelle Insel- und Küstenschutz 26: 85–120

Jørgensen BB (1980) Seasonal oxygen depletion in the bottom waters of a Danish fjord and its effect on the benthic community. Oikos 34: 68–76

Josefson (1990) Increase of benthic biomass in the Skagerrak–Kattegat during the 1970s and 1980s – effects of organic enrichment? Mar Ecol Prog Ser 66: 117–130

Koeman R (1975) Die Makroflora der Watten, Strände und Riffe um den hohen Knechtsand in der Wesermündung. Jahresber 1974, Forschungsstelle Insel und Küstenschutz 26: 41–52

Kolbe K, Kaminski E, Michaelis H, Obert B, Rahmel J (1995) Macroalgal mass development in the Wadden Sea:first experience with a monitoring system. Helgol Meeresunters 49: 519–528

Kornmann P (1952) Die Algenvegetation von List auf Sylt. Helgol Wiss Meeresunters 4: 55–61

Linke O (1939) Die Biota des Jadebusenwattes. Helgol Wiss Meeresunters 1: 201–348

Madsen PB, Jensen K (1987) Population dynamics of *Macoma balthica* in the Danish Wadden Sea in an organically–enriched area. Ophelia 27: 197–208

Martens P (1989) Inorganic phytoplankton nutrients in the Wadden Sea areas off Schleswig–Holstein. I. Dissolved inorganic nitrogen. Helgol Meeresunters 43: 77–85

Mathiesen H, Mathiesen L (1976) Fastvoksende vegetation i Randers Fjord og Grund Fjord.-Kap 2: 16–22 Gudenaundersøgelsen. Randers Fjord. Enviroplan Aj5 København.

Michaelis H (1969) Makrofauna und Vegetation der Knechtsandwatten. Jahresber Forschungsstelle Insel–und Küstenschutz Norderney 1967 (19): 147–173

Michaelis H (1970) Biologische Untersuchung der Watten und Landgewinnungsfelder bei Schillighörn. Jahresber 1968 Forschungsstelle Insel– und Küstenschutz 20: 61–76

Michaelis H (1978) Recent biological phenomena in the German Wadden Sea. Rapp Réun Cons Int Explor Mer 172: 276–277

Michaelis H (1987) Bestandsaufnahme des eulitoralen Makrobenthos im Jadebusen in Verbindung mit einer Luftbild–Analyse. Jahresber 1986 Forschungsstelle Küste Norderney 38: 1–97

Michaelis H, Kolbe K, Thiessen A (1992) The "black spot disease" (anaerobic surface sediments) of the Wadden Sea. ICES Statutory Meet, Rostock, September 1992. Code–Nr E: 36 (unpubl rep)

Nienburg W (1925) Eine eigenartige Lebensgemeinschaft zwischen *Fucus* und *Mytilus*. Ber Deutsch Bot Ges 43: 292–298

Nienburg W (1927) Zur Oekologie der Flora des Wattenmeeres I. Teil. Der Königshafen bei List auf Sylt. Wiss Meeresunters Abt Kiel 20: 146–196

Nienburg W, Kolumbe F (1931) Zur Oekologie der Flora des Wattenmeeres. II. Teil. Das Neufelder Watt im Elbmündungsgebiet. Wiss Meeresunters Abt Kiel 21: 1–112

Nienhuis PH (1983) Temporal and spatial patterns of eelgrass (*Zostera marina* L.) in a former estuary in the Netherlands, dominated by human activities. Mar Technol Soc J 17: 69–77

Nienhuis PH (1992) Eutrophication, water management and the functioning of Dutch estuaries and coastal lagoons. Estuaries 15: 538–548

Nienhuis PH (1993) Nutrient cycling and foodwebs in Dutch estuaries. Hydrobiologia 265: 15–44

Ostenfeld CH (1908) On the growth and distribution of eelgrass (*Zostera marina*) in Danish waters (In Danish). Rep Dan Biol Stn 16: 1–62

Paris Commission (1992) Nutrients in the convention area. Part B: Eutrophication Symptoms and Problem Areas.

Petersen CGJ (1891) Fiskenes biologiske forhold i Holbak fjord 1890–(91). Berentn Minist Landbr Fisk Dan Biol Stn 1: 121–184

Petersen CGJ (1913) Om Baendeltangens (*Zostera marina*) Aars–Produktion i de danske Farvande. Mindeskr Japetus Steenstrups Fods 9: 1–20, GEC Gad København

Petersen CGJ (1915) A preliminary result of the investigations on the valuation of the sea. Rep Dan Biol Stn 23: 29–32

Petersen CGJ (1918) The sea bottom and its production of fish–food. A survey of the work done in connection with the valuation of the Danish waters from 1883–1917. Rep Dan Biol Stn 26: 1–62

Petersen CGJ, Jensen PB (1911) Valuation of the sea I: Animal life of the sea bottom, its food and quantity. Rep Dan Biol Stn 20: 1–79

Philippart CJM (1995) Seasonal Variation in growth and biomass of an intertidal *Zostera noltii* stand in the Dutch Wadden Sea. Neth J Sea Res 33: 205–218

Philippart CJM, Brinkman B (1992) Herkomst zwarte vlekken in de Waddenzee Zwartkijkers? Waddenbulletin 27: 90–93

Philippart CJM, Dijkema KS, van der Meer J (1992) Wadden Sea seagrasses: where and why? NIOZ Publ 20: 177–191

Pinnerup SP (1980) Leaf production of *Zostera marina* L. at different salinities. Ophelia, (Suppl 1): 219–224

Rachor E (1990) Changes in sublittoral zoobenthos in the German Bight with regard to eutrophication. Neth J Sea Res 25(1/2): 209–714

Radach G, Berg J (1986) Trends in den Konzentration der Nährstoffe und des Phytoplanktons in der Helgoländer Bucht (Helgoland Reede Daten). Ber Biol Anst Helgol 2: 1–63

Rasmussen E (1973) Systematics and ecology of the Isefjord marine fauna (Denmark). With a survey of the eelgrass (*Zostera*) vegetation and its communities. Ophelia 11: 81–507

Rasmussen TE (1977) The wasting disease of eelgrass (*Zostera marina*) and its effects on environmental factors and fauna. In: McRoy CP, Helfferich C (eds) Mar Sci 4: 1–15

Reise K (1983) Long-term changes in the macrobenthic invertebrate fauna of the Wadden Sea: are polychaetes about to take over? Neth J Sea Res 16: 29–36

Reise K (1985) Tidal flat ecology. Springer, Berlin Heidelberg New York, 191pp

Reise K, Schubert A (1987) Macrobenthic turnover in the subtidal Wadden Sea: the Norderaue revisited after 60 years. Helgolander Meeresunters 41: 69–82

Reise K, Heere E, Sturm K (1989) Historical changes in the benthos of the Wadden Sea around island of Sylt in the North Sea. Helgol Meeresunters 43: 417–433

Riesen W, Reise K (1982) Macrobenthos of the subtidal Wadden Sea: revisited after 55 years. Helgol Meeresunters 35: 409–423

Rosenberg R (1985) Eutrophication – the future marine coastal nuisance? Mar Poll Bull 16: 227–231

Rosenberg R, Elmgren R, Fleischer S, Jonsson P, Persson G, Dahlin H (1990) Marine eutrophication case studies in Sweden. Ambio 19(3): 102–108

Sand–Jensen K (1975) Biomass, net production and growth dynamics in an eelgrass (*Zostera marina* L.) population in Vellerup Vig, Denmark. Ophelia 14: 185–201

Setchell WA (1929) Morphological and phenological notes on *Zostera marina* L. Univ Calif Publs Bot 14: 389–452

Short FI, Ibelings BW, den Hartog C (1988) Comparison of a current eelgrass disease to the wasting disease of the 1930s. Aquat Bot 30: 295–304

Steffensen DA (1976) The effect of nutrient enrichment and temperature on the growth in culture of *Ulva lactuca* L. Aquat Bot 2: 33–351

Van den Hoek C, Admiraal W, Colijn F, De Jonge VN (1979) The role of algae and seagrasses in the ecosystem of the Wadden Sea: a review. In: Wolff WJ (ed) Flora and vegetation of the Wadden Sea. (Rep Wadden Sea Working Group 3) Balkema, Rotterdam, pp 9–118

Van der Veer HW, van Raaphorst W, Bergman MJN (1988) Eutrophication of the Dutch Wadden Sea – external nutrient loadings of the Marsdiep and Vliestroom basins. In: EON–project group (ed) The ecosystem of the western Wadden Sea: field research and mathematical modelling. EMOWAD II – NIOZ Rep 1988: 113–122

Van Goor ACJ (1921) Die *Zostera*–Assoziation des holländischen Wattenmeeres. Rec Trav Bot Neerl 18: 103–123

Van Lent F, Verschuure JM (1994) Intraspecific variability of *Zostera marina* L. (eelgrass) in estuaries and lagoons of Southwestern Netherlands II. Relation with environmental factors. Aquat Bot 48: 59–75

Veldhuis MJW (1987) The ecophysiology of the colonial alga *Phaeocystis pouchetti*. PhD Thesis, Univ Groningen, Van Denderen Groningen, 127pp

Wetsteijn LPMJ, Peeters JCH, Duin RNM, Vegter F, de Visscher RPM(1990) Phytoplankton primary production and nutrients in the Oosterschelde (The Netherlands) during the pre–barrier period 1980–1984. Hydrobiologia 195: 163–177

Wium–Andersen S, Borum J (1980) Biomass and production of eelgrass (*Zostera marina* L.) in the O in the Øresund, Denmark, Ophelia (Suppl 1): 49–55

Wohlenberg E (1935) Beobachtungen über das Seegras, *Zostera marina*, und seine Erkrankung im nordfriesischen Wattenmeer. Nordelbingen 11: 1–19

Wohlenberg E (1937) Die Wattenmeer-Lebensgemeinschaften im Königshafen von Sylt. Helgol Wiss Meeresunters 1: 1–92

Ziegelmeier E (1977) Ein Naturexperiment im Watt. Beobachtungen an einer Miesmuschelbank. Natur Mus Frankfurt 107: 239–243

8 The British Isles

R.L. Fletcher

8.1 Introduction

Eutrophication of coastal waters by anthropogenic sources of nutrients is a worldwide phenomenon which has important ecological and environmental implications. With its extensive coastline and the dramatic increase in industrialization, urbanization and intense farming practices that have taken place on its coastal plains from the latter part of the 19th century, it is perhaps not surprising that the coastal waters of the British Isles have been subjected to increased loading of pollutants including nutrients. Certainly, there have been many reports of increased nutrient levels in coastal regions associated with population increases. These increased nutrient levels have been attributed to increased sewage input and increased land fertilization rates (Hull 1987; Raffaelli et al. 1989). This has been particularly well documented for estuaries, such as the Tyne (James and Head 1972) and the Thames (Graham 1938), which are usually the primary pathway for the disposal of many wastes to the sea in the British Isles (Edwards 1972). Very large quantities of sewage, for example, are discharged directly through pipes into coastal waters around the British Isles, often in close proximity to shores (Wood 1988). Whilst this increased loading of nutrients into British coastal waters has generally been considered beneficial by enhancing pelagic and benthic productivity and improving fish catches, for example, James and Head (1972) reported increased phytoplankton activity near the mouth of the river Tyne, they can also be detrimental. In this respect, there is particular concern about their influence on marine benthic algal communities, notably the formation and accumulation of excessive growths (termed "green tides") of a relatively small number of green macroalgae.

The Marine Laboratory, School of Biological Sciences, University of Portsmouth, Ferry Road, Hayling Island, Hampshire, PO 11 ODG, UK

Ecological Studies, Vol. 123
Schramm/Nienhuis (eds) Marine Benthic Vegetation
© Springer-Verlag Berlin Heidelberg 1996

The present paper examines the phenomenon and environmental impact of eutrophication on benthic algal communities around the British Isles.

8.2   Sources of Nutrients

In general, primary producers rely on the natural annual cycle of nutrients to complete their biological processes. This primarily revolves around the pelagic activities associated with the spring bloom of phytoplankton. Locally it can involve the decay of accumulations of rotting seaweed (Thrush 1986). Anthropogenic sources of nutrients can, however, be important and have been found to exert a major influence on both pelagic and benthic communities. Around the British Isles, the three main sources of these nutrients are industrial wastes, sewage discharges and agricultural runoff.

The relative proportions of these sources entering the marine environment will vary for different localities. Not surprisingly, estuaries that are heavily industrialized, such as the Tyne on the NE coast of England, will be a source of both domestic and trade waste nutrient discharges, whilst those which are less populated and support only light industries, for example the Wear, also in NE England, and the Ythan in NE Scotland, are more likely to receive a higher proportion of nutrients derived from agriculture and sewage. The source will also determine the chemical form of the nutrients, for example a large amount of ammoniacal nitrogen was reported to be released into Southampton Water from industrial sources (Wright 1980). Tables 8.1–8.3 show the pollution characteristics of three contrasting eutrophicated sites in the British Isles; Langstone Harbour in which the nutrients are derived almost entirely from sewage discharges, Dublin Bay in which the main source of nitrogen is agriculture but which also receives a high input from sewage,

Table 8.1. Sources of pollution load in Langstone Harbour, south coast of England (expressed as kg day^{-1}). (After Wright 1980)

	Ammonium N	Nitrate N	Phosphate P
Streams	3.01	3.24	3.64
Industrial	–	–	–
Sewage	3038	886	1344
Total	3041	889	1347

Table 8.2. Sources of pollution load discharged in Southampton Water (expressed as kg day^{-1}). (After Wright 1980)

	Ammonium N	Nitrate N	Phosphate P
Streams	198	8445	300
Industrial	2675	645	82.4
Sewage	1803	542	517

Table 8.3. Sources of input of nitrogen in Dublin Bay, Ireland. (After Wilson et al. 1990)

River Input	71.6%
Domestic Sewage	24.4%
Atmosphere	4.1%
Natural Nitrogen Fixation	0.2%

and Southampton Water which has a relatively high agricultural input of nitrogen but also receives a substantial amount from local industries. Note, however, that although in some localities like the Ythan, which have a relatively high agricultural input of nutrients, sewage-derived nutrients can still exert an important local effect (Raffaelli et al. 1989).

8.3   Distribution of "Eutrophicated" Localities

Table 8.4 lists the localities in the British Isles that have experienced changes in marine benthic algal communities as a result of eutrophication. It can be seen that eutrophication phenomena have been reported at widely distributed sites around the British Isles including the east and west coasts of Scotland, the south coast of England and the east coast of Ireland. Sites affected include bays, harbours, estuaries, loughs and lagoons. It can also be seen that some reports originate in the late 19th and early 20th century (see Baily 1886; Cotton 1910, 1911; Letts and Richards 1911), indicating that eutrophication is not a new phenomenon. The majority of reports, however, are more recent, notably during the past 20–30 years and this might well be as a result of greater public awareness, the rapid population growth around the coasts and increased recreational pressure (Wilson et al. 1990). It can also be seen that some localities have received very little attention whilst others have been the subject of intense scientific study; the latter include Dublin Bay in Ireland, the Ythan Estuary in Scotland and Langstone Harbour on the

Table 8.4. Bibliographic index of "eutrophic" sites in the British Isles where qualitative and/or quantitative changes have been reported to occur in the marine benthic flora

ENGLAND

Langstone Harbour
  Dunn (1972); Southgate (1972); Martin (1973); Anonymous (1976a); Tubbs (1977); Coulson and Budd (1979); Coulson et al. (1980); Montgomery and Soulsby (1980); Tubbs (1980a,b); Tubbs and Tubbs (1980, 1982, 1983); Wright (1980); Nicholls et al. (1981); Haynes and Coulson (1982); Soulsby et al. (1982); Ford et al. (1983); Lowthion et al. (1985); Montgomery et al. (1985); Soulsby et al. (1985); den Hartog (1994)
Portsmouth Harbour
  Anonymous (1976a); Tubbs (1980a,b); Soulsby et al. (1978, 1985)
Chichester Harbour
  Anonymous (1976a); Coulson et al. (1980); Tubbs (1980a,b)
Pagham Harbour
  Anonymous (1976a)
Holes Bay, Poole Harbour
  Dunn (1972); Southgate (1972); Holes Bay Steering Committee (1974); Anonymous (1976a)
Southampton Water
  Cotton (1910, 1911); Tubbs (1980a)
Weymouth Harbour (Backwater area)
  Cotton (1910, 1911); Letts and Richards (1911)
Medway Estuary
  Wharfe (1977)
Thanet
  Fletcher (1974)
Lynher Estuary
  Warwick et al. (1982)
Hamford Water
  Anonymous (1976a)
Blackwater Estuary
  Anonymous (1976a)
Newton, Isle of Wight
  Anonymous (1976a)
Stour Estuary
  Anonymous (1976a)
Tyne Estuary
  Edwards (1972, 1973, 1975)
Mersey Estuary
  Fraser (1932/33)

SCOTLAND

Firth of Forth
  Smyth (1968); Johnston (1971/72)
Firth of Clyde
  Perkins and Abbott (1972); Clokie and Boney (1979)
Eden Estuary
  Dunn (1939); Owens et al. (1979); Owens and Stewart (1983)
Loch Craiglin
  Raymont (1947)
Ythan Estuary
  Green (1977); Hull (1987); Raffaelli et al. (1989, 1991)

Table 8.4. (*Contd.*)

Loch Forfar (freshwater)
  Coleman and Stewart (1979)
Shetland Isles
  Powell (1963); Russell (1974)

IRELAND

Belfast Lough
  Cotton (1911); Letts and Richards (1911)
Dublin Bay
  Baily (1886); Adeney (1908); Letts and Richards (1911); Pitkin (1977); O'Donovan (1981); Madden (1984); Walsh (1988); Jeffrey et al. (1992)
Rogerstown Estuary
  Fahy et al. (1975); Goodwillie et al. (1970, unpubl.)

south coast of England, the last two localities being recognized by the Paris Commission, 1992 as being possibly eutrophic and requiring further study.

8.4   Effect of "Eutrophication" on Marine Benthic Algal Communities

The effect of eutrophication on marine algae was established during the early part of this century when excess growths in Belfast Lough were attributed to the high levels of nutrient salts in the water originating from sewage (Letts and Richards 1911). The relationship between nutrients and excess algal growths was also established shortly after for several sites (Weymouth, Southampton Water) on the south coast of England (Letts and Richards 1911; Cotton 1910, 1911). Since that time numerous authors have suggested a relationship between increased nutrient input and local changes in algal populations. For example, Hull (1987) referred to increased levels of nutrients influencing algal populations in the Ythan Estuary, Scotland; he reported that an increase in macroalgal distribution had occurred in the Estuary since earlier reports by Green (1977). In a later study, Raffaelli et al. (1989) referred to a two- to three fold increase in nitrogen entering the estuary over a 25-year period which they held responsible for increased algal growths. They attributed this increase to a comparable increase in the local population and changes in agricultural practices, notably the decline in oat farming and an increase in barley/wheat farming which requires more intensive use of nitrogen fertilizers. Increased urbanization, and associated sewage, was also held responsible for both qualitative and quantitative changes observed in the flora and fauna of the inner parts of the Firth of Forth Estuary (Johnston 1971/92) and the Firth of Clyde Estuary (Clokie and Boney 1979); all these authors reported a reduced flora compared with that recorded in the 19th century. Fletcher (1974) also suggested that perhaps the increased incidences of algal deposits on beaches in Thanet (SE England) were the result of increased sewage discharges in local bays. In one locality (Langstone Harbour, south coast of England), increased nutrient inputs (12 to 36 ml/day over a 10-year period in the 1960s to 1970s) from a newly built sewage works, were held responsible for increased growths of green algae on the surface of the mudflats (Dunn 1972; Anonymous 1976a; Montgomery and Soulsby 1980; Tubbs 1980a). There was, for example, little evidence of green algal growths on the mud flats in 1961 but they were fairly widespread in 1966 and 1967 (Dunn 1972; Anonymous 1976a). Dunn (1972) also noted that Perraton (1953) did not mention any obvious colonization of the Langstone Harbour mudflats by green algae except for some plants associated with *Spartina* root hummocks. Later, Soulsby et al. (1978) also concluded that sewage input was mainly responsible for the large algal mats in Langstone Harbour. No notable increase, however, has been reported in algal cover of the mud flats in recent years (Montgomery and Soulsby 1980; Soulsby et al. 1985). Soulsby et al. (1985), for example, recorded no trends in algal cover from 1974 to 1983 in Portsmouth Harbour and from 1973 to 1983 in Langstone Harbour, although it is likely that nutrient input had increased.

Although high levels of nutrients are generally recognised as the prima facie reason for many of the observed changes in benthic algal populations, other environmental factors can also play a major role. There are, for example, many regions in the British Isles, such as the Thames Estuary, which have very high nutrient concentrations but are not associated with reports of excessive blooms of algae (Southgate 1972). Probably one of the most important environmental factors that determine the response of algal communities to excess nutrient loading is the degree of "exposure" at the locality. A characteristic of many of the affected areas, for example, is their relatively enclosed and sheltered aspect. These include, for example, sheltered loughs such as Belfast Lough in Ireland, sheltered estuaries like the Rogerstown in Ireland (Fahy et al. 1975), the quiet Backwater of Weymouth Harbour, the very enclosed, shallow harbour systems of Poole, Langstone and Portsmouth on the south coast of England and large, but enclosed bays, such as Dublin Bay. Even within the large harbours and bays, the algal growth is often restricted to the quieter, inner regions. In some situations, these quiet regions are natural and involve tidal movements which swing currents, carrying the nutrients or sewage, onto shores (Smyth 1968). In other situations, they are produced artificially. For example, the Backwater represents a section of Weymouth Harbour impounded by a dam (Cotton 1910). The situation was then worsened when the construction of a railway bridge reduced the middle current and formed an additional backwater where green algae could flourish. Johnston (1971/72) also reported that coastal developments which increased shelter may have contributed to the floristic changes observed in the Firth of Forth whilst Fahy et al. (1975) reported that the construction of a causeway in Rogerstown, Ireland reduced the current flow and led to the development of mudflats containing dense algal mats. In all these examples, restricted tidal flow and water exchange occurs, the absence of strong currents and breakers allowing nutrients to accumulate in the water column. This was clearly shown by Parsons and Fisher (1977) using radioactive tracer surveys in Langstone Harbour; they reported that the hydrography of the Harbour permitted a very slow turnover and that most of the water re-enters the Harbour on the flood tide permitting a build-up in effluent concentration at the north end. In addition, low current activity will allow the build-up of floating and/or loose-lying mats of algae in the sheltered regions of many of the affected sites. Such a build-up is reported to be a particular problem in Dublin Bay, Langstone Harbour and Belfast Lough. Montgomery and Soulsby (1980), for example, reported that the initial build-up of green algal mats in Langstone Harbour during the 1960s occurred largely in the sheltered mudflat areas. In general, it is likely that in situations where good water circulation occurs, pollution will have only a minimal effect on marine plant and animal communities around the British Isles (Edwards 1975).

In contrast to the above findings, however, have been reports of excessive algal deposits occurring on more open coasts in the south of England. For example, very large annual deposits of macroalgae (ca. 50 000 to 60 000 tonnes in recent years) have occurred on relatively exposed beaches in the Worthing area as far back as 1850 and have produced nauseating smells, encouraged large numbers of seaweed flies (*Coelopa frigida* and *Coelopa pilipes*) and necessitated expensive clearance operations (Anonymous 1987). In a study by Binnie and Partners (Anonymous 1987, 1988), however, it was concluded that the problem was unlikely to relate to the release of nutrients in wastewaters and that the major cast weed deposits, largely comprising the kelp *Laminaria saccharina*, originated from offshore beds and were caused by particular combinations of wind, wave and tides.

A different beach cast flora was, however, identified by Fletcher (1974) in sheltered bays on the Thanet Coast, with *Ulva lactuca* being a main constituent. Fletcher suggested the deposits might be related to increased sewage discharges in the vicinity of the bays. Although the bays are relatively exposed to the English Channel, the situation might be similar to that described for several bays on the Brittany coast. In the gently sloping Bay of Brieuc, for example, large deposits of green algae (*Ulva* sp.), stimulated by anthopogenic sources of nutrients, become trapped in small residual bodies of water between the tides and are periodically swept up onto the shores as large beach deposits (Piriou et al. 1991; Piriou and Menesguen 1992).

Given the right hydrological conditions for the formation of excessive growths of algae, other influential factors include the availability of suitable substrata and the salinity of the water. Cotton (1910, 1911), for example, suggested that mussels played an important role in the development of algal populations in Southampton Water, Weymouth Harbour and Belfast Lough; they provided a firm anchorage for the plants, retaining vegetative fragments by their byssus threads, thereby, allowing the algae to overwinter until new growth was initiated in spring. Jeffrey et al. (1992) suggested that the tubeworm *Lanice* played an important role in the build-up of *Ectocarpus* populations in Dublin Bay by providing a solid substratum and liberating nitrogen and phosphorus as excretory products. Lowthion et al. (1985) reported that the increased development of green algal mats in Langstone Harbour coincided with a decline in *Spartina* beds (Haynes and Coulson 1982) which increased the amount of available mud flat for the algae to colonize. Finally, the ability of *Ulva lactuca* (identified as var. *latissima*) to thrive in brackish water was considered to be an important factor in its successful colonization of localities such as Belfast Lough, Southampton Water and Weymouth Harbour (Cotton 1911).

Whilst many of the above described floristic changes have been attributed to excess nutrient loadings from sewage, some have been attributed to secondary pollutants, notably the high volume of suspended matter discharged with the effluents. Johnston (1971/72), for example, concluded that the depauperate flora observed on the sewage-contaminated shores at Joppa in the Firth of Forth, Scotland, was probably largely the result of the high levels of suspended matter in the water column which reduced light penetration and interfered with the settlement and growth of germlings. It also increased the quantity of mussels which competed for space with the benthic algae. Other reported detrimental effects of suspended matter on plant and animal communities around the British Isles include a reduction in net annual production and depth range of the kelp *Laminaria hyperborea* in heavily polluted sites on the northeast coast of England compared to non-polluted localities in southeast Scotland and northeast England (Bellamy and Whittick 1968; Bellamy et al. 1969, 1972). These authors also showed that polluted ecosystems are dominated by suspension feeders with a great reduction in all other trophic groups. Polluted/silted waters have also been held responsible for smaller kelp blades (Burrows and Pybus 1970, 1971), inhibiting germling development in several algae (Burrows and Pybus 1971; Moss et al. 1973; Wilkinson and Tittley 1979) and the increased abundance of some turf-forming algae such as *Rhodochorton floridulum* (Edwards 1975).

8.5 Marine Benthic Algae
Associated with "Eutrophicated" Waters

Whilst a small increase in the nutrient loading of coastal waters can probably be considered beneficial, and indeed is likely to pass unnoticed in many situations (e.g. Edwards 1972, 1973 reported very little difference in the marine benthic flora of the nutrient-polluted Tyne Estuary and the relatively unpolluted Wear Estuary), when in excess and given the right hydrological conditions, the nutrients can cause quite marked ecological changes. They can, for example, considerably increase the productivity of algae and this was particularly well demonstrated by Raymont (1947) when he artificially fertilized a sea loch in Scotland (Loch Craiglin) during the summer and produced heavy growths of seaweed and *Zostera*. More notably, however, the excess inputs of anthropogenic sources of nutrients have been attributed with both an overall reduction in species diversity and an enlarged productivity of a relatively small number of species. Table 8.5 lists examples of marine benthic algae reported to be characteristic of eutrophicated waters, especially those contaminated by sewage, in the British Isles. In agreement with Edwards (1975), an ideal way to determine pollution effects is to compare the flora prior to and following the advent of pollution. Such historical comparisons like those of Johnston (1971/72), Price and Tittley (1972), Edwards (1975) and Wilkinson and Tittley (1979) are either rare or lacking even for a relatively well-known flora like that of the British Isles and, like the other examples given, do not always distinguish between the different sources of pollutants. However, Johnston's study does relate to sewage pollution and is worth mentioning here. Johnston compared the marine flora and its distribution in the Firth of Forth, Scotland, with data obtained by Traill (1886) in the late 19th century. He recorded a 41.6% reduction in species at Joppa in the inner regions of

Table 8.5. Marine benthic macroalgae, as named by the authors, associated with eutrophicated waters in the British Isles

Species	Locality	Reference
CHLOROPHYTA		
Chaetomorpha linum	Ythan Estuary, Scotland	Green (1977); Hull (1987); Raffaelli et al. (1989)
	Dublin Bay, Ireland	Jeffrey et al. (1992)
Chaetomorpha sp.	Firth of Forth, Scotland	Johnston (1971/72)
Cladophora gracilis	Weymouth, England	Cotton (1911)
Cladophora sericea	Belfast Lough, Ireland	Cotton (1911)
	Southampton Water, England	Cotton (1911)
	Giants Causeway, Ireland	Cotton (1911)
Cladophora sp.	Strangford Lough, Ireland	Cotton (1911)
	Ythan Estuary, Scotland	Hull (1987)
Enteromorpha clathrata	Southampton Water, England	Cotton (1911)
	Weymouth, England	Cotton (1911)
Enteromorpha compressa	Dublin Bay, Ireland	Jeffrey et al. (1992)
Enteromorpha flexuosa	Dublin Bay, Ireland	Jeffrey et al. (1992)
Enteromorpha intestinalis	Belfast Lough, Ireland	Cotton (1911)
	Giants Causeway, Ireland	Cotton (1911)
	Dublin Bay, Ireland	Pitkin (1977); Jeffrey et al. (1992)
	Eden Estuary, Scotland	Owens et al. (1979); Owens and Stewart (1983)
	Ythan Estuary, Scotland	Hull (1987)
Enteromorpha linza	Dublin Bay, Ireland	Jeffrey et al. (1992)
Enteromorpha prolifera	Belfast Lough, Ireland	Cotton (1911)
	Dublin Bay, Ireland	Pitkin (1977); Jeffrey et al. (1992)
	Eden Estuary, Scotland	Owens et al. (1979); Owens and Stewart (1983); Dunn (1939)
	Ythan Estuary, Scotland	Hull (987)
	Loch Forfar, Scotland (freshwater)	Ho (1979)
	Lynher Estuary, England	Warwick et al. (1982)
Enteromorpha radiata	Langstone Harbour, England	den Hartog (1994)
Enteromorpha ramulosa	Dublin Bay, Ireland	Pitkin (1977); Jeffrey et al. (1992)
Enteromorpha sp.	Strangford Lough, Ireland	Cottton (1911)
	Rogerstown Estuary, Ireland	Fahy et al. (1975)
	Dublin Bay, Ireland	Wilson et al. (1990)
	Firth of Clyde, Scotland	Perkins and Abbot (1972)
	Ythan Estuary, Scotland	Green (1977); Raffaelli et al. (1989)
	Firth of Forth, Scotland	Smyth (1968)
	Langstone Harbour, England	Anonymous (1976a)
	Portsmouth Harbour, England	Soulsby et al. (1978)

Table 8.5. *(Contd.)*

Species	Locality	Reference
	Chichester Harbour, England	Anonymous (1976a)
	Holes Bay, England	Anonymous (1976a)
	Blackwater Estuary, England	Anonymous (1976a)
	Hamford Water, England	Anonymous (1976a)
Monostroma fuscum	Belfast Lough, Ireland	Cotton (1911)
Monostroma grevillei	Belfast Lough, Ireland	Cotton (1911)
Percursaria percursa	Dublin Bay, Ireland	Jeffrey et al. (1992)
Prasiola stipitata	Tyne Estuary, England	Edward (1972, 1973)
Rhizoclonium riparium	Dublin Bay, Ireland	Jeffrey et al. (1992)
Ulva curvata	Langstone Harbour, England	den Hartog (1994)
Ulva lactuca	Belfast Lough, Ireland	Cotton (1911);
(often named *U. latissima*)		Letts and Richards (1911)
	Dublin Bay, Ireland	Jeffrey et al. (1992)
	Giants Causeway, Ireland	Cotton (1911)
	Southampton Water, England	Cotton (1910, 1911)
	Thanet, England	Fletcher (1974)
	Weymouth, England	Cotton (1910, 1911)
	Firth of Forth, Scotland	Smyth (1968)
	Ythan Estuary, Scotland	Johnston (1971/72); Hull (1987)
Ulva sp.	Weymouth, England	Cotton (1911)
	Holes Bay, England	Anonymous (1976a); Southgate (1972)
	Langstone Harbour, England	Anonymous (1976a)
	Chichester Harbour, England	Anonymous (1976a)
	Portsmouth Harbour, England	Soulsby et al. (1978)
	Pagham Harbour, England	Anonymous (1976a)
	Dublin Bay, Ireland	Wilson et al. (1990)
	Firth of Clyde, Scotland	Perkins and Abbott (1972)
	Ythan Estuary, Scotland	Green (1977); Raffaelli et al. (1989)
PHAEOPHYCEAE		
Ectocarpus siliculosus	Dublin Bay, Ireland	Wilson et al. (1990); Jeffrey et al. (1992)
Fucus disticus ssp. *edentatus*	Shetland Isles	Powell (1963); Russell (1974)
Ralfsia sp.	Firth of Clyde, Scotland	Clokie and Boney (1979)
RHODOPHYTA		
Ceramium rubrum	Firth of Forth, Scotland	Johnston (1971/72);
	Thanet, England	Fletcher (1974)

Table 8.5. *(Contd.)*

Species	Locality	Reference
Chondrus crispus	Firth of Forth, Scotland	Johnston (1971/72)
Corallina officinalis	Firth of Forth, Scotland	Johnston (1971/72)
Dumontia incrassata	Giants Causeway, Ireland	Cotton (1911)
	Firth of Forth, Scotland	Johnston (1972)
	Firth of Clyde, Scotland	Clokie and Boney (1979)
Gigartina stellata	Firth of Forth, Scotland	Smyth (1968); Johnston (1971/72)
	Firth of Clyde, Scotland	Clokie and Boney (1979)
Griffithsia flosculosa	Thanet, England	Fletcher (1974)
L aurencia caespitosa	Giants Causeway, Ireland	Cotton (1911)
Plocamium cartilageum	Thanet, England	Fletcher (1974)
Polysiphonia nigrescens	Firth of Forth, Scotland	Johnston (1971/72)
Porphyra sp.	Firth of Forth, Scotland	Smyth (1968)
Rhodochorton rothii	Firth of Forth, Scotland	Smyth (1968)

the Firth which he attributed to the effects of sewage discharge associated with increased urbanization; a similar reduction in numbers of species at Joppa was recorded by Wilkinson and Tittley (1979). In the more polluted regions, the flora was very restricted and was comprised mainly of often stunted growths of such algae as *Ulva lactuca, Chondrus crispus, Gigartina stellata, Ceramium rubrum, Polysiphonia nigrescens, Chaetomorpha* sp., *Hildenbrandia rubra, Dumontia incrassata* and *Corallina officinalis*. A similar reduced flora, comprising species of the genera *Dumontia, Gigartina, Hildenbrandia* and *Ralfsia* was also reported for some nutrient-enriched regions in the Firth of Clyde by Clokie and Boney (1979). These algae survived because their basal crustose parts were grazer-resistant and they were able to tolerate the higher than normal littorinid population present. They attributed the latter to the high nitrate levels supporting a heavy cover of diatoms during late winter. Note that Fletcher (1974) also recorded a very restricted flora, largely comprising *Ulva lactuca* and *Ceramium rubrum*, in beach deposits associated with suspected sewage polluted bays in Kent.

In extreme cases of pollution, such as reported for the Mersey Estuary, macroalgae are often lacking and usually only microalgae such as diatoms and the euglenoid *Euglena limosa* are recorded (Fraser 1931/32). In less seriously polluted environments and localities characterized by excess loadings of nutrients, there appears to be a general reduction in the diversity and occurrence of brown and red algae and a corresponding increase in green algae (as reported, for example, in Belfast Lough by Letts and Richards (1911) and Cotton (1911)). In this respect the two most commonly recorded green algal genera are *Enteromorpha*,

(see especially Smyth 1968; Dunn 1972; Perkins and Abbott 1972; Southgate 1972; Anonymous 1976a; Soulsby et al. 1978, 1985; Montgomery and Soulsby 1980; Nicholls et al., 1981; Owens and Stewart 1983; Hull 1987; Raffaelli et al. 1989; Wilson et al. 1990; Jeffrey et al. 1992; den Hartog 1994) and *Ulva* (see especially Cotton 1910, 1911; Letts and Richards 1911; Smyth 1968; Perkins and Abbott 1972; Southgate 1972; Fletcher 1974; Soulsby et al. 1978, 1985; Holme and Bishop 1980; Nicholls et al. 1981; Jeffrey et al. 1992; den Hartog 1994). In much of the literature, only *Enteromorpha* sp. is cited; however, species identified include *E. prolifera* (Cotton 1911; Dunn 1939; Ho 1979; Owens et al. 1979; Owens and Stewart 1983; Jeffrey et al. 1992), *E. intestinalis* (Cotton 1911; Owens et al. 1979; Owens and Stewart 1983; Jeffrey et al. 1992), *E. clathrata* (Cotton 1911), *E. compressa* (Jeffrey et al. 1992), *E. flexuosa* (Jeffrey et al. 1992), *E. linza* (Jeffrey et al. 1992), *E. radiata* (den Hartog 1994) and *E. ramulosa* (Jeffrey et al. 1992). For the genus *Ulva*, either no specific epithet is given or the material is identified as *U. lactuca*, *U. lactuca* var. *latissima* or *U. latissima* (Cotton 1910, 1911; Smyth 1968; Fletcher 1974); more recently, and in-keeping with recent taxonomic treatments of this genus (and *Enteromorpha*; see Koeman and van den Hoek 1980, 1982a,b, 1984), *U. curvata* has been identified (den Hartog 1994). Other green algal genera reported to thrive in sewage-contaminated waters in the British Isles include *Cladophora* spp. [notably *C. sericea* (Cotton 1911) and *C. gracilis* (Cotton 1911)], *Monostroma* spp. [e.g. *M. grevillei* and *M. fuscum* (Cotton 1911), *Prasiola stipitata* (Edwards 1972, 1973)], and *Rhizoclonium* spp. (Ho 1979; Jeffrey et al. 1992).

These reports of excess growths of *Enteromorpha* and *Ulva* spp. in eutrophicated waters find support in a number of laboratory-based experimental studies which have demonstrated that growth of these algae is stimulated by additions of the sewage effluent being discharged around the British Isles (see especially Letts and Richards 1911; Dunn 1972; Southgate 1972; Anonymous 1976a; Ford et al. 1983). There have also been reports that ammonia is preferred to nitrate (Letts and Richards 1911) and that both *Enteromorpha* and *Ulva* spp. have a high capacity for nutrient absorption (Letts and Richards 1911; Anonymous 1976a; Jeffrey et al. 1992). It is interesting to note, however, that Anonymous (1976a) concluded that the inorganic nutrient content alone was not responsible for the growth response and that the effluent contained some additional growth-stimulating compound. Using an *Ulva* disc growth technique, Burrows (1971) concluded that sewage-polluted Dublin mud similarly contained a growth-promoting ingredient. Such findings might be of considerable importance when considering nutrient removal from effluent as a means of controlling algal growth (Ford et al. 1983).

Two notable exceptions to the above generalization are the brown algae *Fucus disticus* subsp. *edentatus* and *Ectocarpus siliculosus* which have been reported to thrive in nutrient-enriched waters. The former has been reported to occur most abundantly in close proximity to sewage outflows in the Shetland Isles, off the north coast of Scotland (Powell 1963; Russell 1974) whilst the latter has been reported to form dense blooms in Dublin Bay, Ireland (Wilson et al. 1990; Jeffrey et al. 1992).

Whilst some of the above reports of increased algal growths refer to attached populations, usually in the vicinity of sewage outfalls, many refer to unattached, loose-lying mats of algae which accumulate on the surface of sheltered mudflats. In Belfast Lough they occurred as a flannelly or ropey mass known as "May Fog" or "Flannel Weed" and comprised a mixture of species (Cotton 1911). In Langstone Harbour, they were referred to as either "blanket-weed" (Southgate 1972) or "slob" (Anonymous 1976a). These loose-lying mats of weed, which probably originated either from attached sporelings (Anonymous 1976a; Jeffrey et al. 1992) or from vegetative plant portions buried within the sediments (Anonymous 1976a), can be quite extensive. Cotton (1911), for example, refers to wide stretches of the Belfast Lough "sloblands", many hundreds of acres in extent, being covered with a green sward of almost pure *Ulva*, whilst Perkins and Abbott (1972) refer to *Enteromorpha* mats occupying substantial areas (e.g. 21%) of some shores in the Firth of Clyde, Scotland. Using aerial surveys, it was estimated that in 1983, 64% of the intertidal mudflats in Portsmouth Harbour and in 1982, 43% of the intertidal mudflats in Langstone Harbour were covered by algal mats (Coulson and Budd 1979; Soulsby et al. 1985). In Langstone, the mats were described as particularly extensive in the northern, more sheltered areas of the harbour, not too distant from the sewage discharge pipe, forming a deep continuous cover of green algae 10–12 cm deep (Montgomery and Soulsby 1980; Nicholls et al. 1981; Lowthion et al. 1985) (Fig. 8.1). Comparable thick mats of *Enteromorpha*, reaching 15–20 cm thick, were also reported for the Firth of Clyde by Perkins and Abbott (1972). Maximum biomass figures for these mats include $3 \, kg \, m^{-2}$ fresh weight (Ythan Estuary, Scotland; Hull 1987), $68 \, g \, m^{-2}$ dry weight (Langstone Harbour, England; Montgomery and Soulsby 1980), $49 \, g \, m^{-2}$ dry weight (Portsmouth Harbour, England; Soulsby et al. 1978; Lowthion et al. 1985), $150 \, g \, m^{-2}$ dry weight (Loch Forfar, Scotland; Coleman and Stewart 1979), $2 \, kg \, m^{-2}$ fresh weight (Ythan Estuary, Scotland; Raffaelli et al. 1989), $> 500 \, g \, m^{-2}$ dry weight (Jeffrey et al. 1992) and $113 \, g \, m^{-2}$ dry weight (Eden Estuary, Scotland; Owens et al. 1979).

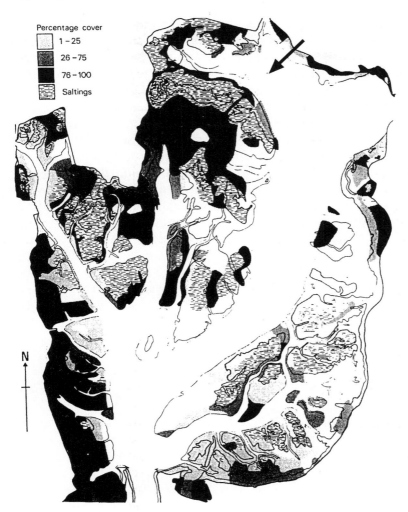

Fig. 8.1. *Enteromorpha* cover in Langstone Harbour, August 1979. Note extensive algal growth over the intertidal mud flats and the particularly high percentage cover west of the Budds Farm sewage outfall (arrowed) and on the western side of the harbour. (After Montgomery and Soulsby 1980)

Whilst in some regions the macroalgal mats can be found throughout most of the year (Montgomery and Soulsby 1980; Nicholls et al. 1981), most authors report a seasonal distribution in their occurrence, with maximum cover (and usually biomass) observed during the summer and minimum cover observed during the winter e.g. in Langstone Harbour (Anonymous 1976a; Lowthion et al. 1985), Eden Estuary (Owens et al.

1979; (Fig. 8.2), the Wash (Anonymous 1976b), Dublin Bay (Jeffrey et al. 1992; Fig. 8.3) and the Firth of Clyde (Perkins and Abbott 1972). Owens et al. (1979) reported that a similar summer bloom of green algae occurs in freshwater lochs in Scotland; see also Ho (1979). Usually, *Enteromorpha* begins development in April/May, reaches maximum development in July/August, begins to decline in September and is usually absent in November (Cotton 1911; Anonymous 1976a, b; Nicholls et al. 1981; Owens and Stewart 1983). There are also some indications of a sequential change occurring in the floristic component of the algal mats in some regions, for example, the occurrence of an annual spring growth of *Ulothrix flacca* preceding the *Enteromorpha* in the Wash, England (Anonymous 1976b) and a seasonal shift from *Enteromorpha* to *Ulva* in Dublin Bay, Ireland (Walsh 1988; Jeffrey et al. 1992). *Ulva* also appears to be less tolerant than *Enteromorpha* to aerial exposure on the mudflats and will only extend above high tide level in pools, streams and shallow water (Cotton 1911). Jeffrey et al. (1992) also reported *Ulva* to occur in greatest quantites in the permanent chan-

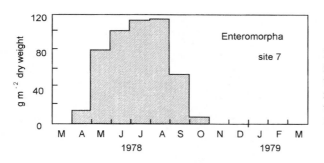

Fig. 8.2. Seasonal variation in *Enteromorpha* spp. in the Eden Estuary, Scotland measured during 1978. (After Owens and Stewart 1983)

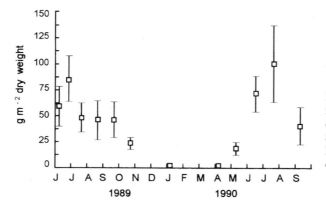

Fig. 8.3. Seasonal variation in green algal biomass during 1989/1990 in the South Lagoon of Dublin Bay, Ireland. (After Jeffrey et al. 1992)

nels of Dublin Bay whilst Anonymous (1976a) reported *Ulva* to show its maximum growth in completely sheltered regions of Langstone Harbour.

Although high nutrient loadings in association with sheltered localities are generally the cause of the abundance and perennial nature of these large algal mats (Soulsby et al. 1978; previously they were often described as more annual and restricted in their occurrence; see Nicholls et al. 1981) and there are reports that in sites receiving more effluent the algal growths occurred much earlier and became much denser (Anonymous 1976a), other environmental factors also play an important role. For example, some authors have noted that the onset of rapid growth can vary by up to a month from year to year (Montgomery and Soulsby 1980) and that large annual fluctuations in algal cover can also occur in some localities, both of which bear no relationship to ambient nutrient levels (Soulsby et al. 1978; Montgomery and Soulsby 1980; Lowthion et al. 1985; Montgomery et al. 1985; Table 8.6). Similarly, there are sometimes distinct spatial distribution and abundance patterns of algal mats which do not appear to relate to nutrient levels either in the water column (Anonymous 1976a; Montgomery and Soulsby 1980; Montgomery et al. 1985; Soulsby et al. 1985) or in the sediments (Montgomery et al. 1985). Particularly interesting was the report by Soulsby et al. (1985) that stopping the release of sewage into Portsmouth Harbour did not result in any overall decrease in the amount of algal cover. All the above evidence points towards nutrients being in excess and that climatic factors such as temperature, sunlight etc. exert a major influence on the response of the algae to the nutrients (Anonymous 1976a; Montgomery and Soulsby 1980; Lowthion et al. 1985).

Pertinent to these findings is Lowthion et al. 's (1985) conclusion that the lowest biomass cover recorded for Langstone Harbour (in 1976) was

Table 8.6. Areas of macroalgal cover present in Langstone Harbour, south coast of England, in August. Note annual variations and lack of continuous trend. Least weed was present in 1976, a year of exceptionally high temperatures and sunshine. (After Montgomery and Soulsby 1980)

Cover Category %	Area covered by each scale (hectares)					
	1973	1974	1976	1977	1878	1979
0	227	384	643	491	484	417
1–25	344	320	264	249	156	206
26–75	319	245	136	231	170	110
76–100	233	168	68	140	300	377
Total	1123	1117	1111	1111	1110	1110

attributed to the exceptional high temperatures and sunshine levels recorded in the summer months. Note, however, that under less extreme conditions (dry conditions, light winds) dense mats of algae were reported for the same harbour (den Hartog 1994). Anonymous (1976a) also concluded that the "optimal" conditions of 1973 (previous winter warmer than average, steady temperature rise during February, March and April, calm conditions etc. all providing low turbidity and high irradiance) were responsible for the excellent growth of *Enteromorpha* below low water neap tide level in Langstone Harbour. Jeffrey et al. (1992) also drew attention to the important role played by climatic conditions in controlling algal growths in Dublin Bay; they suggested that a prolonged high summer temperature would result in a lower biomass. Based on studies in Langstone Harbour, Anonymous (1976a) and Montgomery and Soulsby (1980) further suggested that another important factor that determines algal cover is the direction and speed of prevailing winds during the growing season. Such studies led Soulsby et al. (1985) to the conclusion that the distribution and biomass of algae in Langstone Harbour was controlled by the availability of habitats, physical/climatic factors and grazing activity; they suggested, therefore, that any further increase in the nutrient loading would go undetected in the harbour.

Particularly significant to this discussion, however, is the recent work of Jeffrey et al. (1992) in Dublin Bay. They refute the importance of metereological conditions and draw attention to the sediments underlying the algal mats as a significant source of nutrients; they suggested that the decline in the algal crop in August bore no relationship with environmental factors such as temperature and light (which were still favourable) but, more notably, corresponded with a comparatively low nutrient release rate from the sediments. They also suggested that nutrients released from the sediments play a major role in determining spatial variations in algal biomass in Dublin Bay (Fig. 8.4).

8.6   Effect of Excessive Growth of Algae

Whilst some authors conclude that the extensive green algal mats can sometimes be beneficial by insulating the intertidal invertebrates from the heat of the sun, that they cause no major ecological or environmental problems, or that the effects are not acute enough to warrent corrective action (Montgomery et al. 1985), more commonly the mats are considered a nuisance and that they lower the quality of the ecosystem. For

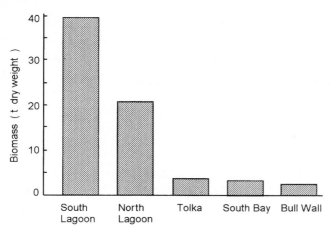

Fig. 8.4. Total biomass (kg dry weight) of green algae in each of the main zones of Dublin Bay, Ireland, recorded in late June 1989. Note the spatial variation and high biomass recorded for the South Lagoon area compared with the North Lagoon despite the area of the latter being four times greater. This was attributed to the plentiful nutrient supply and sheltered conditions in the South Lagoon. (After Jeffrey et al. 1992)

example, dense mats of *Enteromorpha* were reported to have covered and suffocated plants of *Zostera* in Langstone Harbour (den Hartog 1994); by shading the *Zostera* plants they reduced their photosynthetic activity and caused hypoxic to anoxic conditions by preventing water exchange. The loss of *Zostera* beds has considerable ecological implications as the plants are reported to support a rich fauna (Anonymous 1976a) and the leaves are an important food for brent geese (*Branta bernicla bernicla*) and wigeon (Tubbs 1980a,b).

During decomposition, the algal mats produce anaerobic conditions at the mud surface (Perkins and Abbott 1972; Anonymous 1976a) and this is frequently reflected in an infauna that is reduced in diversity and density (Anonymous 1976a; Jeffrey et al. 1992; Fig. 8.5). For example, den Hartog (1994) observed a marked reduction in the numbers of the cockle *Cerastoderma edule* under mats of *Enteromorpha radiata* compared to those under *Zostera* plants in Langstone Harbour. A similar decline in the numbers of *Cerastoderma* was also noted by Perkins and Abbott (1972), Anonymous (1976a) and Raffaelli et al. (1989) under anaerobic mats of green algae in the Firth of Clyde, Langstone Harbour and the Ythan Estuary respectively. In order to escape the anaerobic conditions

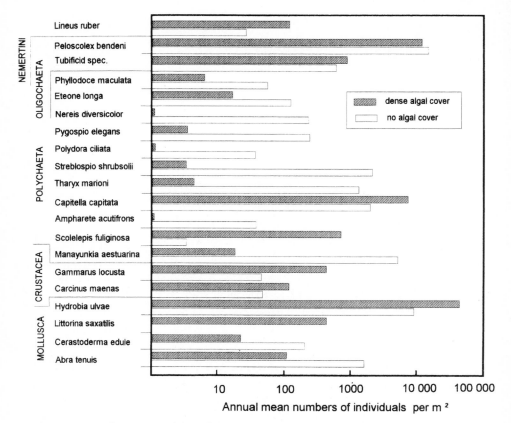

Fig. 8.5. Annual mean numbers of the main species of invertebrates at site 1 (*no algal cover*) and site 2 (*dense algal cover*) in Langstone Harbour, south coast of England. Note that although the species composition of the two sites is very similar, the relative abundance of individual species differs considerably. For example, typical estuarine polychaetes such as *Nereis diversicolor* and *Ampharete acutifrons* occur in low numbers at site 2 where the dominant species are *Capitella capitata* and *Scolelepis fuliginosa*. (After Nicholls et al. 1981)

the *Cerastoderma* moved up to the mud surface and either died and/or were predated by wildfowl. Other invertebrates reported in reduced numbers in areas with high algal cover include *Streblospio shrubsolii* (Soulsby et al. 1982), *Macoma baltica* (Anonymous 1976a; Perkins and Abbott 1972; Hull 1987) and *Corophium volutator* (Hull 1987; Raffaelli et al. 1989).

Some invertebrates, however, are reported to have successfully exploited these polluted mudflats and increased in numbers; these include

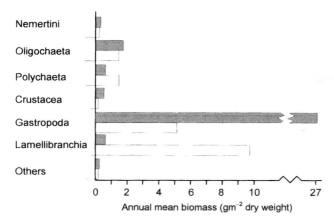

Fig. 8.6. Annual mean biomass (g m^{-2} dry weight) of the various taxa from site 1 (no algal cover, *open bar*) and site 2 (dense algal cover, *shaded area*) in Langstone Harbour, south coast of England. Note that molluscs contribute most of the biomass at both sites; bivalves (Lamellibranchia) are major contributors in open muds, whereas epibenthic Gastropoda dominate the algal site (primarily *Hydrobia ulvae*). (After Nicholls et al. 1981)

infauna such as *Peloscolex benedini, Capitella capitata, Arenicola marina, Nereis diversicolor* and epifauna such as *Hydrobia ulvae* and *Littorina littorea* (Goodwillie et al. 1970, unpubl.; Perkins and Abbott 1972; Fahy et al. 1975; Anonymous 1976a; Soulsby et al. 1978; Montgomery and Soulsby 1980; Nicholls et al. 1981; Soulsby et al. 1982; Hull 1987; Jeffrey et al. 1992; den Hartog 1994; Figs. 8.5, 8.6). The numbers of epifauna, and in particular *Hydrobia*, associated with the algal mats are reported to be particularly high, quite significantly increasing the faunal biomass at many of the sites (Anonymous 1976a; Montgomery and Soulsby 1980; Soulsby et al. 1982; Montgomery et al. 1985; Fig. 8.6). Nicholls et al. (1981), for example, found that 89% of the invertebrate biomass on weed-covered areas comprised epifauna whereas 71% of the biomass on weed-free sites comprised infauna. They also found a tendency for bivalves to dominate open mudflats whereas gastropoda (notably *Hydrobia*) dominated weed-covered areas (Fig. 8.6). It has been suggested that this increase in both algal and faunal biomass can provide an important food source for birds such as brent geese (Montgomery and Soulsby 1980; Soulsby et al. 1982; Montgomery et al. 1985; Jeffrey et al. 1992). However, Tubbs (1977) and Tubbs and Tubbs (1983) concluded that the algal mats are generally detrimental to bird populations with some species avoiding areas with high algal cover (see also Raffaelli et al. 1989).

Of particular ecological importance, however, is the role of the algal mats as a nutrient sink. Soulsby et al. (1985) revealed *Enteromorpha*, in both Langstone and Portsmouth Harbours, to contain very high, luxury levels of nitrogen and phosphorus with the tissue content higher in plants collected from the upper more polluted regions of each harbour. A similar relationship was shown for *Ulva* plants collected from polluted and unpolluted sites in Belfast Lough (Letts and Richards 1911). Certainly, it is well known that *Enteromorpha* and *Ulva* have a marked capacity to absorb ammonia and nitrates compared to many other algae (Letts and Richards 1911; Owens et al. 1979; Jeffrey et al. 1992) and that the species exhibit marked seasonal patterns in nutrient concentrations which reflect aspects of their growth rate and ambient concentrations (Jeffrey et al. 1992; Fig. 8.7). Indeed, Owens et al. (1979) calculated that at the time of maximum production, *Enteromorpha* nitrogen accounted for 95% of the total plant nitrogen in the Eden Estuary. Significantly, the *Enteromorpha* contributes new nitrogen by assimilating it from the water column (Owens et al. 1979). During subsequent decomposition, usually in autumn, the nutrients are then returned to the water column and sediments (Coleman and Stewart 1979). Incubation experiments have revealed that the *Enteromorpha* nitrogen is also released back into the water column in the readily available form of ammonia (Owens et al. 1979). The *Enteromorpha* mats do, therefore, play an important

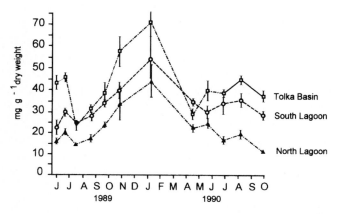

Fig. 8.7. Seasonal variation in nitrogen concentration (mg g^{-1} dry weight) of green algae (1989–1990) for different regions of Dublin Bay, Ireland. It was suggested that the moderate to high nutrient concentrations in spring reflect the actively growing young plants, the lowest concentration in summer reflects the period of high biomass, and the increased concentration in autumn and maximum concentration in winter reflect cessation of growth. (After Jeffrey et al. 1992)

central role in the nitrogen cycle, providing a transfer mechanism for nitrogen and carbon from the water column onto the mudflats.

Additional problems are also encountered when the algal mats become detached from the surface of the mudflats. For example, sometimes they occur as large drifting mats (Anonymous 1976a) which can interfere with boating activities (Montgomery et al. 1985) and when subsequently cast upon the shore as the so-called "green tides" usually during autumn gales, they can accumulate in large quantities (Cotton 1911; Jeffrey et al. 1992) lowering the aesthetic quality of bathing beaches (Wilson et al. 1990; Jeffrey et al. 1992). Noxious smells caused by their decomposition have also been reported, both locally in the bays/harbours (Cotton 1910, 1911; Letts and Richards 1911; Perkins and Abbott 1972; Southgate 1972; Fletcher 1974; Montgomery and Soulsby 1980; Montgomery et al. 1985; den Hartog 1994), and in the vicinity of refuge dumps where the beach casts are sometimes deposited (Fletcher 1974). There have also been reports of algal mats blocking power station intake ducts (Southgate 1972) and, during decomposition on the beaches, encouraging swarms of flies (Anonymous 1987; 1988) and blackening the lead-based paints on boats and houses due to the presence of hydrogen sulphide (Cotton 1910, 1911).

The main "treatment" method usually involves removal of troublesome deposits, especially from recreational beaches (Fletcher 1974; Anonymous 1987, 1988; Jeffrey et al. 1972); at Worthing, on the south coast of England, it was suggested that assistance be given to the "natural" removal of seaweed, involving modifications to groynes and pushing the seaweed down the shingle in reach of the tides (Anonymous 1987, 1988). In some situations, the effluent load is either diverted away (Soulsby et al. 1985) or longer outfall pipes are used (Anonymous 1976a; Ford et al. 1983). Previous considerations given to the chemical treatment of the algae have not been implemented on environmental grounds (Cotton 1911; Anonymous 1976a). Whilst it is widely considered that long term treatment will require a reduction in the levels of nutrients being discharged into nearshore waters and this is already receiving attention at some sites (e.g partial nutrient removal is underway in Langstone Harbour; Ford et al. 1983), this is expensive and would be difficult in the case of diffuse sources via river runoff (Wilson et al. 1990). There is also growing evidence that any changes made in the concentrations of nutrients being released might not necessarily be reflected in seaweed growth patterns (Anonymous 1976a; Soulsby et al. 1985). Clearly, there is an urgent need for more detailed studies to be carried out on the relationship between nutrient inputs and the occurrence of excess algal growths at the various sites affected and such needs are reflected in recent financial assistance from the European Community under the auspices of the NORSAP Programme (for the Dublin Bay studies, see Jeffrey et al. 1992) and the ENVIRONMENT Programme (for Langstone Harbour studies).

References

Adeney WE (1908) Effect of the new drainage on Dublin Harbour. In: Handbook to the Dublin District British Association, British Association. Dublin, 287 pp

Anonymous (1976a) Langstone Harbour Study – the effect of sewage effluent on the ecology of the harbour. Rep to Southern Water Authority, Portsmouth Polytechnic, Portsmouth, 356 pp

Anonymous (1976b) The Wash water storage scheme feasibility study, a report on the ecological studies. NERC Publ Ser C 15, Natural Environment Research Council, London, 36 pp

Anonymous (1987) Initial study of seaweed problem at Worthing and other related matters. Rep to Worthing Borough Council. Binnie and Partners, Redhill, Surrey

Anonymous (1988) Seaweed Problem. Stage 2 studies. Volume 1. Main Report. Volume 2. Appendices. Rep to Worthing Borough Council. Binnie and Partners, Redhill, Surrey

Baily (1886) Rambles on the Irish Coast, 1. Dublin to Howth. Carraig Book Reprints Dublin 2: 8–10

Bellamy DJ, Whittick A (1968) Problems in the assessment of the effects of pollution on inshore marine ecosystems dominated by attached macrophytes. Field Stud 2 (Suppl): 49–54

Bellamy DJ, Whittick A, Jones DJ (1969) How to live with pollution. Spectrum 62: 8–11

Bellamy DJ, John DM, Jones DJ, Starke A, Whittick A (1972) The place of ecological monitoring in the study of pollution of the marine environment. In: Ruivo M (ed) Marine pollution and sea life. Fishing News (Books), Surrey, pp 421–425

Burrows EM (1971) Assessment of pollution effects by the use of algae. Proc R Soc Lond B 177: 295–306

Burrows EM, Pybus C (1970) Culture of Laminaria in relation to problems of marine pollution. Proc Challenger Soc 4: 80–81

Burrows EM, Pybus C (1971) Laminaria saccharina and marine pollution in North-East England. Mar Pollut Bull 2: 53–56

Clokie JJP, Boney AD (1979) David Landsborough (1779–1854): an assessment of his contribution to algal studies in the Firth of Clyde. Glasgow Nat 19: 443–462

Coleman NV, Stewart WDP (1979) Enteromorpha prolifera in a polyeutrophic loch in Scotland. Br Phycol J 14: 121

Cotton AD (1910) On the growth of Ulva latissima in water polluted by sewage. Bull Misc Inform Bot Gard Kew, pp 15–19

Cotton AD (1911) On the growth of Ulva latissima in excessive quantity, with special reference to the Ulva nuisance in Belfast Lough. Royal Commission on sewage disposal, 7th Rep II (Appendix IV), HMSO, London, pp 121–142

Coulson MG, Budd JIC (1979) Weed survey of Langstone Harbour. Rep to Southern Water Authority, Portsmouth Polytechnic, Portsmouth, 16 pp

Coulson MG, Budd JTC, Withers RG, Nichols DJ (1980) Remote sensing and field samp-
    ling of mudflat organisms in Langstone and Chichester harbours, Southern
    England. In: Price JH, Irvine DEG, Farnham WF (eds) The shore environment, vol 1.
    Methods. Syst Assoc Spec Vol 17(a). Academic Press, London, pp 241-263
den Hartog C (1994) Suffocation of a littoral *Zostera* bed by *Enteromorpha radiata*. Aquat
    Bot 47: 21-28
Dunn J (1972) A general survey of Langstone Harbour with particular reference to the
    effects of sewage. Rep to Hampshire River Authority and Hampshire County Council,
    Portsmouth Polytechnic, Portsmouth, 79 pp
Dunn MD (1939) The marine algae of St. Andrews Bay. Trans Bot Soc Edinb 32: 488-501
Edwards P (1972) Benthic algae in polluted estuaries. Mar Pollut Bull 3: 55-60
Edwards P (1973) The benthic marine algae of polluted estuaries in County Durham.
    Proc Challenger Soc 4: 161-162
Edwards P (1975) An assesment of possible pollution effects over a century on the
    benthic marine algae of Co. Durham, England. Bot J Linn Soc 70: 269-305
Fahy E, Goodwillie R, Rochford J, Kelly D (1975) Eutrophication of a partially enclosed
    estuarine mudflat. Mar Pollut Bull 6: 29-31
Fletcher RL (1974) *Ulva* problem in Kent. Mar Pollut Bull 5: 21
Ford GS, Rees RLG, Soulsby PG, Lowthion D (1983) Nutrient removal trials and bioassay
    evaluation at Budds Farm Sewage Treatment Works, Havant. Wat Pollut Contrib 82:
    381-392
Fraser JH (1932-33) Observations on the fauna and constituents of an estuarine mud in a
    polluted area. J Mar Biol Assoc UK 18: 69-85
Goodwillie R, Goodwillie O, Brandt E (1970) Ecological survey of Bull Island mud flat.
    Unpublished
Graham M (1938) Phytoplankton and the Herring. Part III. Distribution of phosphate in
    1934-1936. Fish Invest, London Ser II 16: 3
Green AJ (1977) The production and utilization of *Enteromorpha* spp. in the Ythan
    Estuary, Aberdeenshire. PhD Thesis, Aberdeen Univ
Haynes FN, Coulson MG (1982) The decline of *Spartina* in Langstone Harbour,
    Hampshire. Proc Hampsh Field Club Archaeol Soc 38: 15-18
Ho YB (1979) Inorganic mineral nutrient level studies on *Potamogeton pectinatus* L. and
    Enteromorpha prolifera in Forfar Loch, Scotland. Hydrobiologia 62: 7-15
Holes Bay Steering Committee (1974) Report on *Ulva* growth in Holes Bay. Unpublished
    Report, 11 pp
Holme NA, Bishop GM (1980) Survey of the littoral zone of the coast of Great Britain. 5.
    Report on the sediment shores of Dorset, Hampshire and the Isle of Wight. Rep to
    Nature Conservancy Council, SMBA/MBA Intertidal Survey Unit, 88 pp
Hull S (1987) Macroalgal mats and species abundance: a field experiment. Estuar Coast
    Shelf Sci 25: 519-532
James A, Head PC (1972) The discharge of nutrients from estuaries and their effect on
    primary productivity. In: Ruivo M (ed) Marine pollution and sea life. Fishing News
    (Books), Surrey, pp 163-165
Jeffrey DW, Madden B, Rafferty B, Dwyer R, Wilson J, Allott N (1992) Dublin Bay - Water
    quality management plan. Tech Rep 7. Algal growths and foreshore quality. Environ-
    mental Research Unit, Dublin, 168 pp
Johnston CS (1971/72) Macroalgae and their environment. Proc R Soc Edinb 71: 195-207
Koeman RPT, van den Hoek C (1980) The taxonomy of *Ulva* (Chlorophyceae) in the
    Netherlands. Br Phycol J 16: 9-53
Koeman RPT, van den Hoek C (1982a) The taxonomy of *Enteromorpha* Link (Chloro-
    phyceae) in the Netherlands. I. The section *Enteromorpha*. Arch Hydrobiol (Suppl) 63:
    279-330

Koeman RPT, van den Hoek C (1982b) The taxonomy of *Enteromorpha* (Chlorophyceae) in the Netherlands. II. The section *Proliferae*. Crypt Algol 3: 37–70

Koeman RPT, van den Hoek C (1984) The taxonomy of *Enteromorpha* (Chlorophyceae) in the Netherlands. II. The sections *Flexuosae* and *Clathratae* and an addition to the section *Proliferae*. Crypt Algol 5: 21–61

Letts EA, Richards EH (1911) Report on green seaweeds (and especially *Ulva latissima*) in relation to the pollution of the waters in which they occur. R Comm on sewage disposal 7th Rep vol 11, Appendix III, HMSO, London

Lowthion D, Soulsby PG, Houston MCM (1985) Investigation of a eutrophic tidal basin. 1. Factors affecting the distribution and biomass of macroalgae. Mar Envion Res 15: 263–284

Madden B (1984) The nitrogen and phosphorus turnover of the *Salicornia* flat North Bull Island, Dublin Bay. Thesis, Trinity College, Dublin (unpubl)

Martin GHG (1973) Ecology and conservation in Langstone Harbour. PhD Thesis, Portsmouth Polytechnic

Montgomery HA, Soulsby PG (1980) Effects of eutrophication on the intertidal ecology of Langstone Harbour, UK, and proposed control measures. Prog Water Technol 13: 287–294

Montgomery HAC, Soulsby PG, Hart IC, Wright SL (1985) Investigations of a eutrophic tidal basin: part 2–nutrients and environmental aspects. Mar Environ Res 15: 285–302

Moss B, Mercer S, Sheader A (1973) Factors affecting the distribution of *Himanthalia elongata* (L.) S. F. Gray on the northeast coast of England. Estuar Coast Mar Sci 1: 233–243

Nicholls DJ, Tubbs CR, Haynes FN (1981) The effect of green algal mats on intertidal macrobenthic communities and their predators. Kieler Meeresforsch Sonderh 5: 511–520

O'Donovan G (1981) The role of the green algae in the phosphorus cycle of the intertidal mudflat at Bull Island, Dublin Bay. Thesis, Trinity College, Dublin (unpubl)

Owens NJP, Stewart WDP (1983) *Enteromorpha* and the cycling of nitrogen in a small estuary. Estuar Coast Shelf Sci 17: 287–296

Owens NJP, Christofi N, Stewart WD (1979) Primary production and nitrogen cycling in an estuarine environment. In: Naylor E, Hartnoll RG (eds) Cyclic phenomenon in marine plants and animals. pp 249–258

Parsons TV, Fisher RA (1977) Experience with radioisotope tracing in local tidal waters. Water Pollut Control 76: 59–64

Perkins EJ, Abbott OJ (1972) Nutrient enrichment and sand-flat fauna. Mar Pollut Bull 3: 70–72

Perraton C (1953) Salt marshes of the Hampshire/Sussex border. J Ecol 41: 240–247

Piriou JY, Ménesguen A (1992) Environmental factors controlling the *Ulva* sp. blooms in Brittany (France). In: Colombo G, Ferrari I, Ceccherelli VU, Rossi R (eds) Marine eutrophication and population dynamics. 25th Eur Mar Biol Symp, Olsen and Olsen, Fredensborg, Denmark, pp 111–115

Piriou JY, Ménesguen A, Salomon JC (1991) Les Marées vertes à Ulves: conditions nécessaires, évolution et comparaison de sites. In: Elliot M, Ducrotoy JP (eds) Estuaries and coasts: spatial and temporal intercomparisons. Proc ECSA 19 Symp, Olsen and Olsen, Fredensborg, Denmark, pp 117–122

Pitkin PH (1977) Distribution and biology of algae. In: Jeffrey DW (ed) North Bull Island, Dublin Bay. R Dublin Soc, Dublin, pp 32–37

Powell HT (1963) New records of *Fucus distichus* subspecies for the Shetland and Orkney Islands. Br Phycol Bull 2: 247–254

Price JH, Tittley I (1972) The marine flora of the county of Kent and its distribution, 1597–1972. Br Phycol J 7: 282–283

Raffaelli D, Hull S, Milne H (1989) Long-term changes in nutrients, weed mats and shorebirds in an estuarine system. Cah Biol Mar 30: 259–270

Raffaelli D, Limia J, Hull S, Pont S (1991) Interactions between the amphipod *Corophium volutator* and macroalgal mats on estuarine mud flats. J Mar Biol Assoc UK 71: 899–908

Raymont JEG (1947) A fish farming experiment in Scottish sea lochs. J Mar Res 67: 219–227

Russell G (1974) *Fucus distichus* Communities in Shetland. J Appl Ecol 11: 679–684

Smyth JC (1968) The fauna of a polluted shore in the Firth of Forth. Helgol Wiss Meeresunters 17: 216–223

Soulsby PG, Lowthion D, Houston MCM (1978) Observations on the effects of sewage discharged into a tidal harbour. Mar Pollut Bull 9: 242–245

Soulsby PG, Lowthion D, Houston M (1982) Effects of macroalgal mats on the ecology of intertidal mudflats. Mar. Pollut Bull 13: 182–186

Soulsby PG, Lowthion D, Houston M, Montgomery HAC (1985) The role of sewage effluent in the accumulation of macroalgal mats on intertidal mud flats in two basins in southern England. Neth J Sea Res 16: 257–263

Southgate BA (1972) Langstone Harbour study. Rep to Hampshire River Authority and the Hamsphire County Council, The Counties Public Health Laboratories, London, 20 pp

Thrush SF (1986) The sublittoral macrobenthic community structure of an Irish sea-lough: effect of decomposing accumulations of seaweed. J Exp Mar Biol Ecol 96: 199–212

Traill GW (1886) The marine algae of Joppa in the county of Mid-Lothian. Trans Proc Bot Soc Edinb 16: 395–402

Tubbs CR (1977) Wildfowl and waders in Langstone Harbour. Br Birds 70: 177–199

Tubbs CR (1980a) Processes and impacts in the Solent. In: The Solent estuarine system: an assessment of present knowledge. NERC Publ Ser C 22. Natural Environment Research Council, London, pp 1–5

Tubbs CR (1980b) Bird populations in the Solent, 1951–77. In: The Solent estuarine system: an assessment of present knowledge. NERC Publ Ser C 22. Natural Environment Research Council, London, pp 92–100

Tubbs CR, Tubbs JM (1980) Wader and Shelduck feeding distribution in Langstone Harbour, Hampshire. Bird Study 27: 239–248

Tubbs CR, Tubbs JM (1982) Brent geese *Branta bernicla bernicla* and their food in the Solent, southern England. Biol Conserv 23: 33–54

Tubbs CR, Tubbs JM (1983) Macroalgal mats in Langstone Harbour, Hampshire, England. Mar Pollut Bull 4: 148–149

Walsh T (1988) An ecological study of the green algae (Chlorophyceae) on the intertidal mudflats of North Bull Island: their biomass, nitrogen relations, distribution and species composition. Thesis, Trinity College, Dublin (unpubl)

Warwick RM, Davey JT, Gee JM, George CL (1982) Faunistic control of *Enteromorpha* blooms: a field experiment. J Exp Mar Biol Ecol 56: 23–31

Wharfe JR (1977) The intertidal sediment habitats of the lower Medway Estuary, Kent. Environ Pollut 13: 79–91

Wilkinson M, Tittley (1979) The marine algae of Elie, Scotland: a reassessment. Bot Mar 22: 249–256

Wilson JG, Jeffrey DW, Madden B, Rafferty (1990) Algal mats and eutrophication in Dublin Bay, Ireland. Presentation at the 25th Mar Biol Symp, Sept 10–15, 1990, Ferrara, Italy

Wood E (eds) (1988) Sea life of Britain and Ireland. IMMEL, London, 240 pp

Wright SL (1980) The pollution load entering Southampton Water and the Solent. In: The Solent estuarine system – an assessment of present knowledge. NERC Publ Ser C 22. Natural Environment Research Council, London, pp 62–63

9 The French Atlantic Coasts

P. Dion and S. Le Bozec

9.1 Introduction

Summer mass blooms of macroalgae have been occurring increasingly for two decades along the French Atlantic coastline (Fig. 9.1). The coasts of Brittany are the most affected, considering the nearly 100 000 m³ of green seaweeds collected yearly from the beaches. One member of the genus *Ulva* is particularly involved in these mass blooms, although some ectocarpales, for example *Pilayella littoralis*, recently appeared in some localities.

The enclosed Bay of Arcachon is the second most affected area with about 6000 m³ of green algae removed yearly. Eutrophication also occurs, at least as a natural phenomenon, in the 30 000 ha of salt marshes of the French Atlantic coast.

The present knowledge of the eutrophication mechanisms producing the green tides on the Brittany coasts is presented here. Possible fighting means (curative as well as preventive measures) are also reviewed.

9.2 Green Tides of the Brittany Coasts

The term "green tides" in Brittany generally refers to concentrations of unattached forms of *Ulva* sp., occurring in summer in the nearshore waters (Fig. 9.2), and leading locally to massive beaching (Fig. 9.3).

Green tides are not in themselves a hazard to human health, and a negative impact of decomposing deposits on the ecosystem has not yet

CEVA (Centre d'Etude et de Valorisation des Algues), BP 3, 22610 Pleubian, France

Ecological Studies, Vol. 123
Schramm/Nienhuis (eds) Marine Benthic Vegetation
© Springer-Verlag Berlin Heidelberg 1996

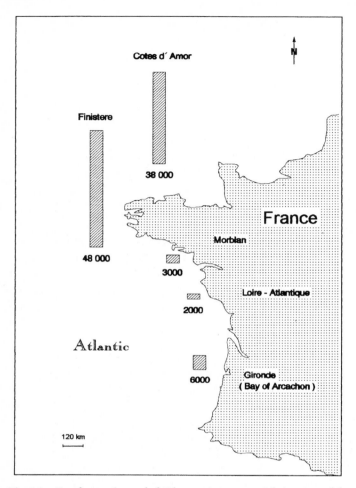

Fig. 9.1. Beach cast (stranded *Ulva* sp. in t wet weight) removed from beaches along the coast of France in 1991

been demonstrated. However, by making the beaches unattractive (visual and odoriferous pollution) during the summer period, they force coastal townships and municipalities to undertake a costly removal scheme in recreational or living areas. Moreover, immeasurable costs probably originate with a decrease in a general degradation of the reputation of the Brittany coastline. It should also be added that massive development of seaweeds can be a physical hindrance for small coastal navigation or mariculture, for example, shellfish cultivation. The hydrodynamic situation along the exposed Brittany coasts with strong swells and tidal water movements prevents degradation of the algal masses in the water

Fig. 9.2. Unattached forms of green tide *Ulva* sp. forming a distinct floating belt at low tide along the beach, Bay of Lannion, Brittany. The belt can be 20 to 200 m wide. Note the breaking swell concentrating the seaweed in the very shallow nearshore waters (0 to 1 m depth). Aerial infrared photograph, IFREMER

Fig. 9.3. Stranded biomass of green tide forming *Ulva* sp

column, such degradation could cause a dystrophic crisis (i.e. oxygen consumption and depletion) at the end of the growth period, as has been observed in shallow closed lagoons (e.g. Mediterranean pools, Venice lagoon: see Chap. 15, this vol). In green tide-infested areas of the "open type", as is the case for the coasts of Brittany, seaweed degradation only takes place after beaching.

The existence of green tides along the Brittany coastlines can already be detected in the Bay of St. Brieuc, Lannion and Douarnenez from IGN aerial photographs taken in the 1950s (cf. Piriou et al. 1991). In the late 1970s, the phenomenon of green tides became evident, and reached an intolerable point only in its recent evolution, during the 1980s. The extent of these green tides now seems to be stabilized in these main sites, but it has been observed that each year an increasing number of coastal communities and townships are reported to be affected (50% increase between 1983 and 1991; Fig. 9.4).

During the summer of 1991, about 25% of these localities were affected by green tides and some of them, particularly seaside resorts, had to undertake cleanup operations. Thus, 98 000 m³ were collected from the coasts of the four Breton departments. It should be pointed out that not all of the stranded seaweeds were collected and that there is no simple relationship between beaching levels and total production.

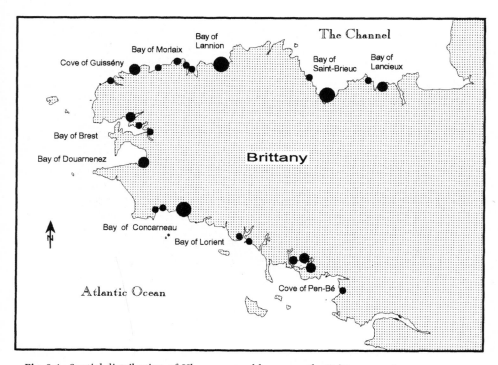

Fig. 9.4. Spatial distribution of *Ulva* sp. mass blooms on the Brittany coast

9.2.1   Production Mechanism of Green Tides

The term green tide generally refers to drifting *Ulva* concentrated along sandy portions of the shoreline. The species involved occur also attached to rocky and gravel substrates of the lower intertidal zone. These attached populations are the primary source of green tides on the coasts of Brittany, but there is now evidence that they are not of preponderant importance in the annual formation of the phenomenon. In fact, attached populations of these species are not particularly abundant around affected areas. In addition, no whole individuals or basal parts with their holdfast can be found in the floating green tide material, which would indicate a recent removal from substrate and direct transport. It is likely that material transfer from attached to free-living populations is small. The bulk of the green tide production occurs in the floating phase, through vegetative reproduction. In spring, the source of the green tide is most probably a remnant of *Ulva* from the preceding year.

Mass development of green seaweeds in other coastal areas of Europe and all over the world is generally attributed to an increase in the nutrient concentrations in coastal waters. In most cases, green tides can be directly linked to that of waters collected by drainage basins. The most responsible nutrients are nitrogen and phosphorus compounds.

On the Brittany coast, enclosed areas, such as the inner parts of bays, where water exchange with the sea is slow, are particularly sensitive to green tide outbreaks (Menesguen and Salomon 1988; Piriou et al. 1991, 1993). Despite the high tidal amplitude of 6–9 m, residual drift of the waterbody between tides is small, as shown in the Bay of St. Brieuc (Menesguen and Salomon 1988). Drifting seaweed and nutrients are therefore trapped together in these areas. Breaking swell, which usually occurs in large intertidal areas with a low slope, probably takes part in concentrating the seaweeds in shallow nearshore coastal waters. Keeping the seaweeds in suspension in this enriched, mixed, easily warmed up and well lighted environment favours rapid growth and vegetative multiplication through fragmentation.

Out of the eight species of *Ulva* that it is possible to identify around the coast of Brittany (Hoeksema and Van den Hoek 1983), only one is involved in large-scale green tide processes. The systematic position of this species is not clear, and a new taxonomic nomenclature will probably have to be put forward for it. Its predisposition to react to eutrophicating conditions at the end of bays seems to be based on particular biological and physical characteristics (Dion 1990).

Ulva species in general, and the *Ulva* form producing mass blooms on Brittany coasts in particular, have a high photosynthetic potential, allowing efficient use of the summer light conditions for growth (Table 9.1).

Table 9.1. Comparison of photosynthetic potentials at saturating light ($2000 \, \mu E \, cm^{-1} \, s^{-1}$) and $15°C$ between green tide *Ulva* sp. and other seaweed life forms

Species	Photosynthetic potential ($mg \, O_2 \, g \, dry \, wt^{-1} \, h^{-1}$)
Ulva sp. (green tides)[a]	60–80
Ulva rigida[a]	30–40
Enteromorpha compressa[b]	25–35
Fucus vesiculosus[b]	7–13
Petrocelis cruenta[c]	1–2

[a]Data from CEVA.
[b]Data from Levavasseur (1986).
[c]Data from Dion (1988).

Compared with other seaweeds (Table 9.2), the green tide *Ulva* species shows high nutrient uptake performances. The ratio V_{max}/km, which is often used to characterize the affinity of seaweeds for nutrients (e.g. Wallentinus 1984), is very high for the green tide *Ulva* sp., demonstrating its opportunistic behaviour in the use of nutrients.

The wide temperature optimum for growth, ranging from $17–23°C$ as determined in laboratory experiments (CEVA, unpubl. data), shows that this species is well adapted to the temperature conditions of shallow waters in flat sandy bays, which easily warm up. Water temperatures close to the beach may be as high as $28°C$ during the flood period. (The open sea temperature never exceeds $18°C$).

Table 9.2. Kinetic parameters of the green tide *Ulva* sp. under batch and laboratory conditions (data from CEVA, unpubl. and for other seaweeds data from Wallentinus 1984)
A. Phosphorus uptake

Species	V_{max} ($\mu mol \, PO_4^{-3} \, g \, dry \, wt^{-1} \, h^{-1}$)	V_{max}/km
Ulva sp.(green tides)	8–35	4–10
Enteromorpha spp.	2–8	2–6
Fucus vesiculosus	0.6–1	0.2–0.4

B. Nitrate uptake

Species	V_{max} ($\mu mol \, NO_3^- \, g \, dry \, wt^{-1} \, h^{-1}$)	V_{max}/km
Ulva sp.(green tides)	60–230	12–37
Enteromorpha spp.	27–130	8–16
Fucus virsoides	5–20	0.5–1.3

Survival and, consequently, mass production of the green tide *Ulva*, would not be possible if it could not maintain itself suspended in the mixed water column and multiply easily by fragmentation. The thinness (30–40 µm) and fragility of the thallus are probably conditional to these capabilities. Generally, the seaweeds involved in eutrophication seem, in the same way, able to colonize the water column physically.

As far as other species are concerned with regard to eutrophication on Brittany coasts, it should be mentioned that agglomerations of unattached mixed *Porphyra* species have been found irregularly in small quantities since the first reports of the green tide phenomenon. They have very rarely led to quantitatively significant beachings.

More important is the brown alga *Pilayella littoralis* (Ectocarpales), which has been recently noted to produce mass blooms in some localities of North Brittany. Such blooms occur irregularly from one year to another, sometimes in alternation with *Ulva* sp. Mass development of *Pilayella littoralis* has also been observed along the coasts of other countries, e.g. USA (Massachussetts), and the Baltic Sea (Breuer and Schramm 1988; Pregnall and Miller 1988). In the same way, the filamentous red alga *Falkenbergia rufalonosa* (sporophyte of *Asparagospsis armata*) has been recently shown to produce mixed mass blooms with the green tides of the Bay of Douarnenez (South Brittany), indicating trends in diversification or succession for species involved in local coastal eutrophication (CEVA, unpubl. data).

9.2.2   Fighting Means and Associated Research Programs

What methods could be used for fighting green tides on the coasts of Brittany? In this context, the relationships between attached and free-floating populations must be considered. The attached populations of the *Ulva* species are scattered and not easily distinguishable from each other. In practice, it is difficult, even un-feasible to operate selectively on these populations. Furthermore, this action on the whole would probably be useless, since sessile populations are not a major cause of the annual formation of green tides, as was pointed out earlier.

As to the control of floating populations, curative and preventive means may be distinguished.

Curative means include collecting the seaweeds and removal, or the direct use of growth inhibitors or toxic compounds. Collecting the beached biomass is commonly used to clean up recreational areas, although this method has its limitations and is often expensive (3 million FF in 1992 for the coasts of Brittany CEVA, unpublished data). Another possibility would be to collect enough seaweeds from those already in the water at the start of the proliferation period to prevent an outbreak of green tides. However, this does not appear to be very feasible from a technical and economic point or view. In addition, storage of collected seaweed masses is an environmental problem in itself. Treatments such as methanization or composting are now technically perfected (Briand 1989), but it is difficult to reach an industrial breakthrough.

At the present state of scientific progress, direct application of any chemical or microbiological method, which would lead to growth inhibition or destruction of the green tide-forming plants, is not ecologically acceptable.

After curative means, preventive means have to be considered. Of the primary environmental factors controlling growth and mass development of the seaweeds, light and temperature, of course, cannot be altered. Undoubtedly, the only realistic preventive means to reduce green tides is to lower nutrient levels in sensitive coastal areas. Three different aproaches are possible, at least theoretically.

1. *Physical Means.* Collecting or rerouting nutrient-loaded waste water has already been used on different scales to prevent coastal zone pollution. This solution would be theoretically possible in some locations of Brittany (e.g. Bay of Lannion), where river discharge is comparatively small and could be canalized or piped out of the sensitive zone. In the Bay of St. Brieuc, rerouting of tidal channels would probably help decrease green tides to a certain extent (Piriou et al. 1991).

   Generally speaking, physical solutions have the disadvantage that they can only be applied in particular cases. In addition, they require major investments, and they do not decrease, but only transfer the pollution problem (even if diluted) from one area to another, with unpredictable long-term effects.

2. *Preventive Means of the Biological Type.* These could, for example, include the cultivation of economically valuable seaweeds that would compete with *Ulva* for nutrients, or the settlement of macrophyte swamps or wetlands, the purifying effect of which is well known (cf. Reddy and Smith 1987). Their disadvantage is that they would require large land or sea areas, which are rarely available along the Brittany coast.

The possibility of using wetlands in Brittany was first considered (in 1990) for the Fremur river (Departement of Côtes d'Armor), which has at least a potential of 80 ha for the wetland settlement. The estimated cost for such a project, however, was discouraging for the local

authorities. A smaller and more experimental project was started recently in the Bay of Douarnenez (Piriou 1993). The introduction of
seaweed cultures as nutrient competitors for the green tides (see
Chap.4, this vol.) was considered in some favourable situations but was
not followed up because there was no demand by the local seaweed
industry for such quantities of seaweeds which would be necessary to
substitute or outcompete the green algae.

3. *Preventive Means of the "Chemical" Type.* These concern decreasing
   the input of nutrients already upstream in the drainage basin by
   efficient effluent treatment and a preventive policy.

In this context, research programs have been undertaken at CEVA and
IFREMER to study the physiological ecology of *Ulva* nutritient requirements and to develop ecological models.

Seasonal variations of N and P content in *Ulva* tissues clearly indicated that external nitrogen N was the major limiting nutrient for
green tide growth (Dion 1988; Piriou and Menesguen 1992; Fig. 9.5).

Ecophysiological laboratory investigations also contributed to a nu-
merical model for the ecological production of *Ulva* (Menesguen and

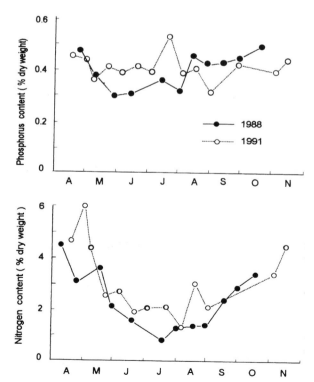

Fig. 9.5. Seasonal variations
in internal N and P contents
in green tide *Ulva* sp.
during mass bloom in the
Bay of St. Brieuc, 1988 and
1991 (CEVA, unpubl. data)

Salomon 1988; Menesguen 1990; Piriou and Menesguen 1992). With the help of this model it was possible to quantify the limiting role of N for growth, and to predict relationships between N and P flows and green tide development in the Bay of St. Brieuc.

Reducing the nitrogen input at the river basin level is undoubtedly the most suitable way (if not the only one available) to prevent green tides in the long term. This solution, however, introduces serious short-term economical problems, in particular when intensive agriculture is the main source of nutrients, as is the case in Brittany. The effects of phosphate removal in wastewater treatment plants have been monitored as an alternative solution in the Bay of St. Brieuc over a period of 5 years. Despite a 50% reduction in the total phosphorus load and a considerable reduction (60%) in tissue phosphorus levels in *Ulva* sp., mass development of green algae continued to occur (CEVA, unpubl.).

9.3   Other Cases of Eutrophication
Along the French Coast

9.3.1   The Bay of Arcachon

The Bay of Arcachon in southwestern France is an example of a relatively closed marine bay, where mass proliferation of green macroalgae occurs. The area of the bay (150 km^2) is divided at low tide into three NE–SW oriented main channels, which are connected with the Atlantic Ocean through a strait (Fig. 9.6). The total low tide area of the main and secondary channels is about one-third of the total Bay area (cf. Auby et al. 1994).

The Bay of Arcachon is well known for its attractive landscape, favouring an important seasonal touristic economy. The bay is urbanized, and crowded in summer. The other basis of the local economy is mariculture, in particular oyster cultivation, producing 15 000 to 20 000 t of oysters per year.

The general topographic conditions in the bay and the tidal regime allow oyster production in the intertidal zone and the presence of 7000 ha of seagrass meadows (*Zostera noltii*) whies colonize mobile sand and the mud substratum.

Macroalgal mass blooms have been occurring in the Bay of Arcachon for many years, but they have become conspicuous only recently, because of increased production of green algae in the water body, and the

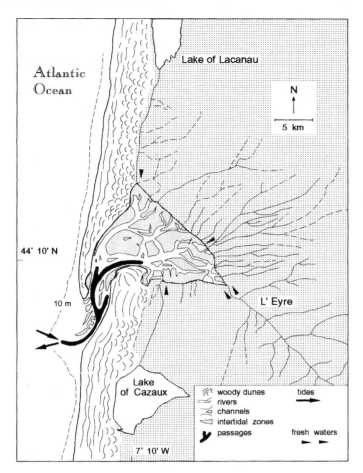

Fig. 9.6. The Bay of Arcachon

strandings of these seaweeds in the large muddy intertidal zone and on tourist beaches. During spring tides and windy periods, large amounts of green algae mixed with drifting *Zostera* are frequently deposited on the muddy and sandy strands, where they decompose rapidly. Because it is a nuisance for fishery and tourism, the decomposing material has to be removed mechanically.

The collected beach cast increased from about 4000 m³ in 1990 to 30 000 m³ in 1992, the latter containing about 6 000 m³ of *Monostroma*, indicating a rapid increase in the green tide phenomenon in this region.

According Auby et al. (1994), mass production of *Enteromorpha* sp. first appeared in the Bay of Arcachon in 1982. This species had been noted earlier at this site but in acceptable amounts. Seasonally, the development of *Enteromorpha* sp. is rapid in June in the intertidal zone, where populations are attached to the substratum, partly in the *Zostera* beds. During this period, the size of the plants increases, until they are torn off from their substrata, then invading the water body where they continue to grow free-floating. This phenomenon usually ends in October. Such proliferation can lead to various detrimental effects including anoxia and fish mortality.

Another green alga has become the dominant species in the bay since 1989; it was identified as *Monostroma obscurum* (Auby *et al.* 1994). The period of maximum biomass development is between January and April, i.e. before the bloom or *Enteromorpha* sp. The vegetative propagation of the unattached thalli is probably the principal way of producing large biomasses, in the subtidal channels and in the shallow water of the intertidal zone. During other periods, significant quantities of algae can be maintained in the water, and they do not completely disappear in autumn and winter. For a few years, the filamentous red alga *Centroceras clavatum* was also observed to produce blooms sporadically in the Bay of Arcachon. Concerning *Monostroma* at least, nitrogen is considered the limiting nutrient.

9.3.2   Eutrophication and Salt Marshes

Salt marshes of the French Atlantic coast cover about 30000 ha from the Gulf of Morbihan (southern Brittany) to the Bay of Arcachon (Aquitaine) and extend mainly in the Régions Pays de Loire and Poitou-Charentes (Fig. 9.7). They consist of complex hydraulic systems of channels and ponds, which used to be exploited by the salt industry in the past and are partly utilized for shellfish aquaculture today.

The natural productivity of these systems is very high and macroalgal blooms occur frequently in the ponds. *Ulva* spp. and *Enteromorpha* spp. are the main species responsible for the mass blooms, but other green seaweeds (*Cladophora* spp., *Ulothrix* sp., *Chaetomorpha* spp.), as well as brown algae (*Colpomenia peregrina,* some members of *Ectocarpaceae*) or red seaweeds (*Gracilaria verrucosa*) also invade the ponds depending on the localities, the seasons and the ponds' management procedures. Such macroalgal blooms can cause dystrophic conditions in the ponds, but also may compete for nutrients with phytoplankton, the food base for shellfish aquaculture. The removal of the seaweeds uses up a considerable proportion of the working time and efforts of the aquaculturists.

The development of eutrophication and pollution by trace elements has recently received attention, particularly in areas where intensive agriculture has spread to the buffer wet zone besides the salt marshes.

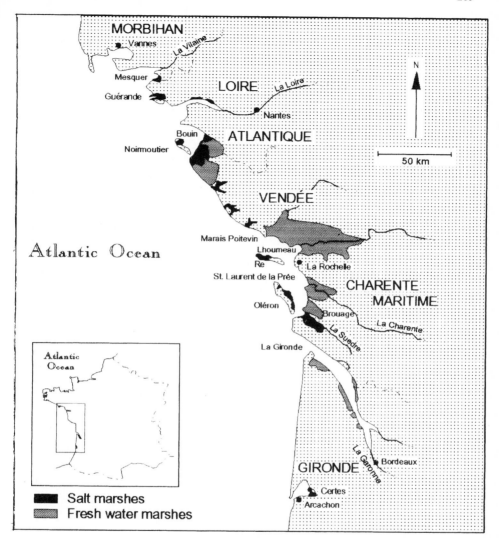

Fig. 9.7. Localization of the French Atlantic salt marshes. (From Clement 1991)

References

Auby I, Manaud F, Maurer D, Trut G (1994) Etude de la prolifération des algues vertes dans le bassin d'Arcachon. Etude IFREMER–CEMAGREF–SSA–SARBAC, 146 pp
Breuer G, Schramm W (1988) Changes in macroalgal vegetation of Kiel Bight (western Baltic Sea) during the past 20 years. Kiel Meeresforsch 6: 241–255

Briand X (1989) Prolifération de l'algue verte *Ulva* sp. en baie de Lannion (France): etude d'une nuisance et de son traitement par fermentation anaérobie. Thése de 3° cyle, Biologie et Physiologie Végétale. Lille I, France (7), 210 pp

Clement O (1991) Typologie aquacole des marais salants de la côte atlantique. Etude Ressources en Eau, n 3. CEMAGREF, National Public Institutes, 232 pp

Dion P (1988) Etude de certains aspects écophysiologiques des marées vertes. Rapport de contrat IFREMER/CEVA R37 National Public Institutes, 35 pp

Dion P (1990) Les marées vertes. Penn Ar Bed 137(21): 63–67

Hoeksema BW, Van Den Hoek C (1983) The taxonomy of *Ulva* (Chlorophyceae) from the coastal region of Roscoff (Brittany, France). Bot Mar 26: 65–86

Levavasseur G (1986) Plasticité de l'appareil pigmentaire des algues marines macrophytes, régulations en fonction de l'environnement. Thése de Doctorat d'Etat es Sciences Naturalles présentée à l'Université Pierre et Marie Curie-Paris VI, 185 pp

Menesguen A (1990) Modélisation des "marées vertes" littorales et ses applications. Houille Blanche 3/4: 237–242

Menesguen A, Salomon JC (1988) Eutrophication modelling as a tool for fighting against *Ulva* coastal mass blooms. In: Schrefler BA, Zienkiewicz OC (eds) Proc Int Conf on Computer modelling in ocean engineering. Balkema, Rotterdam, pp 443–450

Piriou JY (1993) Bilan de l'action de remise en eau du marais de Kervigen (littoral de la baie de Douarnenez) dans le cadre d'une lutte contre la marée verte. Rapport IFREMER DEL/93 21/Brest, 5 pp

Piriou JY, Menesguen A (1992) Environmental factors controlling the *Ulva* sp. blooms in Brittany (France). In: Colombo G, Ferrari I, Ceccherelli VU, Rossi R (eds) Marine eutrophication and population dynamics. 25th Eur Mar Biol Symp. Oben & Olsen, Milwaukee, pp 111–115

Piriou JY, Menesguen A, Salomon JC (1991) Les marées vertes á Ulves: conditions nécessaires, évolution et comparaison de sites. In: Elliott M, Ducrotoy JP (eds) Estuaries and coasts: spatial and temporal intercomparisons. ECSA 19 Symp. Olsen & Olsen, Milwaukee, pp 117–122

Piriou JY, Merot P, Jégou AM, Garreau P, Yoni C, Watremez P, Urvois M, Hallegouët B, Aurousseau P, Monbet Y, Cann C (1993) Cartographie des zones sensibles á l'eutrophisation: cas des côtes bretonnes. Rapport de synthése IFREMER DEL 93/25/Brest, 80 pp

Pregnall AM, Miller SL (1988) Flow of ammonium from surf-zone and nearshore sediment in Nahan's Bay, Massachusetts, USA, in relation to free-living *Pilayella littoralis*. Mar Ecol Prog Sar 50 (1–2): 161–167

Reddy KR, Smith WH (1987) Aquatic plants for water treatment and resource recovery. Proc. Conf on Research and applications of aquatic plants for water treatment and resource recovery, 20–24 July 1986, Orlando, Florida. Magnolia Publ, Orlando, 1024 pp

Wallentinus I (1984) Comparisons of nutrients uptake rates for Baltic macroalgae with different thallus morphologies. Mar Biol 80: 215–225

10   Spanish Atlantic Coasts

F. X. Niell[1], C. Fernández[3], F. L. Figueroa[1], F. G. Figueiras[4], J. M. Fuentes[5],
J. L. Pérez-Llorens[6], M. J. Garcia-Sánchez[2], I. Hernández[6],
J. A. Fernández[2], M. Espejo[8] , J. Buela[7], M. C. Garcia-Jiménez[1],
V. Clavero[1], and C. Jiménez[1]

10.1   Introduction

Spanish Atlantic coasts extend over 1000 km between 36°01' N to 43°40'
N and 01°40' W to 09°20' W in the north and from 05°30' to 07°20' W in
the south. They belong biogeographically to the Lusitanian province of
the Mediterranean-Atlantic European region (van den Hoek and Donze
1967; van den Hoek 1975). The algal flora of this coast is similar to that
of the coast extending from southern Ireland and Great Britain to the Rio
de Oro in the northwestern coast of Africa.

The Canary Islands must be considered part of an unique Macaronesian
region together with the Azores, Sauvages, Madeira and Cabo Verde archipelagos (Gil-Rodriguez and Wilpret 1980). Van den Hoek (1975) included
these islands, in a different province, into the same Mediterranean-Atlantic
region in which the peninsular waters are included. The coasts of these
islands are of volcanic origin, practically lacking a littoral platform, and
are submitted to strong hydrodynamic forces. For these reasons, proliferation of green algae should be unexpected along these open coasts, in
which the sea shore bottom is very steep. Accumulations of seaweeds of
the *Ulva* type are not frequent, but populations of Ulvaceae with low
biomass can be found in restricted areas (Afonso et al. 1979).

Along the peninsular coast, three regions can be delimited by their
hydrographic and biogeographic features (Fig. 10.1).

[1]Department of Ecology, University of Málaga, Campus de Teatinos, 29071 Málaga, Spain
[2]Department of Plant Physiology, University of Málaga, 29071 Málaga, Spain
[3]Department of Ecology, University of Oviedo, Oviedo, Spain
[4]Instituto de Ciencias Marinas, Vigo, Spain
[5]Centro de Investigaciones Marisqueras de Galicia, Vilaxoan, Pontevedra, Spain
[6]Department of Ecology, University of Cádiz, Spain
[7]ENCE. Lourizan, Pontevedra, Spain
[8]Conselleria de Pesca, Autonomic Government of Galicia, Santiago, Spain

Ecological Studies, Vol. 123
Schramm/Nienhuis (eds) Marine Benthic Vegetation
© Springer-Verlag Berlin Heidelberg 1996

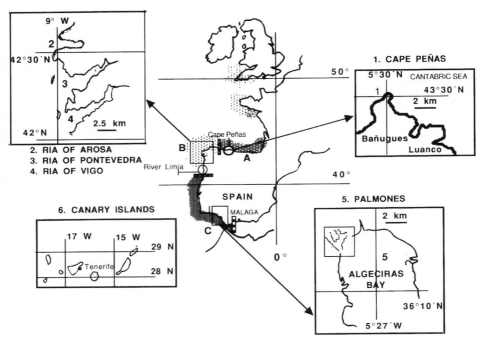

Fig. 10.1. Biogeographical regions in the Spanish Atlantic coast. **A** Cantabric region (*hatched area*); **B** northwestern region (*dotted area*); **C** South Atlantic regions (*dashed area*). Details of the sampling sites are given in the *small framed maps*. *1* Cape Peñas (Fernández 1980); *2* Ria of Arosa (Fuentes 1986); *3* Ria of Pontevedra (Neill and Buela 1976; Niell and Pazó 1978; Figueiras et al. 1985); *4* Ria of Vigo (Niell 1976); *5* Palmones river; *6* Tenerife (Canary Islands) (Afonso et al. 1979)

The first biogeographical area is the northern coast of Spain, the Cantabric Sea, which extends from the French border to the central part of the Asturias region (Anadón and Niell 1981). The platform along this coast is very narrow, and its marine flora similar to that found along the Portuguese coast (Ardré 1970, 1971). Its flora is related to the one of the northwestern corner of Spain, but presents clear meridional characteristics (Fisher-Piette and Duperier 1960, 1961, 1963, 1965, 1966).

The estuaries of this area are small and their water masses are renewed by tidal action almost daily. Despite the existence of high industrial and urban pollution in some areas, no accumulation or mass development of marine algae has been reported.

The second region is the northwestern corner of Spain, which coincides roughly with the region of Galicia. The coast is very rugged, with highly productive positive estuaries (the so-called Rías Bajas) in its western part, which typically show a gradient of vegetation from the open marine coast to the inner shallow estuarine waters. In this region there are relics of some seaweeds very abundant in Northern European waters, such as *Ascophyllum nodosum, Fucus serratus, Chorda filum* and dwarf forms of *Fucus vesiculosus* (Niell et al. 1980a). The estuaries in the northern coast of this region are smaller than the Rías Bajas, and are named Rías Altas. Despite this coincidence in their denomination, no similarities in their hydrographical features can be found between these northern and western estuaries. Rías Altas look like the estuaries of the Cantabric region, but the dominant species of their flora are different.

The northwestern area was studied in recurrent yearly surveys by some classical phycologists. Among them, the works of Fisher-Piette (1955a,b, 1956, 1957a, b, 1958, 1959a, 1963) are very interesting because he followed faunistic and floristic changes from the end of World War II to the seventies. Fisher-Piette (op. cit.) stated that the organisms living in this "corner" are similar to those found on the coasts of French Britanny and used the term "septentrionalization" to refer to the boreal aspect of the algal communities at the biogeographical discontinuity found in Galicia, in contrast to the Portuguese and Cantabric floras. The southern frontier of this "septentrionalized" area is located 40 Km south of the geographical border between Spain and Portugal (Ardré 1970, 1971), in the river Limia (Fig. 10.1).

The Rías are, in some ways, unique systems. Their geological formation has been recurrently studied since the early works of Scheu (1913) to those of Brongersma and Pannekoek (1966) and Pannekoek (1966 a,b). The water circulation in the Rías has also been studied by a plethora of oceanographers. Among them, Saiz and coworkers (1961a,b) and Mouriño et al. (1984) studied the Ría of Vigo, Otto (1975) the Ría of Arousa, and Figueiras et al. (1985) the Ría of Pontevedra. All these works are fundamental for the understanding of the circulation in the Rías, which is not simple, because the estuarine system is influenced by external coastal water movements and sometimes by the Central North Atlantic water, running offshore southwards. Inflow of water into the Rías causes fertilization processes. In addition, a cold upwelling occurs during summer which has also effects on nutrient supply and productivity (Fraga 1979) and is related to the floristic discontinuity found here with abundant presence of the northern flora and fauna species (Niell et al. 1980b; Anadón and Niell 1981).

Ardré and coworkers (1958) made the first attempt to relate the distribution of benthic organisms to the environmental variables in the Rías. Since that first study, numerous investigations have been made on different communities and organisms. The dynamics of the algal communities have been studied by Niell (1977).

The southern Atlantic region is a continuation of the Portuguese coast, extending into the Mediterranean to the mid coast of Málaga. Conde (1981) gives evidence for the delimitation of this region. Among other aspects, it is defined by the presence of *Fucus spiralis* on the Mediterranean coast, the existence of slight tidal movements and by the limit of distribution of the seagrass *Posidonia oceanica*, a Mediterranean endemic species, the frontier of which coincides with the previously mentioned eastern distribution of *F. spiralis*. For these reasons, the Strait of Gibraltar is not really a biogeographical frontier, as Fisher-Piette (1959b) pointed out in his classical and interesting descriptions. The political frontier between Spain and Portugal is also biogeographically irrelevant. This is specially evident due to the presence of Atlantic surface water that enters the Alboran Sea.

In this area there are interesting salt marshes (Doñana, Cadiz Bay, Barbate and Palmones-Algeciras), which, despite their characteristic of being natural bird sanctuaries, are very polluted due to discharge of urban and industrial effluents. The Romans, during the colonization, used to catch fish in these areas by means of controlling the outflow of tidal water in special closed constructions ("esteros"), which were opened at high tide to allow fish to enter. The outflowing water was sieved using nets. These structures were connected by means of a complicated system of channels, and are, together with the salt brine ponds, frequent in many parts of the area. All along this coast there are small shallow tidal estuaries.

This region presents clear differences compared to the northern and north-western coasts. Summer air temperature (close to 40°C), as well as irradiance, is high, and irregular input of nutrients related to stormy and unpredictable rains or to very changeable strong winds in the area of the Strait of Gibraltar is typical. The tidal amplitude is very small with a maximal oscillation of 1 m (Niell et al. 1989), and the emersion period is usually long, obviously causing stress for plants and animals.

10.2   Macroalgal Proliferation Along the Spanish Atlantic Coasts

Until now, no important or catastrophic occurrence of algal masses has been found along the Spanish Atlantic coast. Yearly observations in the Ría of Vigo and in the southern Atlantic region detected only local proliferations of green algae, which have never attained the enviornmental importance that they have had in the shallow coastal platforms of Brittany, the Netherlands, Germany or Denmark.

A survey within a group of phycologists working in different places along the Spanish Atlantic coast revealed that macroalgal proliferations are very scarce. Despite this, some local proliferations in the inner part of the Rías during summer have had catastrophic effects on the populations of bivalves of commercial interest.

The main species that cause proliferations in these areas belong to the Ulvaceae. Data are available from studies made on algal community structure and succession. Of these, three studies have been selected to compare the behaviour of the Ulvaceae in the three coastal areas; Fernández (1980) for the Cantabric area (A), Niell (1976) and several subsequent papers published on the Rías Bajas of the northwestern area (B), and Niell et al. (1989) for the definition of some characteristics of the communities of the Strait of Gibraltar (C).

Proliferations of seaweed of the *Ulva* type are associated with increase of nitrogen in the environment. Ulvaceae are typical opportunistic species. The filamentous and ramified species of *Enteromorpha* (*E. compressa* and *E. prolifera*) are replaced in summer by *E. ramulosa, E. multiramosa* and/or *E. clathrata*. They are most resistant to emersion than the species of *Ulva* (*U. rigida, U. curvata* and *U.cf.gigantea*), which are mainly present at low levels of the intertidal system.

In the Cantabric area, the presence of *Enteromorpha* on the rocky shores is very irregular, with sparse populations and very low biomass (Fig 10.2). The biomass of *Ulva* reached maximal values of 30 g dry weight per m^2 in July (Fernández 1980), whereas Fuentes (1986) found a maximal biomass of 24 g dry weight per m^2 of *Ulva* at the level of *Fucus vesiculosus* in Ría of Arousa (northwestern area), values very similar to those obtained by Niell (1976) at the level of *Himanthalia*, 23.4 g dry weight m^{-2}.

A presumed polluted area close to a pulp kraft mill factory was also studied in the Ría of Pontevedra (Niell 1974; Niell and Buela 1976; Niell and Pazó 1978). Around the source of pollution, a nearly abiotic area was found where only some sulfur bacteria lived. Macrophytic vegetation increased with increasing distance from the point of discharge. In this area, the biomass of Ulvaceae was higher compared to other Rías (Fig 10.3). Waste freshwater of industrial and urban origin caused the presence of green patches in the intertidal system, without the typical Fucacean dominance. In this area, the biomass of *Enteromorpha* was higher than the biomass of *Ulva*.

Data are more scarce in the southern Atlantic region of Spain, and most of them are part of studies on recuperation of cleaned surfaces (Espejo 1984). Figures presented in this study are computed from small

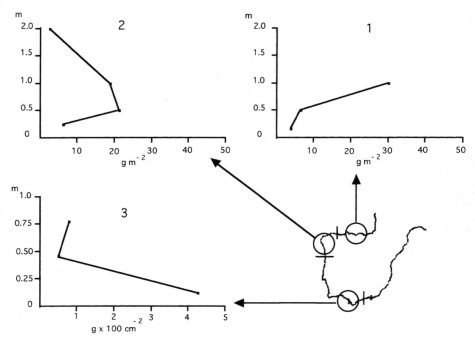

Fig. 10.2. Intertidal profiles of biomass in the three areas mentioned in Fig.1. *1* Cantabric Sea; *2* Rías of Arosa and Vigo; *3* Gibraltar Strait region (note the different scales)

samples of 100 cm² in carpets of continuous Ulvacean vegetation (Fig. 10.2), so data computed for 1 m² are probably risky overestimations, because the cover is not homogeneous. Many of the estimates that can be found in the literature have been obtained by this method, and must be taken carefully. The pattern of the vegetation shows a strong patchiness and all the data computed by extrapolation from small samples are misleading; for this reason, data from Punta Carnero (Espejo 1984), in the south Atlantic region, are expressed per 100 cm² (Fig. 10.2) and cannot be referred to 1 m², because interconversion of data is not realistic or true.

Many data on regeneration of destroyed communities are available (Niell 1977, 1979; Fernández 1980; Fernández et al. 1981; Fernández 1982; Espejo 1984; Niell and Varela 1984; Fuentes 1986; Niell et al 1989). These data, it should be pointed out, are only representative for growth of Ulvaceae in the absence of other macrophytes, although they may reflect the abundance in an unspecific carpet. Table 10.1 shows that in the Spanish Atlantic coast, the highest biomass has been attained in cleaned surfaces in the lowest horizons.

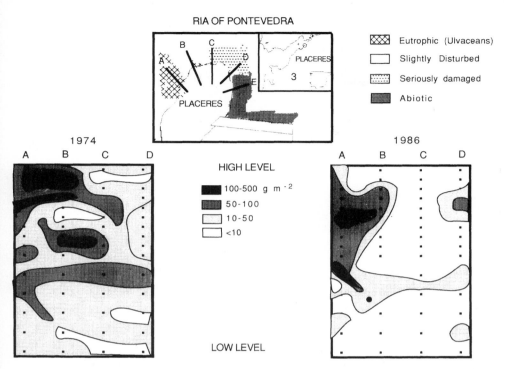

Fig. 10.3. Biomass of seaweeds of the *Ulva* type in 1974 and 1986 in Placeres area (Ría of Pontevedra; site 3 in Fig.10.1)

Calculated P:B values for Ulvaceae in several macroalgal communities ranged from 1 to 4 (Niell 1977, 1979). It has been argued that communities stressed by nutrient inputs show higher P:B ratios than starved ones. The highest P:B values (Table 10.1) have been obtained in polluted and inner estuarine Ulvacean populations, and lowest ones in populations growing on rocky substrata.

The P:B ratio allows an estimate of the potential production expected in the absence of competition. Values of production obtained from actual values of biomass in rocky communities (Table 10.2) are smaller by one order of magnitude than those given by other authors (reviewed Niell 1977) for non-polluted areas. However, populations growing in polluted areas presented values of 0.6 to 0.7 g of carbon m^{-2} day^{-1}, similar to those obtained in the upper horizons dominated by *Fucus*, and slightly lower than a planktonic production of about $0.7\,g\,C\,m^{-2}\,day^{-1}$ as estimated by Fraga (1976) for the Ría of Vigo.

It seems that there is an obvious relationship between algal accumulation and the input of nutrients in water. In the Cantabric coast,

Table 10.1. Estimated values of biomass[a], production and P:B in preinteractive[b] communities with exclusive dominance of Ulvaceans on rocky substrata

Region and Locality	Biomass (g dry wt m^{-2})	Estimated Production (g dry wt m^{-2} year^{-1})	P:B (year^{-1})	Reference
Cantabric Sea				
–Cape Peñas	250	506.62	2.02	Fernández, (1980)
Northwestern Rias				
–*Fucus vesiculosus* horizon	350	1064.34	3.04	Fuentes (1986)
–*Himanthalia* red-algae horizon	750	1140.60	1.52	Niell (1977)
–Polluted area (Placeres) from *Fucus vesiculosus* horizon to upper levels	260 to 560	811.3–2154	3.11–3.84	Niell and Buela (1976); Niell and Pazó (1978); Buela (unpubl.)
Southern coast				
–Higher horizons	827	1424.9	1.72	Espejo (1984)
–Lower horizons	450–750	775–1303	1.72	Espejo (1984)

[a]All of them must be considered as overestimations because ulvaceans never presented 1 m^2 of homogeneous cover and patchiness; competition with other species reduced the biomass 1/5 to 1/10 of the original values.

[b]These figures have been obtained after total cleaning of rocky surfaces in the intertidal system.

Table 10.2. Tentative values of production in Ulvaceans assuming that P:B is similar in non-competitive and in multispecific communities

Region and locality	Biomass[a] (g dry wt m^{-2})	Production[b] (mg C m^{-2} day^{-1})
Cantabric Sea (Cape Peñas)[c]		
-High intertidal level	68.6	40
-Mid intertidal level	13.3	8
-Low intertidal level	13.2	8
Northwestern Rias		
-*Fucus vesiculosus'* horizon[c]	70.0	46
-*Himanthalia*-red algae' horizon[c]	33.4	21
-Polluted area (Placeres) from *Fucus vesiculosus'* horizon to upper levels	933.4–1152[d]	613
Southern coast		
-High intertidal level	102	67
-Low intertidal level	66.04	43
Planktonic production in the Rias[e]		71

[a]Maximal values. Values per square meter were calculated from samples in which the pattern of distribution of Ulvaceans allowed for extrapolations.
[b]Values obtained taking into account P:B values from Table 10.1.
[c]Values are real estimations and were obtained taking into account the patchiness of Ulvaceans.
[d]Intertidal communities exclusively dominated by Ulvaceans in a very eutrophic area.
[e]From Fraga (1976), estimated by means of the ^{14}C method.

maximal concentration of nitrate and phosphate is found in winter. followed by a summer depletion, as is also found in the mediterranean or in the Baltic Sea. (J.A. Fernandez, unpubl; Schramm et al. 1988). Nitrate concentration in the water ranged from 0 to 9 µM and phosphate from 0 to 2 µM.

The seasonal picture is different in the northwestern Rías. Data from Figueiras and coworkers (1985) from the Ría of Pontevedra state that the main allochthonous sources of nutrients are the river Lerez during rainy seasons, and the Central North Atlantic water during summer. In addition, there is an internal source that arises from the mineralization of the nutrients in the Rías. Maximal nutrient concentrations in this region are found in autumn and winter, with a peak in January. The rest of the year the concentration of nutrients is quite low (around 1 µM), but in the Ría of Vigo a second peak sometimes occurs in May. In the inner part of the Rías the concentration of nitrogen and phosphorus is higher and fluctuating. Dissolved oxygen in this area is always close to saturation. Under these conditions no strong eutrophication can be expected.

A totally different situation is found on the Atlantic coast of southern Spain, with two clear peaks of nutrient concentration, one in summer and one in winter with a higher range of variation in nitrate and phosphate concentrations compared to other areas of the Spanish coast (Niell et al. 1989). The only locations in which an accumulation of *Ulva* has been detected are some small estuaries of this area close to the Strait of Gibraltar (e.g. Palmones river estuary) and in some salt–brine ponds, for which an average of 30% coverage by Ulvaceae has recently been reported (Drake and Arias 1993).

The data from Fernández (1980) show that in the northern region the Ulvaceae settle when nutrients increase and days become longer, that is, from March/April to the summer. Some populations remained in the low intertidal system during summer and became larger. When nutrients increased again in September and days became shorter, plants settled again. Some photoperiodical mechanisms may be involved, but no concrete data exist on this subject. The fact is that the combination of several variables, such as high nutrient concentration, sufficient light and high temperature, favours the proliferation of Ulvaceae.

From the existing data, no significant differences in settlement behaviour could be derived between the three above-mentioned areas. It could be stated that spring is the preferred season for settlement and proliferation of green algae with a second peak in autumn. It could also be concluded that the season of settlement of *Entermorpha* is later than that of *Ulva*, and that filamentous *Enteromorpha* occurs before ramified forms.

10.3   The Ulvaceae of Palmones River Estuary: A Case Study of Green Algal Accumulation

The Palmones river (Fig. 10.1) forms a small estuary of 2.5 km^2 at its confluence with the sea and an increasing input of nutrients has been detected in this estuary in recent years (Clavero 1992; Clavero et al. 1991, 1992). The net inflow of carbon and organic phosphorus to the sediment, promoting anaerobic activity and net denitrification, suggests that this estuary could be filled up by sediment deposition in a very short time. In the last 2 years, the proliferation of *Ulva* has been evident. This increase in biomass led to the disappearance of seagrass beds (*Zostera noltii*) and *Gracilaria* mats that previously were very abundant in the estuary (Pérez-Llorens 1991). The estuary is very shallow and water masses are completely mixed daily. It is very difficult to describe the dynamics of its high nutrient levels, because the time course of phosphate and nitrate variations is absolutely chaotic and unpredictable Clavero et al. (1991, 1992) observed a significant trend of organic phosphorus accumulation in the sediment. Very recently, an increase in biomass of *Ulva* was detected, and the sites of accumulation change over time. The cover is sometimes higher than 75% and the blade area index is more than 2 m^2 m^{-2}, sometimes reaching values of $4\,m^2$ m^{-2} or even higher in the floating pleustophytic algal masses.

No seasonal quantitative direct evaluation of standing crop has ever been carried out but using laboratory data, some calculations could be made in order to estimate the potential increase in biomass of Ulvaceae. Garcia-Jiménez (1990) gives data of fresh weight of 2.7 mg cm^{-2} for thalli of *Ulva*. Assuming a cover of 60–100%, a blade area index of 2 to $8\,m^2$ m^{-2}, and a conversion index of wet to dry weight of 25%, the expected biomass would range from 11.8 g m^{-2} in clean areas, to 189.6 g m^{-2} in areas with a large accumulation of algae. Direct measurements made in November 1993 in order to check these estimates gave a biomass of *Ulva* of 97.6 ± 63.5 g dry weight m^{-2}. These values differ considerably from those obtained for the Venice lagoon, in Denmark or Veersemeer in the Netherlands, areas strongly affected by proliferation of green macroalgae (Table 10.3). The data of Meulstee et al. (1986) for the Zandkreek, in Zeeland (Netherlands), are in the middle of our range, but it should be taken into account that our highest estimates have been made just in a few small and localized populations.

Table 10.3. Comparative values of standing crop of Ulvaceans living in sandy eutrophic shallow areas of the European coast

Locality	Standing crop (g m^{-2})	Reference
Zandkreek (NL)	40.8	Meulstee et al. (1986)
Veersemeer (NL)	285	Nienhuis (unpubl.)
Nakehoole (DK)	1200	Anonymous (1991)
Palmones (E)	47–189[a,b]	This work
Venezia (I)	2400	Nienhuis (unpubl.)
Delta of Ebro (E)	250	Martínez Arroyo (1990)
Inner Rias, Placeres (E)	933–1152[c]	Niell (op. cit.)

[a]Theoretical estimations at different blade area index (2–4) and cover (40–100%).
[b]Values obtained in a survey in November 1993 were 97.6 (63.5) g m^{-2}; no seasonal values exist.
[c]Attached populations on rocky substrate in a polluted area.

10.4   Trends, Some Historical References and Comparison of Data

The long-term development of algal production during the past decades can be shown by a few comparative studies. Our impression is that in recent years changes in biomass of Ulvaceae are not significant in non-polluted areas. Perhaps their overall production increased, however, the standing crop remained constant, at least on the rocky shores of Asturias, Galicia and southern Spain. Species richness is decreasing in intertidal communities, but the species that are disappearing are the small ones, so the diversity, computed from biomass values, and the standing crop seem not to change.

A second point of reference is the polluted area in the Ría of Pontevedra, where the first quantitative survey of Ulvaceae in relation to pollution was made in 1974 (Niell and Buela 1976; Niell and Pazó 1978). Ten years later, the biomass has clearly decreased (Fig. 10.3), and maximal values are half those of the past. However, the biomass of the whole community of intertidal Fucaceans has increased (Fig. 10.4). It is difficult to explain these phenomena, but present trends of recuperation of biomass in the area are evident.

Seawards in this study area, there is a zone of continous sediment perturbation due to active fishing of bivalves using spades and similar tools. Although sediment turbation by women's shell-collecting activities releases large amounts of nutrients from the sediment to the water, no floating green algae have been observed.

In the Palmones river, the proliferation of Ulvaceae is recurrent, and fast increases of biomass have been observed year after year (Garcia-Sánchez and Hernández, unpubl.). Southern Spain is a semi-arid area, where rains are scarce, but very heavy (the last heavy rains took place in 1984 and 1989, and in the autumn of 1993). The resulting floods remove all the biomass of macrophytes from the estuary, favouring the dominance of *Ulva*, and causing the disappearance of *Zostera noltii*, the dominant seagrass in the estuary. Turbidity of the water in Palmones is high with an attenuation coefficient of 1.5 m^{-1}, on clear summer days (2500 μE m^{-2} s^{-1}). The irradiance at 2 m water depth for example would be only 15 μE m^{-2} s^{-1}, which is close to the light compensation point of photosynthesis in *Z. noltii* (Jiménez et al. 1987; Pérez–Llorens and Niell 1993). Measurement of attenuation by the laminar thalli of Ulvaceae give an average attenuation coefficient of 1.6 per blade (Niell, unpubl.), which means that three to four thalli of *Ulva* would decrease light to values

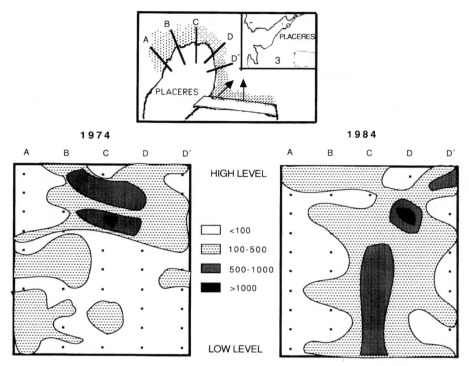

Fig 10.4 Biomass of *Fucus* species (or fucoid seaweed) in 1974 and 1984 in Placeres area. Position of transects see Fig.10.3

below the compensation point of *Zostera noltii*. A proliferation of *Ulva* could therefore lead to the disappearance of *Zostera*. Taking into account that the values of PAR (photosynthetic active radiation) used in this calculation are the highest recorded along the southern coast of Spain, it may be concluded that the presence of *Z. noltii* indicates, at least in our latitudes, the absence of previous recent green algae accumulations.

Despite the uncertainties which lie in extrapolations, *Zostera noltii* and *Z. marina* are abundant in the shallow bays of southern Spain (Cádiz Bay) and in the inner parts of the Rías. The presence of these seagrasses probably shows the relatively small impact of Ulvaceae living as pleustophytes in those eutrophic areas, which is very different from the situation in the Palmones river estuary.

The maintenance and proliferation of conspicuous masses of green tide forming species of the *Ulva* type requires certain environmental conditions: shallow waters, excess of nutrients and favourable conditions for competition with other photoautotrophs. Spanish Atlantic coasts have a very narrow littoral platform, usually with very strong wave action, so that the persistence of green algal blooms can be expected only in the estuaries. Northwestern estuaries have their origin in submersed fluvial valleys which are quite deep, with an acute profile, very sharp margins and steep slopes.

There is evidence that at present in the Rías the phytoplanktonic blooms, basically formed by diatoms, are more frequent, compared to 30 years ago. They are probably the consequence of fertilization processes caused by the input of the Central North Atlantic water, or may be due to mixing processes or terrigenous and fluvial inputs of nutrients, which are utilized faster by diatoms than by green algae. Diatoms sink fast, sequestering nutrients at the bottom in the form of organic matter, followed by mineralization. On the other hand, tidal water remains in the Ría for a very short time (renewal time 8 to 10 days). When nutrients have been exhausted by the diatoms and the levels of released dissolved organic matter increase, dinoflagellates become dominant and can form red tides, if hydrodynamic conditions are stable enough (Jiménez et al. 1992). If hydrodynamic conditions are unstable, big centric diatoms prevail in the nutrient-poor, well-mixed waters. The more frequent occurrence of red tides, together with diatom blooms, points to an anthropogenic origin of nutrients, a result of increasing coastal residential development, local industrial pollution, frequent recurrent fires and increased erosion.Under these conditions, Ulvaceae are not able to increase their biomass and are restricted to shallow areas and rocky shores.

Our impression is that, at present, green tides are not a problem along the Spanish Atlantic coasts, except in some restricted areas, and that the increasing levels of eutrophication in recent years induce more frequent diatom blooms and red tides. The accumulation of green algae is, at present not as important as on the North Atlantic European coasts.

Acknowledgments. We are indebted to Dr. A. Borja (IEO, San Sebastian), Dr. A. Juanés (University of Cantabria), Prof. R. Anadón (University of Oviedo), Prof. A. Carballeira (University of Santiago) and Dr. A. Arias (CSIC, Cádiz), for their comments on the presence of proliferation of Ulvaceae in different areas of the Spanish Atlantic coast. This work has been supported by project AMB93-1211 of the CICYT and the ENCE (Pontevedra).

References

Afonso J, Gil-Rodriguez MC, Wildpret W (1979) Estudio de la vegetación algal de la costa del futuro poligono industrial de Granadilla (Tenerife). Vieraea 8: 201–242

Anadón R, Niell FX (1981) Distribución longitudinal de macrófitos en la costa asturiana (N de España). Invest Pesq 45: 143–156

Anonymous (1991) Eutrophication of coastal waters. Coastal water quality management in the country of Funen, Denmark, 1976–1990. Funen County Council, 288 pp

André F (1970) Contribution á l'étude des algues marines du Portugal. I La Flore. Port Acta Bot (D) x (1–4): 415

André F (1971) Contribution á l'étude des algues marines du Portugal. II Ecologie et chorologie. Bull Cent Etud Rech Sci Biarritz 8: 359–575

André F, Cabañas-Riescas F, Fischer-Piette E, Seoane J (1958) Petite contribution à une monographie bionomique de la Ria de Vigo. Bull Inst Oceanogr Monaco 1127: 56 pp

Brongersma P, Pannekoek AJ (1966) Investigations in and around the Ria de Arosa, North-West Spain, 1962–1964. Leidse Geol Medelingen 37: 1–5

Clavero V (1992) Estudio experimental y modelo de intercambio de fostato en la interfase sedimento–agua en el estuario del rio Palmones (Algeciras, Cádiz). PhD Thesis, Univ Málaga, 206 pp

Clavero V, Niell FX, Fernández JA (1991) Effects of *Nereis diversicolor* O.F. Muller abundance on the dissolved phosphate exchange between sediment and overlying water in Palmones river estuary (southern Spain). Estuar Coast Shelf Sci 33: 193–202

Clavero V, Fernández JA, Niell FX (1992) Bioturbation by *Nereis* sp. and its effects on the phosphate flux across the sediment-water interface in the Palmones River estuary. Hydrobiologia 235/236: 387–392

Conde F (1981) Estudio sobre las algas bentónicas del litoral de la provincia de Málaga. PhD Thesis, Univ Málaga

Drake P, Arias AM (1993) Incidencia de la acuicultura intensiva en salinas sobre los macroinvertebrados bentónicos. Actas IV Congr Nac Acuicult, pp 771–775

Espejo M (1984) Sucesión primaria en un intermareal sometido a perturbaciones térmicas y nutricionales. Ms Thesis, Univ Málaga, 216 pp

Fernández JA (1982) Estudios estructurales del sistema intermareal de Punta Carnero (Algeciras). Ms Thesis, Univ Málaga, 109 pp

Fernández MC (1980) Estudios estructurales y dinámica del fitobentos intermareal (facies rocosa) de la región de Cabo de Peñas, con especial atención a la biologia de *Saccorhiza polychides* (Le Jol.) Batt. PhD Thesis, Univ Oviedo, 278 pp

Fernández MC, Niell FX, Fuentes JM (1981) Remarks on succession in rocky intertidal systems. In: Levring T (ed) Xth Int Seaweed Symp De Gruyter Berlin, pp 283–288

Figueiras FG, Niell FX, Zapata M (1985) Hidrografia de la ria de Pontevedra (NO de España) con mencion especial del banco de Placeres. Inv Pesq 49: 451–472

Fisher-Piette E (1955a) Sur les déplacements des frontiéres biogéographiques intercotidales, observables en Espagne: situation en 1954–1955. CR Acad Sci Paris 241: 447–449

Fisher-Piette E (1955b) Répartition le long des cotes septentrionales de l'Espagne des principaux espéces peuplant les rochers intercotidaux. Ann Inst Ocean 31: 38–124

Fisher-Piette E (1956) Sur les déplacements des frontiéres biogéographiques intercotidales, actuellement en cours en Espagne: situation en 1956. CR Acad Sci Paris 242: 2782–2784

Fisher-Piette E (1957a) Sur les progrés des espéces septentrionales dans le bios intercotidale ibérique: situation en 1956–1957. CR Acad Sci Paris 245: 373–375

Fisher-Piette E (1957b) Sur les déplacements de frontières biogéographiques, observables au long des cotes ibériques dans le domaine intercotidal. Inst Biol Aplic 26: 35–40

Fisher-Piette E (1958) Sur l'écologie intercotidale ouest-ibérique. CR Acad Sci Paris 245: 373–375

Fisher-Piette E (1959a) *Pelvetia canaliculata* examinée de proche en proche de la Manche au Portugal. Coll Int du CNRS, LXXXI, Ecol Alg Mar Dinard, 20–28 Sept 1957, pp 65–73

Fisher-Piette E (1959b) Contribution a l écologie intercotidale du détroit de Gibraltar. Bull Inst Ocean 1145: 1–32

Fisher-Piette E (1963) La distribution des principaux organismes nord–ibériques en 1954–1955. Ann Inst Ocean 40: 165–311

Fisher-Piette E, Duperier R (1960) Variations des Fucacés de la cote basque, de 1894 à 1959. Bull CERS Biarritz 3:1

Fisher-Piette E, Duperier R (1961) Situation des Fucacés de la cote basque en 1960. Bull CERS Biarritz 3: 3

Fisher-Piette E, Duperier R (1963) Situation des Fucacés de la cote basque en 1961 et 1962. Bull CERS Biarritz 4: 4

Fisher-Piette E, Duperier R (1965) Situation des Fucacés de la cote basque en 1963 et 1964. Bull CERS Biarritz 5: 3

Fisher-Piette E, Duperier R (1966) Situation des Fucacés de la cote basque en 1965. Bull CERS Biarritz 6: 1

Fraga F (1976) Fotosintesis en la Ria de Vigo. Invest Pesq 40: 151–167

Fraga F (1979) Profundidad de visión del disco de Secchi y su relación con las concentraciones de fitoplancton y arcilla. Invest Pesq 43: 519–528

Fuentes JM (1986) Dinámica, estructura y producción de una comunidad fitobentónica intermareal (horizonte de *Fucus vesiculosus*) en las Rias gallegas. PhD Thesis, Univ Santiago de Compostela, 288 pp

Garcia-Jiménez MC (1990) Uso fotosintético de la luz y el carbono en agua y en aire en *Gracilaria bursa-pastoris* (S.G. Gmelin) Silva, *Ulva gigantea* (Kutzing) Bliding y *Ulva rigida* C. Agardh. Ms Thesis, Univ Málaga, 92 pp

Gil-Rodriguez MC, Wilpret W (1980) Contribución al estudio de la vegetación ficológica marina del litoral canario. Encicl canaria, Cabildo Insular de Tenerife, 100 pp

Jiménez C, Niell FX, Algarra P (1987) Photosynthetic adaptation of *Zostera noltii* Hornem. Aquat Bot 29: 217–226

Jiménez C, Niell FX, Figueiras FG, Clavero V, Algarra P, Buela J (1992) Green mass aggregations of *Gyrodinium* cf. *aureolum* Hulburt in the Ria of Pontevedra (north-west Spain). J Plank Res 14: 705–720

Martinez-Arroyo MA (1990) Estudio ecológico de algas efemeroficeas: papel en los flujos de materia y energia en un sistema estuárico (Bahia de los Alfaques), Delta del Ebro, España. PhD Thesis, Univ Barcelona, 166 pp

Meulstee C, Nienhuis PH, van Stokkom HTC (1986) Biomass assessment of estuarine macrophytobenthos using aerial photography. Mar Biol 91: 331–335

Mouriño C, Fraga F, Fernández F (1984) Hidrografia de la Ria de Vigo: 1979–1980. Cuaderno da Area Ciencias Mariñas. Semin Estudos Galegos 1: 91–103

Niell FX (1974) Efectos de los vertidos industriales de una fábrica de pasta de papel sobre la estructura del sistema intermareal. Ciencias 39: 363–371

Niell FX (1976) Estudio sobre la estructura, dinámica y producción del fitobentos intermareal (facies rocosa) de la Ria de Vigo. PhD Thesis, Univ Barcelona, 181 pp

Niell FX (1977) Rocky intertidal benthic systems in temperate seas: a synthesis of their functional performances. Helgol Wiss Meeresunters 30: 315–333

Niell FX (1979) Structure and succession in rocky algal communities of a temperate intertidal system. J Exp Mar Biol Ecol 36: 185–200

Niell FX, Buela J (1976) Incidencia de vertidos industriales en la estructura de poblaciones intermareales. I. Distribución y abundancia de Fucáceas caracteristicas. Inv Pesq 40: 137–149

Niell FX, Pazó JP (1978) Incidencia de vertidos industriales en la estructura de poblaciones intermareales. II. Distribución de la biomasa y de la diversidad especifica de comunidades de macrófitos de facies rocosa. Invest Pesq 42: 213–239

Niell FX, Varela M (1984) Initial colonization stages on rocky coastal substrates. PSZNI Mar Ecol 5: 45–56

Niell FX, Farnham WF, Fuentes JM (1980a) Sobre el limite meridional en las costas europeas de *Isthmoploea sphaerophora* (Carm.) Kjellman. Caracteristicas de los ejemplares de la Ria de Arosa. Invest Pesq 44: 211–216

Niell FX, Miranda A, Pazó JP (1980b) Studies on the morphology of the *Megaecade limicola* of *Fucus vesiculosus* L. with taxonomical comments. Bot Mar 23:303–307

Niell FX, Espejo M, Fernández JA, Algarra P (1989) Performances and use of energy in intertidal systems related to random disturbances and thermal stress. Sci Mar 53:293–299

Nienhuis PH (1992) Ecology of coastal lagoons in the Netherlands. Vie Mil 42:59–72

Otto L (1975) Oceanography of the Ria de Arosa (NW spain). Koninklijk Ned Meteoro Inst Medelingen en Verhandelingen 96:210 pp

Pannekoek AJ (1966a) The Ria problem: the role of antecedence, deep weathering and Pleistocene slope-wash in the formation of the West Galician Rias. Tijdschr Kon Ned Aardr Gen 83:289–297

Pannekoek AJ (1966b) The geomorphology of the surroundings of the Ria de Arosa, Galicia, NW Spain. Leidse Geol Medelingen 37:7–32

Pérez-Llorens JL (1991) Estimaciones de biomasa y contenido interno de nutrientes, ecofisiologia de incorporación de carbono y fostato en *Zostera noltii* Hornem. PhD Thesis, Univ Málaga, 168 pp

Pérez-Llorens JL, Niell FX (1993) Temperature and emergence effects on the net photosynthesis of two *Zostera noltii* Hornem morphotypes. Hydrobiologia 254:53–64

Saiz F, López-Benito M, Anadón E (1961a) Estudio hidrográfico de la Ria Vigo. Invest Pesq 8:3–14

Saiz F, López-Benito M, Anadón E (1961b) Estudio hidrográfico de la Ria Vigo II. Invest Pesq 8:97–33

Scheu E (1913) Die Rias von Galicien. Ihr Werden und Vergehen. Z Ges Erdk Berl 84–114: 193–210

Schramm W, Abele D, Breuer G (1988) Nitrogen and phosphorous nutrition of two community forming seaweeds (*Fucus vesiculosus, Phycodrys rubens)* from the Western Baltic (Kiel Bight) in the light of eutrophication processes. Kieler Meeresforsch 6:221–270

Van den Hoek C (1975) Phytogeographic provinces along the coasts of the northern Atlantic Ocean. Phycologia 14: 317–330

Van den Hoek C, Donze M (1967) Algal phytogeography of the European Atlantic Coast. Blumea 15: 63–89

11 Portugal

J.C. Oliveira and G. Cabeçadas

11.1 Introduction

Portuguese coastal waters, in spite of being subjected to considerable inputs of organic matter and nutrients from different sources, generally present acceptable environmental conditions.

The nutrients, nitrogen and phosphorus reaching the coastal waters are mainly transported by rivers and by untreated or poorly treated urban and industrial effluents from sources located along the coast. A process of upwelling, temporarily occurring in the western coast, also provides nutrients to the biologically productive coastal zone. Due to hydrographic, geomorphological and climatic conditions, the Portuguese coastal waters do not show permanent symptoms of eutrophication.

However, some estuaries and lagoons situated along the coast of Portugal (Fig. 11.1) have very complex topographies, presenting interior zones and enclosed arms, which show seasonal signs of eutrophication, such as poor conditions of oxygenation, increasing nutrient levels and development of large amounts of microalgae. In these areas, localized proliferation of some macroalgae, mainly *Ulva* and *Enteromorpha* spp. has been observed.

In general, they do not constitute a nuisance, except occasionally in some aquaculture units and when close to beaches.

The vascular vegetation of salt-marshes is also disturbed, probably as a result of human activities like waste discharges and dredging, as indicated by changes in the distribution and abundance of the communities of *Zostera* and *Spartina* spp. and Chenopodiaceae.

Another indication for eutrophication or pollution in some localized areas may be the occurrence of red tides along the Portuguese coast during the past years.

Portuguese Institute of Marine Research, AV. Brasilia, 1400 Lisboa, Portugal

Ecological Studies, Vol. 123
Schramm/Nienhuis (eds) Marine Benthic Vegetation
© Springer-Verlag Berlin Heidelberg 1996

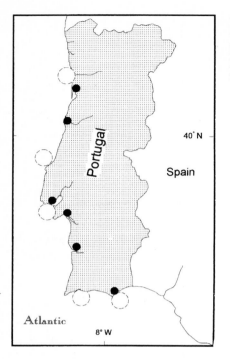

Fig. 11.1. Map of Portugal showing the locations of some of the studied estuaries and lagoons (*filled circles* north to south: Aveiro, Mondego, Tejo, Sado, Mira, Formosa), and five red water-prone regions (*dotted circles*)

11.2 Characterization of Selected Estuaries and Lagoons

11.2.1 Ria de Aveiro Lagoon

The Ria de Aveiro lagoon, covering approximately 45 km^2, is formed by three main branches, a long narrow and shallow channel (Canal de Mira), a wide and deep channel (Canal de S. Jacinto), and a channel, which crosses the town of Aveiro (Canal da Cidade; Fig. 11.2).

The morphology of the lagoon offers favourable conditions for interchange with the sea, particularly because occasional dredging at the entrance keeps water depths relatively stable, although the propagation of the tide slows down considerably with distance from the inlet.

Several rivers discharge into the lagoon. The Vouga river carries 65% of the total freshwater inflow (Hall 1980), equivalent to 23% of the total water volume of the lagoon (Barrosa 1985). The river carries considerable amounts of suspended solids, predominantly formed by clay and organic debris. Some of the enclosed arms of the lagoon show conditions of oxygen depletion and microbiological contamination, especially those

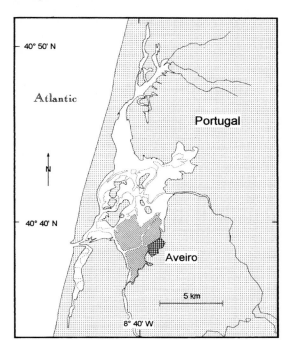

Fig. 11.2. Map of the Ria de Aveiro lagoon

40° 50' N

Atlantic

Portugal

N

40° 40' N

Aveiro

5 km

8° 40' W

receiving urban sewage (Hall 1980). As a result, water quality became inadequate for aquaculture and table salt production, two of the main economical activities in the Ria.

A study undertaken in 1989 (Alcântara et al. 1990), reports comparatively low nutrient levels in the marine-influenced area of the lagoon (NO_3^--N 1.4–2.3 µM, NH_4^+-N 0.9–1.8 µM and PO_4^{3-}-P 0.2–0.3 µM) contrasting with the high levels found in some of the interior sections of the channels (Vouga river: NO_3^--N 20 µM, NH_4^+-N 8 µM and PO_4^{3-}-P 0.7 µM and in Largo da Coroa: NO_3^--N 40 µM, NH_4^+-N 20 µM and PO_4^{3-}-P 4 µM). Chlorophyll a concentrations ranged from 12 to 49 mg m^{-3} in the Vouga river and inner parts of the lagoon, while in most of the more marine areas the values always remained below 7 mg m^{-3}.

Besides fertilizers used in agriculture, other nutrient sources may add to eutrophication. Some industries discharge their effluents to nearby streams, mainly a paper mill (Caima Pulp Company) and a chemical complex located at Estarreja which is not provided with a complete sewage system. The effluents of this industrial unit contain large amounts of ammonia, nitrates, sulphates and metals. The effect of the nutrients discharged is apparent in the extensive infestation of some areas with attached macrophytes (Hall 1980).

11.2.2   Sado Estuary

The Sado estuary, like the Tejo estuary, belongs to the coastal plain lagoon type, dominated by a wide tidal-flat bay and a narrow channel (Fig. 11.3). The Sado estuary has two major sources of freshwater, the Sado river, which contributes 80 to 90% of the freshwater inflow, and the Marateca river which contributes approximately 10%. The Sado river discharges exhibit a pronounced dry/wet regime and large interannual variations. The tidal regime is semi-diurnal, with amplitudes ranging from 1 to 4 m. The upper limit of salt intrusion is located near the town of Alcacer do Sal, and the tidal influence extends 25 km upstream. Salt marshes extend to some parts of the estuary, especially over the Marateca bay. The Sado drainage basin covers an area of 7640 km^2. Major industrial, municipal and aquacultural activities have been established along the northern shore, and agriculture dominates the entire drainage basin. The southern shore is utilized as recreational sites during summer.

Studies on the hydrography of the estuary were conducted by Wollast (1978, 1979) and studies on phytoplankton biomass were undertaken by L. Cabeçadas (1993). In the upper and lower estuary, a monitoring programme has been carried out by the Instituto Hidrografico since 1986.

The assessment of eutrophication conditions of the estuary, based on the data referred to as well as on studies carried out by Cabeçadas and Brogueira (1991) and Oliveira (1992), leads to the conclusion that the system is biologically moderately productive and that the symptoms of

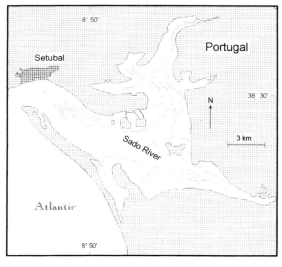

Fig. 11.3. Map of the Sado estuary

eutrophication have not yet been detected. Although there is evidence that nitrogen is the limiting nutrient to phytoplankton productivity along the estuary, the Alcacer zone presents higher nutrient levels in winter (NO_3^--N up to 96 µM, NH_4^+-N up to 7.0 µM, PO_4^{3-}-P up to 1.7 µM and SiO_2 up to 160 µM) and is more productive (chl *a* up to 33.1 mg m^{-3}) compared to the main body and the Marateca bay (NO_3^--N up to 21 µM, NH_4^+-N up to 6.5 µM, PO_4^{3-}-P up to 1.5 µM and SiO_2 up to 52 µM and chl *a* up to 14.3 mg m^{-3}; Cabeçadas and Brogueira 1991).

11.2.3 Ria Formosa Lagoon

The Ria Formosa is a shallow lagoonal system extending more than 50 km along the southern coast of Portugal. The lagoon is permanently connected to the sea through several narrow channels, the most important being the Barra de S. Luiz, Barra de Faro-Olhao, Barra Grande and Barra de Tavira (Fig. 11.4). The tidal amplitude varies between 0.5 and 3.5 m. The lagoon covers an area of 16 000 ha, with approximately 10 000 ha of intertidal areas. Banks with clams occupy approximately 10 km², and large areas of the bottom are covered by vegetation.

There is no significant freshwater input into the lagoon, therefore, salinity remains stable around 36 PSU most of the time. Temperature ranges annually from approximately 12 to 26°C.

Falcáo et al. (1991) report that dissolved oxygen in the water column is most of the time close to or above saturation levels. Nutrients

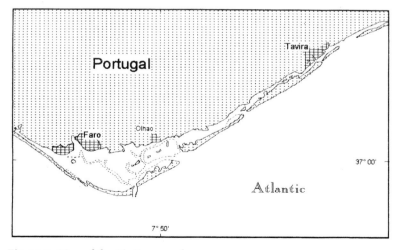

Fig. 11.4. Map of the Ria Formosa lagoon

(Table 11.1) as well as rates of photosynthetic efficiency (21.6 mg C mg^{-1} chl a^{-1} h^{-1}, mean value between July and September) indicate that the lagoon system presents good environmental conditions.

11.3 Symptoms of Eutrophication and Effects on Benthic Vegetation

The most extensive study on the taxonomy and ecology of the Portuguese marine algae is still that of Ardré (1969, 1971), referring to more than 460 species, their distribution and abundance. Succeeding studies have concentrated mainly on the biomass, distribution and growth of *Gelidium sesquipedale,* the major exploited agarophytes (Palminha et al. 1982; Oliveira 1986, 1989; Santos 1993).

Due to geographic and oceanographic conditions, several species have their northern or southern limits of distribution in the Portuguese west coast.

Observed temporal changes in the density of algal communities of brown algae, mainly Laminariales and Fucales, are probably due to natural causes such as variation in solar irradiance levels (Ardré 1971). More recent surveys confirmed these cyclic oscillations, and were also observed in *Porphyra* and *Gelidium* populations, the harvest of which was affected by these temporal variations. On the other hand, the exposure of the open west coast to strong wave action and tidal currents rapidly disperses polluted waters discharged from rivers or coastal towns. Changes in the natural flora on account of pollution have therefore not been noticed, except occasionally near harbours or sewage discharges.

Because of lack of long-term studies on coastal algal vegetation in Portugal, no reliable information on significant changes or modifications is available. Nevertheless, in spring and summer, in some eastern beaches of the south coast, occasional green tides of *Ulva rigida, U. lactuca* and *Enteromorpha* spp. (free floating species) have occured. The

Table 11.1. Nutrient levels

Zones	NO$_3$-N (µM)	PO$_4$-P (µM)	SiO$_2$ (µM)	chl a (mg/m^3)
Faro-Olhao	0.3–4.1	0.05–4.4	0.8–7.2	1.2
Tavira	0.3–10.7	0.05–3.9	0.5–26.1	2.4

settlement of dense communities of the invasive *Sargassum muticum* in large semi-open basins of sand and mud bottoms in the Minho coast (north of Porto) has also been observed.

Changes in coastal benthic vegetation, possibly as a result of eutrophication, pollution or other man-made changes of the environment, have also been observed for marine phanerogams. Several studies on salt-marsh ecosystems have been undertaken in lagoons and estuaries, and settlements of halophilic communities have been observed and described.

Moreira (1986) refers to the most important species found in coastal lagoons and estuaries as *Hamilione, Arthrocnemum* and *Juncus* spp. in upper salt marshes, and *Spartina maritima* in lower salt marshes. In the Sado estuary and Faro lagoon, this last species is replacing the Chenopodiaceae as the sand covers muddy soils after dune erosion provoked by man. Salines, rice fields and aquaculture units, as well as dredging, sand extraction and pollution in several estuaries have disturbed and modified the extension of the highly productive *Spartina* and *Chenopodiaceae* marshes (Moreira 1986). In the Tejo estuary, for example, productivity was estimated as 700 and 350 million kcal m^{-2} year^{-1}, respectively (Catarino and Ramos 1981).

Ferreira and Ramos (1989) estimated an annual net production of 405, 213 and 67 g C m^{-2} year^{-1} for *Fucus vesiculosus, Ulva lactuca* and *Gracilaria verrucosa*, respectively, and a total algal estuarine gross primary production of 5077 t C year^{-1}.

Seagrasses (*Zostera marina* and *Z. noltii*) together with algae (mainly *Ulva, Enteromorpha, Fucus* and *Gracilaria* spp.) colonize muddy surfaces in most estuaries and lagoons. In some restricted interior areas eutrophicated by urban and industrial waste discharge, mass development has occurred. In the most important estuaries and lagoons (Douro, Ria de Aveiro, Tejo, Sado and Ria Formosa), however, the strong tidal currents reduce or prevent serious eutrophication, which may explain the lack of intensive studies on floristic changes in salt marshes in relation to eutrophication.

At present, J.C. Marques and his team from Coimbra University are studying structural dynamic models for estuarine ecosystems, together with qualitative changes in macrophytes due to eutrophication in the Mira and Mondego estuaries. They observed a progressive increase in *Enteromorpha* spp. populations and biomass (4–5 kg m^{-2}), together with the reduction in *Zostera* beds in the southern arm of the Mondego estuary (pers. comm.).

Besides the observed effects of local eutrophication on benthic vegetation in the enclosed areas of some coastal areas, the phenomenon of red tides has increasingly been observed in Portuguese coastal waters, mainly in five restricted areas (Sampayo 1987; Fig. 11.1).

It is generally assumed that development of red tide blooms is associated with adequate amounts of nutrients and specific physical conditions. As the referred coastal zones are influenced by upwelling (Fiuza 1982) and are often located in the vicinity of lagoons or estuaries subjected to river discharges and/or large urban and industrial effluents, it cannot be excluded that eutrophication processes are involved in this phenomenon.

Most red tides which have been detected along the Portuguese coast had no deleterious effects (Sampayo and Cabeçadas 1981; Sampayo and Moita 1984; Brogueira and Sampayo 1983; Cabeçadas et al. 1983). However, the red waters observed in the polluted Obidos lagoon caused PSP, DSP and fish kills (Moita et al. 1984), and the phytoplankton bloom on the north-west coast in 1986, involving the presence of the dinoflagellate *Gymnodinium catenatum*, was responsible for an outbreak of PSP intoxication in bivalves (Sampayo 1987).

11.4   Conclusions

The morphology of the main Portuguese lagoons and estuaries, together with their tidal regimes, favours interchange with the sea and limits eutrophication to localised interior areas, confirmed by several environmental studies.

Quantitative and qualitative changes in benthic vegetation (algae and marine phanerogams) of some lagoonal and estuarine areas as a result of human activities have been observed and studied. Most conspicuous are local proliferations of *Ulva* and *Enteromorpha* spp. in areas disturbed by waste discharges or other sources of nutrients. On the open coast, where strong wave action and tidal currents rapidly disperse urban and industrial effluents, no algal proliferation or significant floristic changes have been observed, except occasionally near coastal towns. However, recent settlements of the invasive *Sargassum muticum*, which forms dense communities in the north, green tides of *Ulva* in the south, as well as red tide blooms in several coastal areas, generally unrelated to eutrophication, have occurred. Long-term changes in the density of observed algal communities of brown and red algae (*Laminaria, Fucus, Porphyra, Gelidium*) are probably due to natural causes such as the pluriannual variation in solar irradiance levels. The absence of serious problems of eutrophication and consequently significant changes in benthic vegetation and biotopes in large areas of the country may partially explain the lack of intensive studies in this connection.

References

Alcántara F, Almeida MA, Nicolau F, Quintaneiro MI, Rua J (1990) Qualidade físico-química e microbiológica da água da Ria de Aveiro. In: 40 Encontro Saneamento Básico, Univ Aveiro, Aveiro

Ardré F (1969) Contribution à l'étude des Algues marines du Portugal I. La Flore. Port Acta Biol B 10(1/4) : 1–423

Ardré F (1971) Contribution à l'étude des Algues marines du Portugal II. Ecologie et chorologie. Bull CERS Biarritz 8(3): 359–574

Barrosa JO (1985) Breve caracterizaçáo da Ria de Aveiro. In: Jornadas da ria de Aveiro, 2. Câmara Municipal Aveiro, Aveiro

Brogueira MJ, Sampayo MA (1983) On a *Noctiluca Scintillans* (Macartney) Ehremb. bloom off the south coast of Algarve, Portugal. ICES/CM E: 35

Cabeçadas G, Brogueira MJ (1991) The Sado Estuary, Portugal: eutrophication conditions in areas of aquacultural activities. In: Nath B, Robinson JP (eds) Environmental pollution, vol. 2. Interscience Enterprises, London, pp 810–820

Cabeçadas G, Cunha ME, Moita MT, Pissarra J, Sampayo MA (1983) Red tide in Cascais Bay, Portugal. Bol Inst Nac Invest Pescas 10: 81–123

Cabeçadas L (1993) Ecologia do Fitoplancton no Estuário do Sado: para uma estratégia de Conservaçáo. Estud Biol Conserv 9 SNPRCN: 1–60

Catarino FM, Ramos MI (1981) Cartografia da vegetaçáo superior halofila do estuario do Tejo. Estrutura e produtividade. Port Acta Biol (B) 2: 13

Falcáo M, Pissarra JL, Cavaco MH (1991) Caracteristicas químico-biológicas da Ria Formosa:análise de um ciclo anual (1985–1986). Bol Inst Nac Invest Pescas 16: 5–21

Ferreira JG, Ramos L (1989) A model for the estimation of annual production rates of macrophyte algae. Aquat Bot 33: 53–70

Fiuza A (1982) The Portuguese coastal upwelling system. In: Seminar on actual problems of oceangraphy in Portugal. JNICT (ed), Lisbon, pp 45–71

Hall A (1980) Water quality problems in Ria de Aveiro. A preliminary assessment. In: Seminar on actual problems of oceangraphy in Portugal. JNICT (ed), Lisbon pp 159–169

Moita MT, Cunha ME, Sampayo MA (1984) Portuguese coastal red waters. ICES/CM B: 9

Moreira ME (1986) Man-made disturbances of Portuguese salt-marshes. Thalassas 4(1): 43–47

Oliveira JC (1986) Distribuiçáo cobertura biomassa e crescimento de *gelidio*. Relatorios de 5 campanhas Jul/Nov 1985. Relat Inst Nac Invest Pescas 70: 1–91

Oliveira JC (1989) A note on the elongation rate of *Gelidium sesquipedale* in in-situ cages in Portugal. In: COST 48 DG XII CEC (ed) Outdoor Seaweed Cultivation. Proc 2nd Worksh COST 48 SGI: 69–75

Oliveira MR (1992) Estado trófico e dinâmico do fitoplankton das zonas superior media e inferior do estuario do Sado. Relat Tec Cient Inst Nac Invest Pescas 59: 1–34

Palminha F, Melo R, Santos R (1982) A existencia de *Gelidium sesquipedale* (Clem) Bornet et Thuret, na costa sul do Algarve I – Distribuicao local. Bol Inst Nac Invest Pescas 8: 93–105

Sampayo MA (1987) Red tides of the Portuguese coast. In: Okaichi T, Anderson DM, Nemoto T (eds) Proc 1st Symp on Red tides. Elsevier, New York, pp 89–92

Sampayo MA, Cabeçadas G (1981) Occurrence of red water of Algarve. Bol Inst Nac Invest Pescas 5: 63–87

Sampayo MA, Moita MT (1984) An *Olisthodiscus luteus* red water. Its dynamics during 24 hours. ICES/CM B: 12

Santos R (1993) A multivariate study of biotic and abiotic relationships in a subtidal algal stand. Mar Ecol Prog Ser 94: 181–190

Wollast R (1978) Rio Sado Campagne de Mesures–Decembre 1978. Rapp Tech, SEA, Lisboa

Wollast R (1979) Rio Sado Campagne de Mesures d'Avril. Rapp. Tech, SEA, Lisboa

**Part B.II
The Mediterranean**

12 The Spanish Mediterranean Coasts

J. Romero[1], F. X. Niell[2], A. Martínez-Arroyo[3], M. Pérez[1], and J. Camp[4]

12.1 Introduction

The Mediterranean coastline of Spain is subjected to high human pressure, caused by both permanent and seasonal populations. For a total
length of coast of about 2200 km, the population living within 5–10 km
of the coast is around 6 million people. In addition, the coastal areas
receive about 6 million visitors annually, most of them in summer. The
extent to which the wastewater from this population is treated before it
returns to the environment is unknown, and very variable, depending on
the area and on the season. These wastes may thus constitute chronic or
seasonal sources of pollution.

Industrial activity is locally intense, especially around large cities
(Barcelona, Tarragona, Valencia, Cartagena), and mining wastes are
dumped into the water in some areas, inducing zones of acute pollution.
Agricultural pollutants have a diffuse, but not significant influence on
coastal waters. Permanent rivers are additional sources of different kinds
of pollutants, although such rivers are scarce. The Ebro is the only large
river in the Spanish Mediterranean coast (Fig. 12.1). In the Archipelago
of the Balearic Islands (Fig. 12.1), with several sites of natural interest,
tourism is the most important source of pollution.

There is a general agreement that high nutrient concentration in the
water is a key variable for algal production. In the unpolluted areas of
the Mediterranean coast, concentrations are low [0–5 (10–25) μmol
dm^{-3} NO_3-N; 0–0.1 (0.3) μmol dm^{-3} PO_4-P; Romero 1985; Ballesteros

[1]Department of Ecology, University of Barcelona, Diagonal 645, 08028 Barcelona, Spain
[2]Department of Ecology, University of Málaga, Campus de Teatinos, 29071 Málaga, Spain
[3]Centro de Ciencias de la Atmósfera, Universidad Nacional Autónoma de México, CP
04510 México DF, Mexico
[4]Instituto de Ciencias del Mar, CSIC, Paseo Nacional s/n, Barcelona, Spain

Ecological Studies, Vol. 123
Schramm/Nienhuis (eds) Marine Benthic Vegetation
© Springer-Verlag Berlin Heidelberg 1996

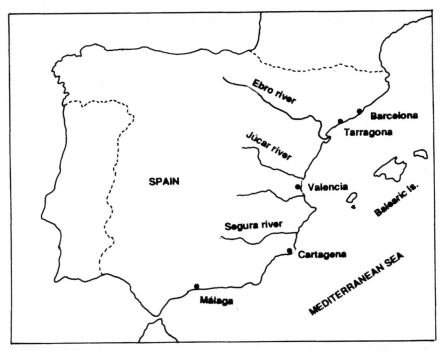

Fig. 12.1. Map of the Iberian Peninsula. Along the Mediterranean coast, the main towns and industrial zones are indicated

1992]. In more than 50% of the samples, values are below analytical detection limits. Thus, recycled production is expected to be more important in sustaining biomass than new production, which is related to external inputs. Marine macrophytes are present throughout the year, and their growth occurs even when the nutrients are undetectable in the water (Ballesteros 1989). In contrast to the Atlantic communities, Mediterranean algal communities never show catastrophic outbursts in biomass, which agrees with the view that primary production at low nutrient concentrations depends on regenerative processes. Algal growth is then supported for most of the time by uptake through high affinity systems, nutrient storage and, eventually, internal recycling, i.e. mobilization for metabolic uses in times of depletion (Ballesteros 1989).

Terrestial inputs provide an important quantity of nutrients to the coastal waters; however, freshwater with high nutrient concentrations is quickly diluted, and thus its availability to seeweeds is local and/or episodic.

12.2   Green Algal Communities and Proliferation

A number of quantitative studies have been carried out on green algal communities and their proliferation mainly on rocky shores. Ballesteros (1992) related these not only to nutrient enrichments, but also to substrate instability. The presence and abundance of green algal communities is rather irregular, varying from site to site and from year to year (Ballesteros 1989). Growth occurs throughout the year, but is frequently and irregularly interrupted by disturbances, as for example storms or persistent low see levels. Biomass values as high as 250 g dry weight m^{-2} for *Enteromorpha compressa*, and 375 g dry weight m^{-2} for *Ulva rigida* have been reported (sample size: 100–200 cm^2; Ballesteros 1992). Their production values are from 0 to 4 g C m^{-2} day^{-1} as estimated by weekly measurements (Pérez 1984; Ballesteros 1992), and the average biomass duplication time is 23 days (annual minimum 5 days). Yearly production for *E. compressa* is around 100 g C m^{-2} year^{-1}, and blade area index can reach 10 (Ballesteros 1992). Such values seem to be high, but it should be taken into account that this biomass accumulation is reached only in small patchy areas (maximum 1 m^2) or in narrow vertical zones of less than 0.5 m, which however, can extend horizontally for some tens of meters. In addition, the communities referred to are on semi-exposed rocky substrates, which means a certain degree of vertical structure, along with movement of the thalli that prevents the self-shading caused by blade accumulation (see Niell et al., Chap. 10, this Vol.).

The distribution of green algae such as *Cladophora* ssp., *Ulva* ssp. and *Enteromorpha* ssp., and other "pioneering" species (*Scytosiphon lomentaria*, *Codium fragile* ssp. *tomentosoides*) was used as a quality indicator in an extensive survey of the NE coast (Catalonia, > 600 km of coastline) in summer 1982 (Ballesteros et al. 1984). No massive green algal proliferations were reported, although in some coastal zones a dominance of pioneering species was evident. Blade indices of ulvaceans were always between 1 and 4. Green algae were also abundant inside some harbours and marinas, but within the same range of biomass.

Macroalgal green tides along the Spanish Mediterranean coast have not been reported during recent years, except for one area which is described below. Even if the number of quantitative data and systematics surveys supporting this contention is still low, it should be remembered that there is a good knowledge of the qualitative aspects of the Mediterranean Spanish Flora (Gallardo et al. 1985; Ribera et al. 1992; Gallardo et al. 1993). A considerable number of phycologists working in this area agree about the absence of green algal proliferation.

The only record of extensive macroalgal accumulation is found in newspapers. Since the 1970s, accumulations of jelly masses of Ceramiaceae (*Spyridia filamentosa, Ceramium tenerrimum*) on beaches, especially in the Valencia Gulf, have been reported every year. These originally attached species are torn off by water movements and accumulate near the beaches in the wave breakfront. This phenomenon has not be quantified, and cannot be considered as an algal proliferation.

Thus, we conclude that green algal masses like those described for Venice lagoon, or northern Europe (see corresponding chapters in this volume) are not a common phenomenon.

12.3   Green Algal Proliferations: A Case Study

The shallow bays on both sides of the Ebro river Delta (Fig. 12.2) are peculiar ecosystems with high nutrient inputs and limited water exchange. The southern Alfacs Bay has been extensively studied (Pérez and Camp 1986; Camp and Delgado 1987; Delgado and Camp 1987; Camp et al. 1992; Vidal et al. 1992), whereas only few data are available for the northern Fangar Bay (Pérez and Camp 1986).

The Alfacs bay, covering a surface of 40 km^2, has a 2-km-wide connection to the sea (Fig. 12.2). The edges of the bay are shallow (0.5 – 1 m; 18 km^2 are less than 1 m deep), while the central zone is deeper (4 – 6 m). The northern part of the shallow platform is silty, with high seagrass cover (80 – 90%), whereas the sandy southern part has a much scarcer macrophyte cover (10%). The central basin is muddy, and mostly without vegetation.

Freshwater discharge channels are located in the northern shore of the bay. The fresh water entering the bay comes from the irrigation system of the rice fields, and amounts to about 275×10^6 m^3 year^{-1}, with yearly averages of 1 – 2 μM PO$_4$-P, 10 – 50 μM NH$_4$-N, 10 – 50 μM NO$_3$-N and 1.5 – 4 μM of particulate phosphorus and 20 – 30 μM of particulate nitrogen. The water discharge is strongly seasonal, depending on the rice culture, and occurs only from April – May to September. A small amount of fresh water enters the bay during the rest of the year through the phreatic system, decreasing the average salinity of surface waters to 35 PSU, slightly less than the Mediterranean seawater (mean salinity 37 PSU).

Because of the absence of significant tidal movements, the water exchange, driven mostly by estuarine circulation and by the wind, is

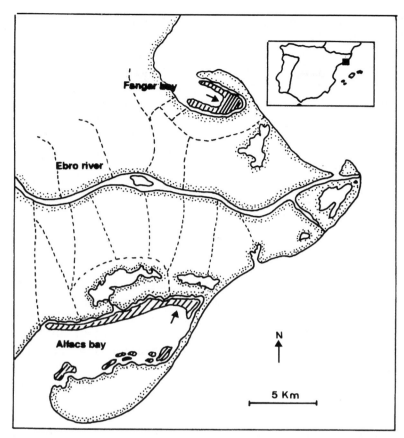

Fig. 12.2. Map of the Ebro delta, showing the Fangar Bay and Alfacs Bay. The distribution of rooted macrophytes is indicated (*hatched*), as well as the zones of preferential macroalgal accumulation (*arrows*). (Perez and Camp 1986)

complex. It has been estimated that the average residence time (ratio between fresh water inputs and fresh water content of the bay) for fresh water in the bay is 15–21 days during summer, and 40–60 days during winter (Camp 1994).

Primary production (8900 t C year^{-1}), is dominated by phytoplankton, which contributes more than 60% of the total. Benthic microalgae account for about 13% and seagrasses for 18%, while macroalgae with 5% play only a secondary role in the total carbon budget of the system (Camp et al. 1992). This is due to the fact that both phanerogams and macroalgae are restricted to a narrow fringe of the shallow water, as a result of shading caused by planktonic growth (average chlorophyll concentration in the centre of the bay: 5 mg m^{-3}, range 1–20 mg m^{-3}).

Plankton together with particulate matter and sediments from the channels induce high turbidity with an average extinction coefficient of 0.4 m^{-1} in the centre of the bay (Pérez 1989). In contrast, in the northern Fangar Bay, with comparable freshwater input and morphological features, but with a different residence time of fresh water (only 2–3 days during spring-summer), primary production is mainly achieved by macrophytes, which cover more than 80% of the bottom, including the central basin. This is because rapid water exchange prevents accumulation of planktonic biomass in the bay, and thus the water is much clearer, so that dense macrophyte beds can develop.

The conditions described above, together with the fact that planktonic growth is not continuous, but occurs in pulses, suggest the possibility of anoxia events in the deep waters of the Alfacs Bay, particularly since a saline stratification is present throughout the year. This possibility has been carefully examined (Camp et al. 1992), both through field measurements and conceptual modelling. Field data report oxygen depletion only in extreme and peculiar cases, usually less than once a year, after phytoplankton blooms when winds do not blow and temperature is high. In general, anoxia is prevented mainly by estuarine circulation, which brings oxygen-saturated offshore waters, together with other processes (diapycnal eddy diffusion, wind-driven vertical mixing).

Macroalgae in the Ebro delta have been studied by Martínez-Arroyo (1990). Usually they are present in the shallow platforms, but with distinct local variations in cover. While in the southern and eastern shores algal cover is low (0–5%), in the northern shore locally it can be as high as 25–50%, with maximum values in the NE part (> 60%). Green algae (i.e. *Ulva* sp.) dominate in the areas closer to the freshwater inputs, while in other parts of the bay other species grow together with *Ulva* (*Chondria tenuissima, Gracilaria verrucosa*; see Ferrer (1993) for a detailed study on *Cystoseira barbata*). In general, young thalli grow attached on small hard substrata (pebbles, mollusc shells etc.), but soon become free-living, mostly lying on the sediment or the seagrasses, and seldom floating on the surface.

Biomass of macroalgae shows a very irregular annual pattern. *Ulva* sp. seems to culminate in winter and late spring, reaching 250 g m^{-2}, dry weight (Fig. 12.3). These values are mostly found in the northern shore, where for an area of about 2 km^2, cover by this species is around 50% during maximal development. However, it is difficult to distinguish the biomass originating from local algal growth from the biomass advected by winds and currents. Such accumulations usually disappear within 1 month, either dispersed by winds or due to decay (decay rates: 0.03–0.08 day^{-1}).

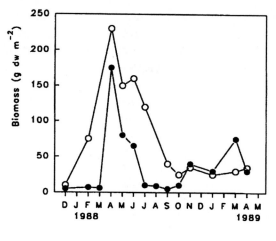

Fig. 12.3. Seasonal evolution of the biomass of *Ulva* sp. in two different stations of the Alfacs Bay. Location I is closer to the freshwater inputs (*open circles*), compared to location II (*solid circles*). (Martinez-Arroyo 1990)

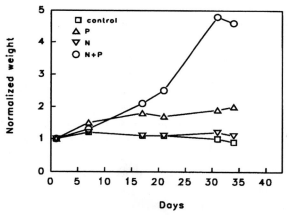

Fig. 12.4. Growth of *Ulva* sp. in laboratory media. *Control* (□) Seawater (< 0.1 μmol dm^{-3} PO_4-P and < 1 μmol dm^{-3} dissolved inorganic nitrogen); P (Δ) $+ 5$ μmol dm^{-3} PO_4-P; N(∇) 10 μmol dm^{-3} NH_4-N $+ 10$ μmol dm^{-3} NO_4-N; $N + P$ (0) both treatments together. After 3 weeks of treatment, all the thalli were transferred to untreated seawater. Data are expressed in fresh weight normalized for the weight at the beginning of the experiment, i.e. the ratio between weight at time t and weight at time 0. (Martínez-Arroyo 1990)

Growth of *Ulva* sp. is relatively slow ($0.011-0.019$ day^{-1}), which means a biomass duplication time of $36-63$ days (Fig. 12.4). This growth seems to be alternatively limited by P and N, however when both nutrients are supplied in excess, growth rates can increase to 0.115 day^{-1}, which corresponds to a duplication time for biomass of 6 days.

This high production potential is supported by efficient phosphorus uptake mechanisms and very high storage capacities. Phosphorus uptake in short-term experiments (30 min) does not show any saturation up to

$6 \mu mol \; dm^{-3}$ of soluble reactive phosphorus, which indicates high uptake rates at unusually high phosphorus concentrations. Nutrient addition after a starvation period can increase the total phosphorus content
of the plant by a factor of 10, while the nitrogen content is only increased
by a factor of 2. It is likely that this species benefits from very short
episodes of high phosphorus concentration (Vidal 1991).

In short, conditions for major green macroalgal accumulation occur
only close to the discharge channels, which provide the nutrients needed
for fast growth, and in the inner part of the bay, where water movement
is more restricted and waters and shallow (< 1 m depth). From observations in the Fanger Bay similar results can be concluded (Fig. 12.2).

However, these proliferations do not appear to be much more frequent or with higher biomasses than in the recent past. We have no
reliable data on the biota of the bay before 1986, but it seems (on the
basis of aerial photographs) that no changes have occured in seagrass
cover during the last 15 years. As stated by Niell et al. (Chap. 10, this
Vol.), algal cover interferes with light adsorption by seagrasses; since
Cymodocea nodosa has a light compensation point of $10-35 \; \mu E \; m^{-2} \; s^{-1}$
(Pérez and Romero 1992), a cover of two *Ulva* blades would reduce the
light intensity at the canopy level enough to cause the disappearance of
the seagrass. Thus, we conclude that the system, which is in equilibrium
between the three main groups of primary producers, has been resistant
to recent changes in nutrient inputs. Such changes have not been dramatic, although the water quality of the Ebro River has deteriorated in
the last 30 years, but the rice culture has remained more or less unchanged since the beginning of the century. However, care must be taken
in not increasing nutrient loads or decreasing the exchange rate between
the bay and the open sea. An imbalance in the present ratio of the
productivity of phytoplankton, rooted macrophytes and macroalgae
could drive the system to severe eutrophic events, either through phytoplankton blooms or mass development of green algae.

12.4   Green Tides in the Mediterranean:
A Conceptual Framework and Future Goals

In summary, excessive growth of green macroalgae does not occur unless
at least three conditions are met: (1) high nutrient loading; (2) limited
freshwater dilution; (3) suitable (hard) substrata. For massive proliferation, two additional conditions are needed; (4) calm waters which, apart

from preventing dilution, allows both biomass accumulation and the settling and developing of algal spores on very small hard substrata; (5) a high surface to volume ratio for the water mass, since macroalgae can outcompete phytoplankton in very shallow waters, but are in turn out-competed in "deep" waters (> 1 m) through the shading effect of plank-tonic biomass.

At first sight, it seems that the Mediterranean is less susceptible to eutrophication events than the North Atlantic, due to its lower nutrient concentrations; but in the zones of high human pressure (i.e. the north-western part, and particularly along the Spanish coast), large areas are probably subjected to the same degree of nutrient levels as the Atlantic. Thus, if dilution of fresh water carrying high nutrient concentrations is prevented or retarded by physical barriers (e.g. harbours, persistent calm weather, closed bays), algal blooms can take place. On open coasts, or in deep harbours and embayments, phytoplankton would dominate these blooms (as described in recent years for the NE part of Spain; J. Camp, unpubl. data), with the greens and other macroalgae being restricted to the hard substrata. Macroalgal blooms would only appear under extremely calm conditions and a high surface to volume ratio (i.e. large areas of shallow waters, as described for the Ebro delta bays). The situ-ation could be even worse than in the Atlantic, especially in natural or man-made closed areas, where the lack of tidal movements limits water exchange.

Despite the fact that no significant problems were caused by green algal proliferations until now, prospects for the near future are somewhat worrying. The development of tourism in Spain is generating increasing nutrient and organic loads to coastal waters, and considerable modifica-tions of the original coastal line (building of new beaches and protection against erosion by breaks, frequent construction of harbours and ma-rinas), which create conditions leading to eutrophic events. It is clear that the development of the seashore must be carefully surveyed in the future.

The productivity of Mediterranean coastal ecosystems is not as low as their biomass and/or nutritional conditions may suggest, and there is always a resilient biomass, even in quite pristine waters, which indicates that the uptake of nutrients depends on high substrate affinities, or that the regeneration of the nutrients is of paramount importance in support-ing biomass. This is a very different situation from the Atlantic coasts, and thus conceptual approaches to pollution problems should be consist-ently different. Future studies should take into account these differences between the Mediterranean and boreal-Atlantic systems, which will lead to differences in both pollution dynamics and in the response of the system to such pollution. In the Mediterranean, there is chronic pollution that is often undetected. This situation is broken irregularly in an unpredictable way by episodic events. The study of such events would be the most effective approach for the future. There is also an urgent need to continuously monitor Mediterranean ecosystems, looking at the disturbances produced by chronically increasing pollution.

Concepts of pollution level and sensitivity of Mediterranean ecosystems are different from the Atlantic coasts, and future efforts should focus on the study of the actual interacting effects between environmental conditions and ecosystem structure in Mediterranean systems. We are convinced that the extrapolation of critical conditions standardized in Atlantic systems are not valid for the Mediterranean system.

Acknowledgments. We would like to thank E. Ballesteros, F. Boisset, F. Conde, T. Gallardo, A. Gómez-Garreta, and M.A. Ribera for the valuable information on green algal proliferations. Part of this work benefited from a grant CICYT CE 89-0017.

References

Anonymous (1991) Eutrophication of coastal waters. Coastal water quality management in the country of Funen, Denmark, 1976–1990. Funen Country Council, 288 pp

Ballesteros E (1989) Production of seaweeds in northwestern Mediterranean marine communities: its relation with environmental factors. In : Ros JD (ed) Topics in marine biology. Sci Mar 53 (2–3): 145–158

Ballesteros E (1992) Els vegetals i la zonació litoral: espècies, comunitats i factors que influeixen en la seva distribució. Inst Estudis Catalans, Arxius de la secció de ciències, Barcelona, 613 pp

Ballesteros E, Pérez M, Zabala M (1984) Approximación al conocimiento de las comunidades algales de la zona infralitoral superior de la costa catalana. Collect Bot 15: 69–100

Camp J (1994) Aproximactiones a la dinámica ecológica de una bahía estuárica mediterránea. PhD Thesis, Univ Barcelona, 245 pp

Camp J, Delgado M (1987) Hidrografía de las bahías del delta del Ebro. Invest Pesq 51(3): 251–369

Camp J, Romero J, Pérez M, Vidal M, Delgado M, Martínez A (1992) Production-consumption in an estuarine bay: how anoxia is prevented in a forced system. Oecol Aquat 10: 145–152

Delgado M, Camp J (1987) Abundancia y distribución de nutrientes inorgánicos disueltos en las bahías del delta del Ebro. Invest Pesq 51(3): 427–441

Fera P (1993) Marine eutrophication along the Brittany coast. Origin and evolution. Eur Water Pollut Control 3(4): 26–32

Ferrer E (1993) Contribució a l'estudi ecofisiològic de *Cystoseira barbata* f. *aurantia* (Kützing) Giaccone. Ms Thesis, Univ Barcelona, 156 pp

Gallardo T, Gómez-Garreta A, Ribera MA, Alvarez M, Conde F (1985) Preliminary checklist of Iberian benthic marine algae. Publ Real Jardín Bot Madrid, 83 pp

Gallardo T, Gómez-Garreta A, Ribera MA, Cormaci M, Furnari G, Giaccone G, Boudouresque CF (1993) Checklist of Mediterranean seaweeds. I. Fucophyceae (Warming 1884). Bot Mar 36: 399–421

Martínez-Arroyo A (1990) Estudio ecológico de algas efemerofiíceas: papel en los flujos de materia y energía en un sistema estuárico (Bahía de los Alfaques, Delta del Ebro, España). PhD Thesis, Univ Barcelona, 166 pp

Pérez M (1984) Estudio ecológico de las comunidades nitrófilas mediolitorales. Collect Bot 15: 351–363

Pérez M (1989) Fanerógamas marinas en sistemas estuáricos: producción, factores limitantes y algunos aspectos del ciclo de nutrientes. PhD Thesis. Univ Barcelona, 244 pp

Pérez M, Camp J (1986) Distribución espacial y biomasa de las fanerógamas marinas de las bahias del delta del Ebro. Invest Pesq 50(4): 519–530

Pérez M, Romero J (1992) Photosynthetic response to light and temperature of the seagrass on *Cymodocea nodosa* and the prediction of its seasonality. Aquat Bot 43: 51–62

Ribera MA, Gómez-Garreta A, Gallardo A, Gallardo T, Cormaci M, Furnari G, Giaccone G (1992) Check list of Mediterranean seaweeds. II. Chlorophyceae (Wille). Bot Mar 35: 109–130

Romero J (1985) Estudio ecológico de las fanerógamas marinas de la costa catalana: Producción primaria de *Posidonia oceanica* (L.) Delile en las islas Medas. PhD Thesis, Univ Barcelona, 261 pp

Vidal M (1991) Estudios sobre la dinámica de nutrientes en sistemas estuáricos: flujos sedimento-agua de nutrientes y oxígeno. PhD Thesis, Univ Barcelona, 305 pp

Vidal M, Romero J, Camp J (1989) Nutrient fluxes accross sediment surface: preliminary results of "in situ" experiments in the Ebro delta. In: Ros JD (ed) Topics in marine biology. Sci Mar 53(2–3): 505–511

Vidal M, Morgul JA, Latasa M, Romero J, Camp J (1992) Factors controlling spatial variability in ammonium release within an estuarine bay (Alfacs bay, Ebro delta, NW Mediterranean). Hydrobiologia 235/236: 519–525

13   France – The Mediterranean Lagoons

M.-L. De Casabianca

13.1   Introduction

The Mediterranean lagoons can be characterized by their salinity regimes, which are determined by the balance between freshwater input by land run-off, or rainfall, mainly from November to May, and seawater input between the periods of rainfall, particularly from May to November. Evaporation in the summer accentuates these variations and may even isolate the lagoon from the sea. The extremes vary from year to year, therefore pluriannual variation data give a better insight into the characteristics of a lagoon (Fig. 13.1).

The range of salinity variations is usually wider in the shallow, littoral lagoons that are more influenced by fresh water input, compared to the deep marine-type lagoons of tectonic origin. The average salinity determines the type of plant communities, whereas the species diversity is inversely proportional to the amplitude of the salinity variations (De Casabianca-Chassany et al. 1973).

Usually the lagoons with pronounced salinity variations are more eutrophicated, compared to those with a balanced salinity regime. There may be two major reasons for this. First of all, a highly physico-chemically variable environment is usually characterized by population fluctuations. This also means production of plant material, accumulation and breakdown of organic matter together with regeneration of nutrients, which favour the development of macrophytic algae in shallow, littoral lagoons. The second reason is the additional supply of nutrients and other materials with the fresh water input.

It is therefore possible to define different degrees of "natural eutrophication" for each type of lagoon, depending on the extent to which it is

Station Méditerranéenne de l'Environment Littoral, Université Montpellier II, 1, Quai de la Daurade, 34200 Séte, France

Ecological Studies, Vol. 123
Schramm/Nienhuis (eds) Marine Benthic Vegetation
© Springer-Verlag Berlin Heidelberg 1996

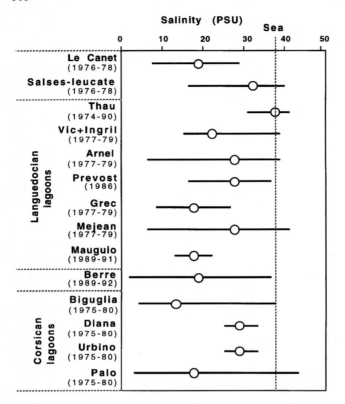

Fig. 13.1. Range and mean of salinity in the main French Mediterranean lagoons

influenced by these factors (Table 13.1). Superimposed on these varying degrees of natural eutrophication may be additional anthropo-genic eutrophication originating from the area neighbouring the lagoon.

In the following, the relationship between: (1) eutrophication and population structure; and (2) macrophytic production and eutrophication will be considered for various lagoonal systems.

13.2 Eutrophication and Macrophyte Community Structure in Corsican Lagoons

Natural eutrophication is an inherent characteristic of Mediterranean brackish zones, due to their high degrees of variability and productivity, as pointed out before. Lagoons are transition zones between land, the

Table 13.1. Nutrient concentration (mean and range for the observation period) in main Corsican and Languedocian lagoons

		N-NH$_4$ (mg/l)	N-NO$_3$ (mg/l)	N-NO$_2$ (mg/l)	P-PO$_4$ (mg/l)	N/P (molar)
Thau	Mean	0.054	0.097	–	0.041	6.05
1986–1987	Min.	0.013	0.027	–	0.013	0.4 (Feb)
	Max.	0.144	0.213	–	0.058	17.7 (May)
Mauguio	Mean	–	0.106	–	0.217	3.1
1990	Min.	–	0.014	–	0.055	0.09 (Sep)
	Max.	–	0.456	–	0.508	14.4 (Dec)
Prevost	Mean	–	0.027	0.01	0.052	2.8
1975	Min.	–	0.013	0.001	0.016	0.3 (Jul)
	Max.	–	0.062	0.035	0.116	10.3 (Jan)
Berre	Mean	0.03	0.06	0.024	0.027	11.87
1989–1992	Min.	0.001	0	0.001	0.002	1.1 (Jun)
	Max.	0.194	0.46	0.04	0.407	35.5 (Apr)
Diana	Mean	0.026	0.092	0.003	0.019	9.9
1981	Min.	0	0	0	0	0.1 (Jun)
	Max.	0.052	0.266	0.026	0.046	27.6 (Apr)
Urbino	Mean	0.019	0.066	0.001	0.011	14.3
1981	Min.	0	0	0	0	0.1 (Jun)
	Max.	0.049	0.319	0.001	0.018	91.2 (Apr)
Biguglia	Mean	0.01	0.138	0.006	0.026	23.67
1981	Min.	0	0	0	0	0.2 (Jun)
	Max.	0.038	0.499	0.043	0.132	176 (Jul)
Mediterranean Sea		0	0.265	0	0.011	16

freshwater environment and the sea. Variability is therefore maximal. Here, the effects of desiccation, temperature and salinity variations lead to large-scale mortalities. The water's edge is therefore a privileged site for the development of floating or attached macrophytes that thrive on this organic matter.

From these characteristic features several questions arise: (1) What is the impact of a given eutrophication event on the systems? (2) What can be considered as man-made eutrophication: pollution from land (agriculture, industry, urbanization) or pollution in the water itself (intensive fish and shellfish farming)? (3) How can seasonal, annual and pluriannual variations inherant in the ecosystem be distinguished from the effects of man-made eutrophication?

To answer these questions, we have chosen two different paths of investigation: (1) comparative studies of diversified and little polluted lagoons of the Corsican coasts which are neither overexploited, nor close to urban or industrialized areas, and which represent various degrees of eutrophication; and (2) studies of the effects of artificially induced eutrophication at one of these sites.

Using this approach, it was expected that species indicative of the degree of eutrophication could be identified. Brackish zones show higher degrees of eutrophication than marine zones, in which the occurrence of indicator species can be used to analyze progressive eutrophication. From a practical point of view, the ability to control the development of these species prior to offshore effluent discharge can be useful in the fight against pollution.

13.2.1   Characteristics of the Lagoons and Study Methods

A lagoon may be characterized by its annual salinity variations which are basically controlled by freshwater inputs and seawater intrusion for a given topography.

From a topographical point of view, two types of lagoons can be distinguished on the east coast of Corsica (De Casabianca-Chassany 1988; Fig. 13.2): (1) Deep and rounded basins of tectonic origin (Diana and Urbino Lagoon, about 10 m deep and 550 and 750 ha surface area, respectively); (2) basins isolated from shallow zones by a spit of land, e.g. the lagoons of Biguglia and Palo (1 m deep, 1600 and 250 ha, respectively). They are shallow, elongated and parallel to the coast line.

Other smaller lagoons can be included in one or the other of these categories. As far as the freshwater input is concerned, the rainy period lasts from October to April (700 to 1100 mm annual rainfall), the dry season from May to October. Rainwater input into the deep basins is low, whereas it is high but intermittent in the two shallow lagoons.

Sea water penetrates into the basins between rainy periods, and especially during the dry period before the lagoons become isolated from the sea. This isolation may occur as the result of a lowering of the water level in the lagoon and the silting up of the connecting canals between the lagoon and the sea.

Due to these factor combinations, the first two lagoons are marine-type brackish lagoons, with little pluriannual variations (salinity variations over a 5-year period for the Diana Lagoon 25–28 PSU, and for the Urbino Lagoon, having the lowest freshwater input, 31–38 PSU which is close to marine conditions).

The second type comprises brackish lagoons with large pluriannual salinity variations (from 2 to 39 PSU at Biguglia, and 1 to 42 PSU at Palo). Palo, the smaller lagoon, reacts with more pronounced extremes compared to Biguglia Lagoon. An increasing gradient of variation (in temperature and salinity) occurs from the centre towards the edges of the lagoon, with an accumulation of detritus around the shores.

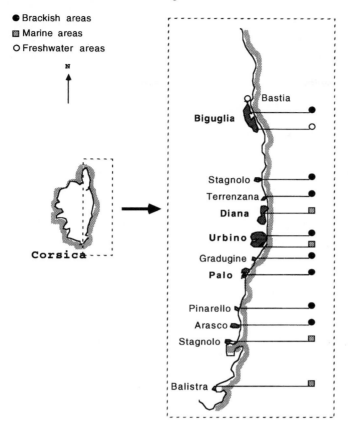

Fig. 13.2. Corsican lagoons and the classification of their littoral zones according to pluriannual salinity regimes

The bottom of the lagoons can be divided into two zones: a bare zone close to the edge, partly subjected to emersion due to variations in the water level of the lagoon. This entire zone may be invaded by floating macrophytes during spring. A second zone is characterized by the presence of seagrass beds and their associated fauna.

This study covers diverse shores from lagoons of various types, with slopes of 3 to 100%, with various substrates and with different degrees of emersion. As the plant populations develop in spring from elements that have survived winter, the extremes of salinity variation encompassing the reproductive period of these species from April to October, i.e. practically the salinity minimum in spring and the maximum in autumn, can be used as a criterion to define such populations and to classify the different biotopes (cf. Fig. 13.1).

For a spatial analysis of the shores, a pluri-annual time scale was chosen because the extremes of the physico-chemical parameters play an important role in determining the nature and composition of the populations.

The populations selected for this study are those with upper salinity tolerance limits close to seawater salinity (cf. Fig. 13.2). We have analyzed the marine and brackish-water areas of the Corsican lagoons, before and after the agricultural development of the western plain which was still surrounded by bush land in 1975.

13.2.2   The Shores of Corsican Lagoons (1975–1980)

The Corsican lagoons were studied during 1975–1980. An investigation of Urbino basin (western shore) followed in 1980–1990 in view of the developing agricultural exploitation.

Figure 13.3 shows some chemical parameters determined for the four large lagoons Biguglia, Diana, Urbino, and Palo during the first investigation period from 1975–1980, and for the western Urbino basin from 1980–1990.

The greatest differences were observed at the brackish sites with a shallow slope (3–10%) and fluctuations in temperature, salinity as well as oxygen levels during summer. In addition, for all of the lagoons with comparable slope, a decrease in the detritic plant material and other particulate matter was noted from the shoreline towards the seagrass beds. Particulate organic C and N decreased from 13 to 4 mg dm^{-3} and from 2 to 0.04 mg dm^{-3}, with an increase in the C/N ratio from 6.2 ± 0.6 to 10.3 ± 0.2, respectively.

The various aquatic nearshore communities studied in 1975–1980 have been classified in relation to the different environmental parameters. Annual salinity variations act on the community structure in two ways: the average salinity level determines the type of community (brackish or marine), while the amplitude or extremes of the salinity variations controls species diversity.

Communities of the marine type correspond to salinities between 26 and 39 PSU (De Casabianca-Chassany 1988). In Corsica this applies mainly to the shores of the large marine-type lagoons Diana and Urbino, as well as Stagnolo sud and Balistra. Brackish water communities correspond to much wider salinity variations (1.8 to 39 PSU) than can be found at the shores of the large brackish lagoons of Biguglia and Palo, the small lagoons of Stagnolo Nord, Gadugine, Terrenzana, Pinarello and Arasco, or at the shores of shallow, marine-type lagoons subject to irregular freshwater input or high degree of emersion (Pompugliani

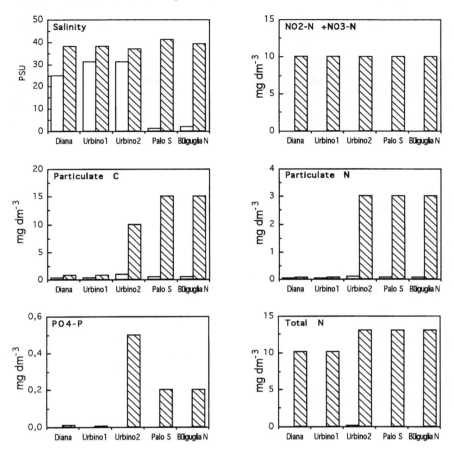

Fig. 13.3. Hydrographical characteristics of the littoral zones of the four Corsican lagoons Biguglia, Diana, Urbino 1 and Palo during the investigation period 1975–1980; Urbino 2 during 1980–1990 after the beginning agricultural exploitation of the adjoining land areas northwest of the Urbino lagoon. Observed minimum and maximum values = *open* and *hatched bars*, respectively

marshes at the Diana lagoon, northwestern shore of Urbino peninsula; De Casabianca et al. 1973).

The study of the nearshore, *marine-type communities* of the large tectonic lagoons, Diana and Urbino, has shown that the community structure can be related to the following major biotopes, characterized by depth, substrate and slope: from 0–0.4 m depth we may distinguish marly cliff with 50–100% slope from pebble or marl shore with 15–35% slope, and sandy shore sloping 10–30%, respectively. From 0.4 to 1 m depth the development of *Cymodocea nodosa* beds depends on sandy or marly substrate (Table 13.2).

Table 13.2. Species composition and distribution in relation to depth, substrate and slope in the two Corsican marine-type lagoons Diana and Urbino

Species	Lagoons		Depth, Substrate, Slope			
			0 – 40 cm		40 – 100 cm	
	Diana S:25–38 PSU	Urbino S:31–38 PSU	Marl cliffs p:35–100%	Pebble p:15–35%	Sand p:10–35%	Sand with seaweed
ATTACHED ALGAE						
Codium fragile	+					
Codium vermilaria	+	+	+			
Gracilaria arcuata	+		+	+		
Alsidium helminthochorton	+	+	+	+		
Alsidium corallinum	+	+	+	+		
Padina pavonia	+	+	+	+		
Cystoseira barbata	+		+	+		
Cystoseira discors	+		+	+		
Enteromorpha clathrata	+	+		+		
Acetabularia acetabulum	+	+		+		
Laurencia papillosa	+	+		+		
Cladophora laetevirens	+	+				+
PHANEROGAMS						
Ruppia spiralis	+	+				+ +
Zostera noltii	+	+				+ + +
Cymodocea nodosa	+	+				
FAUNA						
Ostrea edulis v. *cyrnusi*	+		+			
Mytilus galloprovincialis	+	+	+			
Cerianthus membranaceus	+		+			
Balanus crenatus	+	+	+			
Pholas dactylus	+	+	+			

Species	I	II	III	IV	V	VI
Eriphia spinifrons				+	+	
Ilia nucleus						+
Pachygrapsus marmoratus				+	+	+
Xantho hydrophilus			+	+	+	+
Suberites carnosus			+	+	+	+
Paracentrotus lividus			+	+	+	+
Sphaeroma ghigii			+	+		
Diogenes pugilator		+	+	+	+	+
Asterina gibbosa		+	+	+		
Astropecten bispinosus		+	+	+		
Holothuria polii		+	+	+		
Murex trunculus		+		+		
Macoma tenuis		+		+	+	+
Tapes aureus et decussatus		+		+	+	+
Amphioxus lanceolatus	+	+		+		
Pomatoschistus marmoratus	+			+		
Anemonia sulcata	+			+	+	+
Platynereis massiliensis	+			+	+	+
Idotea viridis	+			+	+	+
Erichthonius brasiliensis	+			+	+	+
Microdeutopus gryllotalpa	+			+	+	+
Carcinus mediterraneus	+			+	+	+
Hippolyte squilla	+			+	+	+
Bittium reticulatum	+			+	+	+
Brachydontes marioni	+			+	+	+
Gibbula adamsoni	+			+	+	+
Rissoa grossa	+			+	+	+
Botryllus shlosseri	+			+	+	+
Gobius paganellus	+			+	+	+
Gobius ophiocephalus	+			+	+	+

The first two biotopes have a number of species in common as a result
of the hard substrate, and both include species associated with the presence of shelter-providing pebbles (*Sphaeroma ghigii*) and/or marly cliffs
(*Cerianthus, Pholas*). Compared to the abrupt cliffs, the shallower slope
of the pebble beaches results in the presence of fine particles. As a
consequence, a belt of specific attached algae and their associated fauna
has developed. The third clear sand biotope includes a zone devoid of
large algae and with infaunal (*Amphioxus, Tapes*) and surface dwelling
animals (*Asterina, Holthuria, Murex, Diogenes*).

The Urbino lagoon, which has a lower annual salinity variation than
the Diana lagoon, has a more diversified population structure. It is noteworthy that the communities have a narrower depth distribution compared to the marine coastal environment.

The *communities of the brackish water type* are independent of substrate type. They occur on slopes from 3 to 30%. They are composed of a
small core of euryhaline species (phanerogams and macroalgae) and a
euryhaline fauna that is mainly benthic and detritus-feeding, associated
with the living plant or with decomposing organic plant material (Fig. 13.4).

From the centre of the basin towards the shore, the following zones
can be distinguished:

1. *Ruppia maritima* beds with associated herbivorous fauna and seagrass dependent species (*Hydrobia acuta, Idotea viridis, Hyppolite
   squilla, Brachydontes marioni*).
2. An intermediate zone, invaded by macroalgae such as the widespread
   Cladophora vagabunda and *Ulva lactuca* which are most abundant at
   the beginning of spring at salinities from 2 to 17 PSU. *Chaetomorpha
   linum* occurs at salinities between 1.8 to 38 PSU, while *Enteromorpha
   intestinalis* seems to prefer a brackish environment with a tendency to
   become hypersaline. The associated fauna is comprised of micropar-
   ticulated detritus feeders (*Corophium insidiosum* and ostracods) and
   primary predator fishes (*Pomatoschistus marmoratus, P. microps*).
3. A third zone close to the edge, where the algae are predominantly
   decomposed, is dominated by a high density of macroparticulate detritus feeders (*Sphaeroma hookeri* and *Gammarus aequicauda*).

From the study of the various biotopes with different salinity regimes
from almost stable marine conditions towards extreme salinity variations
(1.8–45 PSU), the following conclusions can be drawn:

1. There is a shift from a high diversity of communities and species to a
   single community with a low species diversity.
2. A change from stenohaline communities and species, which are finely
   tuned indicators of salinity and of certain types of substrate, to a

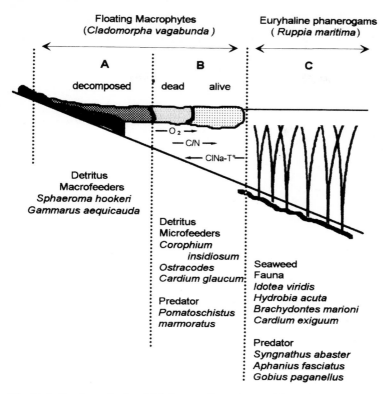

Fig. 13.4. Section of a brackish lagoon littoral zone (slope 3–10%). **A** Zone subjected to immersion; **B** intermediate zone; **C** seagrass zone

community with a high degree of adaptability and which is independent of the salinity and of the substrate type.

3. Communities typical for clean environments are replaced by a community exploiting a great deal of available detritus. The flora is characterized by the appearance of free-floating macroalgae, indicative of eutrophication while the fauna is characterized by a predominance of detritus feeders, mainly utilizing decomposed organic plant material.

13.2.3 The Urbino Basin (Western Shore) with Respect to the Agricultural Exploitation of the Drainage Basins (1980–1990)

Owing to a more diffuse land runoff (leaching) on the southern side of the peninsula, extreme concentrations of mineral nitrogen, up to tenfold, increased particulate C and N levels, and in the extreme case 100-fold higher concentrations of PO_4^{-3}-P are typical for the Urbino basin (Fig. 13.3–13.5).

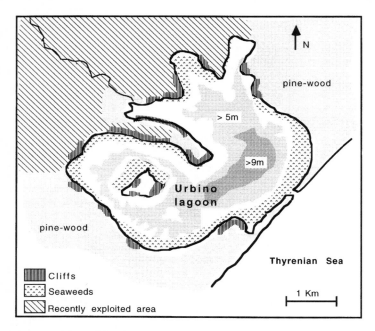

Fig. 13.5. The Urbino lagoon

The inorganic nitrogen concentrations are only slightly different from those found in the lagoons in Languedoc (Thau, Prévost, Ingril, Arnel; Picot et al. 1990). The P-PO$_4^{-3}$ concentrations, on the other hand, are distinctly lower than those of the Palavas lagoons (0.2 – 1.4 mg dm^{-3}) and also of the Thau lagoon (1 – 6 mg dm^{-3}) during the period preceding the red-water phenomena (De Casabianca-Chassany 1979b).

The N:P ratio ranges from 0 to 100 in the Corsican lagoons (peak of P-PO$_4^{-3}$ caused by fertilizers), while in the Prévost and Thau lagoons this ratio is below 1 during the summer period. Nevertheless, the particulate carbon and nitrogen values are lower than those of the Prévost lagoon: 13.56 ± 6.21 mg dm^{-3} (extreme values from 0 to 30 mg dm^{-3}) for carbon and 1.8 ± 0.7 mg dm^{-3} (extreme values from 0 to 4 mg dm^{-3}) for nitrogen, respectively (De Casabianca-Chassany 1992).

Generally, influences from land, in particular leaching, are accentuated to a greater or lesser extent by brush clearing. The influence on *community structure* is similar to the effects of an increase in extreme salinity values and results in the following phenomena as observed from the centre of the Urbino lagoon towards the shore (Fig. 13.6):

1. In the sandy areas, the euryhaline *Ruppia spiralis* grows at about 20 cm depth, whereas *Cymodocea nodosa* beds extend their lower depth distribution by 25 – 30 cm.

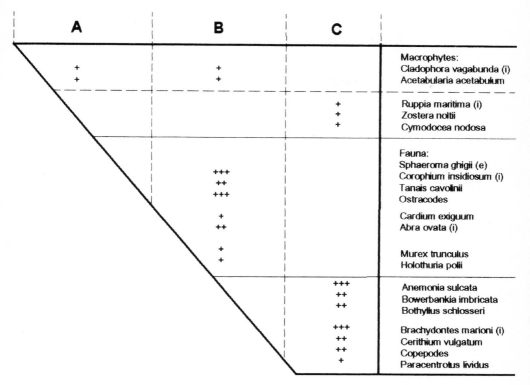

Fig. 13.6. Schematic presentation of the influence of agricultural exploitation of the western side of the Urbino lagoon on the structure of benthic communities south of the Urbino lagoon peninsula. (*e* Species eliminated; *i* species appeared; + + + abundance of species. **A** Zone subjected to immersion; **B** intermediate zone; **C** seagrass zone

2. In the bare zone close to the shoreline an invasion of floating algae (*Cladophora vagabunda*), the more the shallower the slope is. The floating macroalgae are involved in the resorption of nutrients regenerated from surplus organic material.
3. Reduction or disappearance of hypogeal and endogeal species characteristic of zones without macrophytes (*Amphioxus, Asterina, Sphaeroma ghigii*). These are replaced by more euryhaline detritus-feeding species usually restricted to typically brackish zones such as the detritus feeder *Corophium insidiosum* and *Brachydontes marioni* living on the algae.

From these observations we conclude that the influence of the land surrounding a marine lagoon, in this specific case brush clearing of the drainage area, results in a decrease in species diversity through the disappearance of stenohaline species characteristic of clear water. These are replaced by free-floating macroalgae and a detritus-feeding euryhaline fauna. The regression from a typically diverse stenohaline marine community to a euryhaline less diversified community seems to be indicative of increasing eutrophication.

13.3   Macrophytic Production and Eutrophication

Eutrophication of lagoons may come from various sources, as for example shown in Fig. 13.7. External sources include various industrial, agricultural or domestic waste waters that are brought into the lagoon

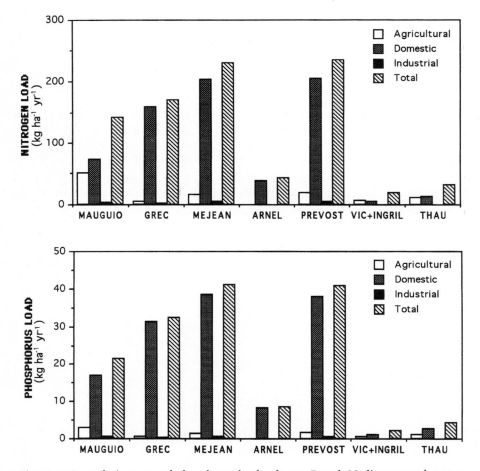

Fig. 13.7. Annual nitrogen and phosphorus loads of some French Mediterranean lagoons. (Agence de L'Eau-Montpellier 1991)

with the land run-off. Particularly affected are shallow lagoons. Internal sources are mainly intensive mariculture systems such as fish and shell-fish farms. While agricultural waste waters contain mainly nitrate and domestic wastes are dominated by phoshorus, intensive mariculture contributes to the ammonia input into the lagoons.

Seasonal changes in the nutrient cycle of a lagoon can be summarized as follows: (1) nitrate levels are mainly controlled by the fresh-water inputs. They show therefore a maximum at the end of periods of heavy rainfall, i.e. at the end of spring in the Mediterranean lagoons (De Casabianca-Chassany et al. 1990). (2) Likewise phosphate levels are greatly controlled by fresh-water inputs and by deposition in the sediments. (3) Ammonium-nitrogen originates from biodeposits on the sediment surface. (4) Both phosphate and ammonia are released at the end of summer when anoxic conditions occur. (5) Due to these seasonal variations, the N:P ratio in the water may change considerably from 16 to 1.

13.3.1   Languedocian Lagoons

Compared to the Corsican lagoons that do not suffer from urban pollution and for which the agricultural exploitation of the surrounding areas has only just begun, the Languedocian lagoons (Fig. 13.8; Table 13.1) are characterized by the influence of domestic discharge (Fig. 13.9). These carry high loads of phosphate which is stored in the sediment and may be released during periods of distrophia causing very high concentrations (> 1 mg dm^{-3} phosphate; De Casabianca-Chassany 1977).

13.3.1.1   Thau Lagoon

In the Languedoc region, the Thau lagoon is a large tectonic, marine lagoon of 7500 ha, stretching over 15 km parallel to the sea (Fig. 13.10). The average depth is 5 m (maximum depth 10m). The lagoon which is connected with the Mediterranean Sea by the Séte and Marseillan canals is the site of intensive oyster farming and is subject to periods of severe distrophia during the summer. There are various zones of eutrophication induced by occasional nutrient inputs due to human activity: (1) shell-fish farming zones covering 1500 ha. Faecal biodeposition from mussels and oysters causes a predominance of NH_4-N. (2) Fresh-water input causes increased nitrate and nitrite levels in the eastern part of the lagoon, whereas increased phosphate concentrations are due to waste-water discharge in the north at Méze. (3) Particulate matter input (construction of the Moure-Blanc port, and industrial zone to the SE).

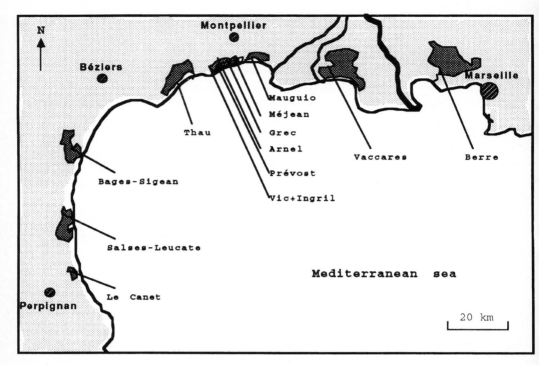

Fig. 13.8. Lagoons in Languedoc-Roussillon (France)

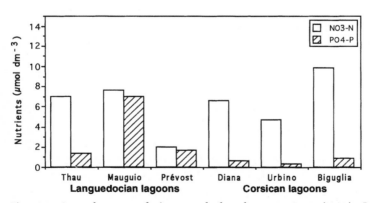

Fig. 13.9. Annual means of nitrate and phosphate concentrations in Languedocian and Corsican lagoons (cf. Table 13.1)

The mean and range of nutrient concentrations is given in Table 13.1. The dominant macrophytes are the seagrass *Zostera marina* (2500 ha, 2000–3000 t dry weight, < 4 m depth), the red algae *Gracilaria verrucosa* and *Gracilaria bursa-pastoris* (2000 ha, 2000–5000 t dry weight), the

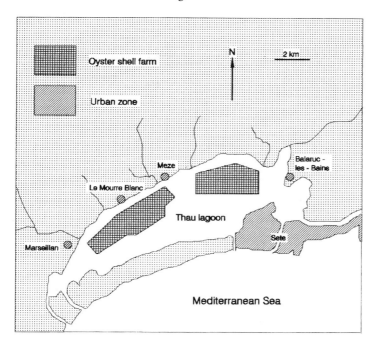

Fig. 13.10. Thau lagoon (Languedoc, France)

green algae *Ulva rigida* and *Ulvaria obscura* (100 ha, 25 t dry weight, and mainly on shellfish culture structures and rocks), the introduced brown algae *Laminaria japonica*, *Sargassum muticum*, and *Undaria pinnatifida* (500 t dry weight; Lauret et al. 1985; Ben Maïz et al. 1987).

Biodeposition, excluding that from shellfish farming, has been determined as 0.26 g m^{-2} day^{-1} carbon and 0.035 g m^{-2} day^{-1} nitrogen. Beneath the shellfish culture structures, the values were 2.6 g m^{-2} day^{-1} carbon and 0.35 g m^{-2} day^{-1} nitrogen for oysters and 2 to 4.5 g m^{-2} day^{-1} carbon and 0.38–0.46 g m^{-2} day^{-1} nitrogen for mussels. Based on these figures, the over all biodeposition for the entire area of the lagoon results in 500 kg year^{-1} nitrogen and 3000 kg year^{-1} carbon, of which 370 kg is nitrogen and 2000 kg carbon from the shellfish cultures, respectively (De Casabianca-Chassany 1977).

Although biodeposition is very high and the contribution by macroalgae is significant, it is far less important than the role which macrophytes play in shallow lagoons. The N passing through the macrophytic compartment is only 0.5% of the total biodeposition.

13.3.1.2 Prévost Lagoon

In the *Prévost lagoon*, the dominant macrophyte is *Ulva rigida* (*rotundata*) with a mean biomass of 4 kg m^{-2} (Riouall 1976; De Casabianca-Chassany 1989). It tends to decompose into particulate matter (De Casabianca-Chassany 1979a; Figs. 13.11, 13.12).

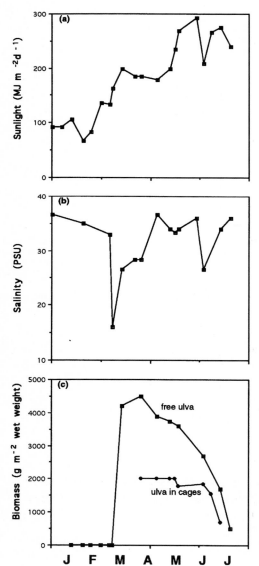

Fig. 13.11. Prevost lagoon. Seasonal variations of *Ulva* biomass in the natural environment and in cage culture. **a** Sunlight: MJ m^{-2} d^{-1}; **b** salinity: PSU; **c** biomass: g m^{-2} wet weight

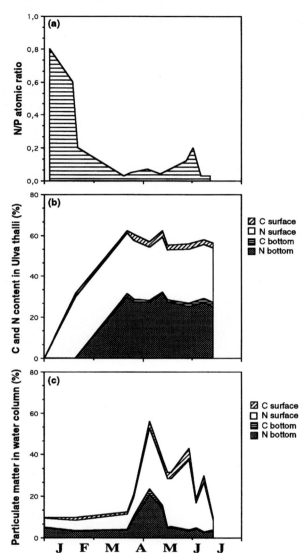

Fig. 13.12. Prevost lagoon. a N/P ratio, and seasonal variation of carbon and nitrogen content (%) in the thalli b or in particulate matter c

The estimation of the macrophytic production shows that the nitrogen passing through macrophytic production averages about 62 kg ha^{-1} year^{-1}, which is 20 times that of Thau lagoon (De Casabianca-Chassany 1992).

Table 13.3. Biguglia lagoon, Corsica: estimates of macrophytic annual production of
Chaetomorpha linum (as $kg\,m^{-2}$ wet weight, carbon or N, respectively), and fraction
carried away by currents or buried in the sediments

	Production			Biomass carried away by streams (kg wet wt. /m²)	Biomass blended with sediment (kg wet wt. /m²)
	Wet weight (kg/m²)	Nitrogen (g/m²)	Carbon (g/m²)		
1969	5–10	0.1–0.2	1–2	3.3	4.2
1970	5–10	0.1–0.2	1–2	1.5	6
1971	3–6	0.06–0.12	0.6–1.2	0.5	4

13.3.2   Example of the Corsican Lagoon Biguglia

The Corsican *Biguglia lagoon* (1600 ha, 1 m depth), an eutrophicated,
shallow lagoon with large variations in salinity (cf. Fig. 13.1), can be
characterized as a very simple system with one dominant macrophyte,
Chaetomorpha linum, mainly supporting populations of the crustacean
Corophium insidiosum that feeds on the detritus derived form the decomposition of this green alga.

Primary production of *Chaetomorpha linum*, as derived from biomass
production during two periods preceding and following the peak of
biomass in summer, varies from 3–10 kg m^{-2} wet weight (200–900 g
m^{-2} dry weight equivalent) to 14–64 g m^{-2} carbon or 3–13.5 g m^{-2}
nitrogen, respectively. Based on the difference in biomass between autumn and succeeding spring, we estimate that one-ninth to one-half of
the biomass produced is carried to the sea by currents. The remaining
portion is deposited in the sediments of the lagoon (De Casabianca-
Chassany 1974, 1979a; Table 13.3). In general, it can be concluded that
the macrophytic production is 10 to 50 times greater in the shallow
lagoons than in the seawater lagoons. The nutrient concentrations seem
to be positively correlated with the varying annual macrophytic production levels as well as with the secondary biomass (*Corophium insidiosum*;
Fig. 13.13). An explanation could be the rapid turnover of organic material and regeneration of nutrients.

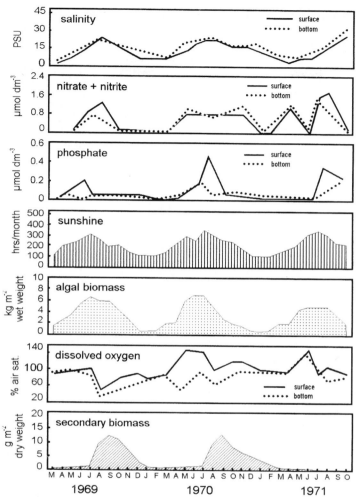

Fig. 13.13. Interannual variations of macrophytic primary production (*Chaetomorpha linum*), secondary production (*Corophium insidiosum*) and some environmental parameters in Biguglia lagoon, Corsica, over 3 years (1969–1971)

13.4 General Conclusion

The following relationships have been examined:

1. Eutrophication and community structure. The example of the Corsican lagoons prior to their exploitation clearly demonstrates various degrees of natural eutrophication, increasing from the most marine-type

lagoon to the most brackish one and leading to a decrease in species diversity that is as significant as the growth of macrophytic algae.

2. Eutrophication and macrophyic primary production. In the two types of lagoon, periods of dystrophia may occur. Macrophytic production plays a key role, which is generally more pronounced in the brackish lagoons, and may allow some control over eutrophication in the same manner as sea/lagoon interchange. Knowledge of the role of macrophytes in these environments may allow for their more efficient management and their orientation towards a finalized and adapted production.

Acknowledgments: Many thanks to Service Maritime de Navigation (Marseille), for the data concerning Berre Lagoon.

References

Ben Maïz N, Boudouresque CF, Riouall R, Lauret M (1987) Flore algale de l'étang de Thau (France, Méditerranée): sur la présence d'une Rhodyméniale d'origine japonaise, *Chrysymenia wrightii* (Rhodophyta). Bot Mar 30: 357–364

Coste B (1987) Les sels nutritifs dans le bassin occidental de la Méditerranée. Rapp Comm Int Mer Médit 30(3): 399–410

De Casabianca-Chassany M-L (1974) Dynamique et production d'une population de crustacés en milieu lagunaire (lagune de Biguglia, Corse). Thèse de, doctorat d'Etat, Univ Luminy Marseille 300 pp

De Casabianca-Chassany M-L (1977) Résultats préliminaires des expériences sur la bio-déposition en milieu lagunaire. Rapp Comm Int Mer Médit 24(6): 91–92

De Casabianca-Chassany M-L (1979a) Production macrophytique dans un écosystème lagunaire simple. Rapp Comm Int Mer Medit 25/26(3): 173–174

De Casabianca-Chassany M-L (1979b) Phosphates dans les étangs mediterranéens: hautes teneurs, prévision et déclenchement des "eaux décolorées". Rapp Comm Int Mer Médit 25/26(3): 105–108

De Casabianca-Chassany M-L (1982) Lisières saumâtres corses et leurs indicateurs de fonctionnement. Bull Soc Ecol 13(2): 165–188

De Casabianca-Chassany M-L (1988) Les étangs. In: Camps G (ed) Terrina et le Terrinien: recherche sur le chalcolitithique de la Corse. Ecole Française de Rome, pp 27–47

De Casabianca-Chassany M-L (1989) Dégradation des ulves (*Ulva rotunda*, lagune du Prévost, France). CR Acad Sci Paris 308 (3): 155–160

De Casabianca-Chassany M-L (1992) Method for evaluating algal production and degradation based on nitrogen levels in particulate organic matter. Bioresour Technol 39: 1–7

De Casabianca M-L, Kiener A, Huve H (1973) Biotopes et biocénoses des étangs saumâtres corses: Biguglia, Diana, Urbino et Palo. Vie Milieu 13 (2): 187–227

De Casabianca-Chassany M-L, Boonne C, Semroud R (1990) Relations entre les variables physico-chimiques dans une lagune méditerranéenne par l'analyse en composante principale (lac Mellah, Algérie). CR Acad Sci Paris 310 (3): 397–403

De Casabianca-Chassany M-L, Semroud R, Samson-Kechacha F-L, Boonne C (1991) Etude spatio-temporelle des sels nutritifs et des principales variables hydrobiologiques dans une lagune méditerranéenne: le lac Mellah (Algérie). Mésogée 51: 15–23

Hervé P (1978) Ichthyofaunes comparées de deux étangs littoraux du Roussillon: Salses-Leucate et Canet-Saint Nazaire. Thèse de doctorat, Univ. Perpignan

Lauret M, Riouall R, Dubois A (1985) L'acclimatation et la croissance de *Sargassum muticum* (Yendo) Fenshold (Pheophyceae) dans l'étang de Thau (Languedoc-France). 110eme Congr Nat des Sociétés savantes, Montpellier, sciences, fasc II: 291–295

Perez JM, Picard J (1955) Biotopes et biocénoses de la mediterranée occidentale comparés à ceux de la Manche et de l'Atlantique nord-oriental. Arch Zool Exp Gen 92 (1): 12

Picot B, Pena G, Casellas C, Bondon D, Bontoux J (1990) Interpretation of the seasonal variations of nutrients in a mediterranean lagoon: étang de Thau. Hydrobiologia 207: 105–114

Riouall R (1976) Etude quantitative des algues macrophytes de substrat meuble de l'étang du Prévost (hérault). Naturalia Monspeliensa Sér Bot 26: 73–94

Tournier, H, Hamon PY, Landrein S (1982) Synthèse des observation réalisées par l'ISTPM sur les eaux et le plancton de l'étang de Thau de 1974 à 1980. Rev Trav Inst Pêch Marit 45 (4): 283–318

14   Italy – The Orbetello Lagoon and the Tuscan Coast

V. Bombelli[1] and M. Lenzi[2]

14.1   Introduction

In the past 20 years, the Tyrrhenian coast has suffered from increasing human activities, leading to environmental changes including trophic conditions.

Because of this development, serious environmental problems have affected the Orbetello Lagoon, which is the largest brackish water basin of the Italian west coast (Fig. 14.1). Hypertrophic conditions have had severe consequences: during several months of the year, conditions under which shlphur occurs with development of reducing and toxic substances (H_2S, CO, CH_2OH, CH_4) are common, resulting in a drastic reduction of species numbers and fish kills. With the development of muddy bottoms, murrains occurred.

In the last 15 years, these conditions have led to deep changes in the structure of the benthic macrophyte communities. In the past 4–5 years, macroalgal blooms have become important events, repeatedly resulting in the developing anoxia and dystrophia after decomposition of the plant material.

Risk of eutrophication is also possible on a wider scale in the open coastal Thyrrenian Sea, as a result of offshore discharge of wastewaters from the densely populated region, without tertiary treatment steps for nutrient reduction.

[1]KOBA Company (Biomass Provision and Conversion), via del Bollo 4, 20134 Milano, Italy
[2]Laboratorio die Ecologia Lagunare e Acquacoltura via Leopardi 15, 58015 Orbetello (Gr.), Italy

Ecological Studies, Vol. 123
Schramm/Nienhuis (eds) Marine Benthic Vegetation
© Springer-Verlag Berlin Heidelberg 1996

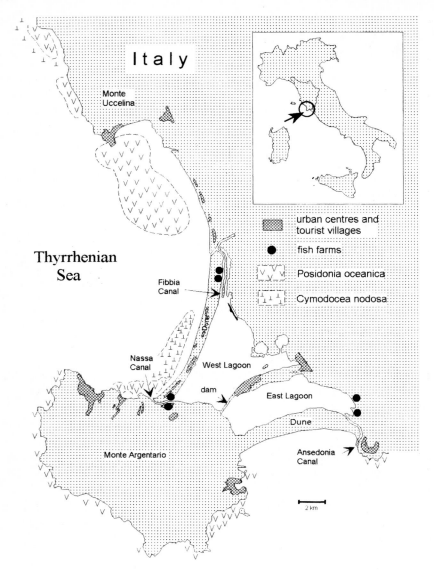

Mediterranean

Fig. 14.1. The Orbetello lagoon and surroundings

14.2 The Orbetello Lagoon

The Orbetello Lagoon, with an area of 2600 ha, is located at the southern coast of Tuscany. It is bounded by two sand dunes which follow the coastline as far as Monte Argentario. The lagoon is divided into two basins (western and eastern) by the Orbetello isthmus and by a dam which connects Orbetello to Argentario (Fig. 14.1). The lagoon is connected to the sea by three canals: the 1.5-km-long Ansedonia canal running east, the 0.5-km-long Nassa canal going west, and the Fibbia canal which leads 3 km to the north.

14.2.1 Physicochemical Parameters

Since 1975, the physicochemical parameters of the Orbetello Lagoon have shown wider and wider fluctuations, even over short periods, while the content of dissolved phosphorus and nitrogen salts have shown a discontinuous progress with irregular variations over recent years (Cognetti et al. 1978; Lenzi and Angelini 1985; Caprioli et al. 1988; Table 14.1).

14.2.2 Economic Activities and Eutrophication

The resident population of about 12 000 people has been stable since 1975, however, tourism results in considerable demographic fluctuations.

Table 14.1. Ranges of some physical and chemical parameters in the central zone of the western basin of the Orbetello lagoon

Parameter	Range
pH	7.25–8.6
Temperature ($^\circ$C)	21–30
Salinity (PSU)	39–50
Oxgen ($cm^3 dm^{-3}$)	2–37
NH^4-N ($\mu mol\ dm^{-3}$)	<0.0001–0.72
NO^2-N ($\mu mol\ dm^{-3}$)	<0.0001–0.42
NO^3-N ($\mu mol\ dm^{-3}$)	<0.0001–22.4
PO^4-N ($\mu mol\ dm^{-3}$)	<0.0001–0.14
N:P atomic ratio	111402–0.22

Sources: Maldura (1935), Cognetti et al. (1978), Lenzi and Angelini (1984, unpubl. data), Caprioli et al. (1988, 1989), TEI (1989), Bassano et al. (1993), Local Sanitary Agency (1993, pers. comm.).

From the 1980s to the 1990s, the annual number of tourists has doubled from 0.5 to 1 million. According to Cossu and coworkers (1992), such data allow the evalutation of the human, in particular, tourist contribution to the lagoon's eutrophication, which was estimated as 54 t year^{-1} nitrogen and 11 t year^{-1} phosphorus for the residents, 65 t year^{-1} nitrogen and 13 t year^{-1} phosphorus for the tourists, respectively (Bombelli et al. 1993).

With regard to other economic activities, only intensive aquaculture has undergone significant development. According to the Local Sanitary Agency and ENEA Agency, the contribution from these activities to the lagoon's eutrophication is about 37 t year^{-1} nitrogen and 1.2 t year^{-1} phosphorus, respectively (Lenzi 1992). Total contribution of nutrients to the lagoon's budget as a result of human activities has been estimated as 68 g m^{-2} year^{-1} nitrogen and 1 gm^{-2} year^{-1} phosphorus, respectively.

14.2.3 Benthic Vegetation

Behveen 1976 and the present, a large reduction of eelgrass beds has been observed, while some seaweed species, typical of this ecosystem in the past (*Valonia aegagropila, Lamprothamnion papulosum*) have disappeared (Cognetti et al. 1978; Lenzi and Angelini 1985; Naviglio et al. 1988; Lenzi unpubl. data 1988). Other species (*Gracilaria verrucosa, Chaetomorpha linum, Cladophora vagabunda*) are now dominating in turn. In the last 4–5 years, the occurrence of these seaweeds has become totally unpredictable, both in quantity and quality.

14.2.4 Seasonal Environmental Events

The development of seaweeds and microphytes are the dominant events for the Orbetello Lagoon. Macroalgal growth occurs between winter and spring. In the latter part of spring, the biomass decays and piles up on the bottom. After the decomposition of the seaweed masses and further remineralization in the spring/summer period, the lagoon becomes subject to frequent distinct blooms of microphytes (Lenzi and Angelini 1985). Despite an sample supply of nutrients from the decaying seaweeds, the availability of additional nutrients from domestic wastes during the tourist season is important for many microalgae, including *Dinophycea*.

Decomposition of the seaweed masses in the summer heat, and the subsequent sulphate reduction processes (Izzo 1988; Lenzi 1988) cause a drastic decrease of dissolved oxygen and development of toxic reducing gases from the bottom, and sometimes fish kills (Lenzi and Salvatori 1986; Lenzi 1992). This has led to a considerable reduction of fish catch from about 200 kg ha^{-1} to 120 kg ha^{-1}.

Besides eutrophication stress, benthic communities also suffer periodically from detrimental effects of low water levels and the consequent decomposition of the dying macroalgae followed by anoxia and development of toxic hydrogen sulphide. Owing to all these causes, the resilience of lagoon communities has greatly decreased: communities are not able to recover during the short-term marine influences.

14.2.5 Adopted Ways of Management

In order to limit further negative effects on the lagoon ecology in the future, some management decisions have been made.

Three pump systems have been set up at the lagoon mouths of Nassa, Ansedomia and Fibbia with pumping capacities of 4000 l s^{-1} in 1982 increased to 8000 l s^{-1} in 1993 at Nassa, 4000 l s^{-1} in 1986 at Ansedomia, and 4000 l s^{-1} in 1994 at Fibbia, respectively. In addition, since 1987 continuous harvesting and removal of 2500–10 000 t wet weight of seaweeds per year have been carried out, employing boats.

Management decisions are aiming at increasing the water circulation when tides are too feeble, and removal of part of the system's excess energy which, in this case constitutes the seaweed biomass (Lenzi 1992).

In recent years, purification by plants as a way of treatment of both domestic and agricultural waste waters has been suggested (Bombelli et al. 1993). However, such solutions have met with difficulties from the local administration which favours engineering solutions. Thus, a project is under development constructing a big sewer trunk-line to collect all wastes from the Orbetello lagoon basin and discharge them offshore.

14.2.6 Marine Coastal Areas Around the Orbetello Lagoon

The marine coastal areas around the Orbetello lagoon are important, because of their high diversity of submerged biocoenoses due to the wide range of substrates which include mud, sand, and rocky bottoms. Wide meadows of *Posidonia oceanica* and *Cymodocea nodosa* can be found near the Parco Naturae Maremma (oof the Uccellina Hills) and along the coast of Monte Argentario (Lenzi and Micarelli 1993; Fig. 14.1). Two of these meadows are *Posidonia* barrier reefs (Lenzi 1987).

The increase in human population during summer has brought about several almost insoluble problems. At present, all wastewaters, treated or untreated, are discharged into the sea at several points along the lagoon or marine coasts. The planned concentrated discharge of all wastes at a few points offshore may not improve the situation, but may have even worse effects as they may damage the seagrass meadows. Already an increase in nutrient levels has been observed along this coastline compared to offshore (Innamorati et al. 1993). Moreover, the presence of mucilaginous aggregates has been reported by divers since 1982 that could be a direct consequence of increased nutrient levels. These aggregates occur at depths of more than 15 m in many localities along the Tuscan coastline and at nearby islands, where they clog the substratum, endangering the benthic communities. Unlike the situation in the Adriatic Sea, such mucilage has not yet floated onto the surface.

According to Sartoni and Sonni (1991), in Thyrrenian benthic mucilaginous aggregates, the most abundant species were *Tribonema marinum* and *Acinetospora criniata*. According to Innamorati et al. (1944) diatoms predominate.

References

Bassano E, Boniforti R, Jacovelli M, Marri P, Nair R, Schirone A (1993) Montio-raggio della laguna di Orbetello. Dati idrochimici, completamento relazione giugno-luiglio 1993, relazione 20 luglio-25 agosto 1993. ENEA, CRAM, La Spezia

Bombelli V, Lenzi M, Mattei N (1993) L'evoluzione del sistema laguno costiero di Orbetello: problemi e soluzione gestionali. Convegno AMBIA, Milano 1: 10-93

Cognetti G, De Angelis CM, Orlando E, Bonvicini Pagliai AM, Varriale AM, Crema R, Mari M, Mauri M, Tongiorgi P, Zunarelli R, De Fraja Frangipane E, Brambati A, Giaccone G, Olivotti R (1978) Risanamento e protezione dell'ambiente idrobiologico delle lagune di Orbetello. Comune di Orbetello, Regione Toscana, Firenze

Cognetti G, Ghiara C, Minguzzi C, Orlandi C (1988) Risanamento ambientale della laguna di Orbetello: ciclo stagionale dei nutrienti e velocita' di accumulo dei sedimenti e caratterische geochimiche delle acque. ENEA RT/PAS/88/9 Casaccia Roma

Cossu R, Andreottola G, Raggazzi M, Casu G (1992) Direct and indirect domestic nutrient load evaluation by mathematic model: the Venice Lagoon case study. Science of the Total Enviroment, suppl. Elsevier, Amsterdam

Innamorati M, Nuccio C, Lazzare L, Mori G, Massi L, De Pol M (1993) Condizioni trofiche, biomass e popolamenti fitoplantonici dell'Alto Tirreno. "Progetto Mare" Regione Toscana, Univ Firenze, pp 103-156

Innamorati M, Melly A, Castelli C, Gorini G, Balzi M (1994) Mucillaggini tirreniche. nutrienti, cenosi fitoplactoniche. Biol Mar Medit 1(1): 189-193

Izzo G (1988) Risanamento ambientale della laguna di Orbetello: i ruoli dell'attività microbica dei sedimenti nella laguna di Orbetello. ENEA RT/ PAS/88/ 12 Casaccia Roma, 14 pp

Izzo G, Creo C, Signorini A, Grosso M, Candirelli M (1989) Studio della dinamica dei processi biogeochimici che determinano le crisi anossiche della laguna di Orbetello. ENEA RT/PAS/89/18 Casaccia Roma

Lenzi M (1984) Indagine sulla distribuzione delle macrofite nella Laguna di Orbetello. Quad Mus Stor Nat Livorno 5:37-55

Lenzi M (1987) Le recif-barriere de Posidonia oceanica (L.) Del. de Santa Liberata (Toscana, Italie): cartographie et biometrie. Giorn Bot Ital 121 (3-4): 155-164

Lenzi M (1988) Cause della morie della fauna ittica e ipotesi operative per una gestione delle risorse nutrizionali nella Laguna di Orbetello. Atti del Convegone "Risanamento della Laguna di Orbetello. Metodologia integrata e soluzioni progettuali", Orbetello 9 aprile 1988, Comune di Orbetello

Lenzi M (1992) Experiences for the management of Orbetello Lagoon: eutrophication and fishing. Science of the Total Environment. Elsevier, Amsterdam, pp 1189-1198

Lenzi M, Angelini M (1985) Indagine sulle condizioni ambientali della laguna di Orbetello. Chimico fisica e carico microfitico. Atti Mus Civ Stor Nat Grosseto 3: 18-30

Lenzi M, Bombelli V (1985) Prime valutazioni della biomassa macrofita della Laguna di Orbetello (GR), in considerazione di uno sfruttamento industriale. Nova Thalassia 7 (Suppl 3): 355-360

Lenzi M, Micarelli P (1993) La vegetazione sommersa di fondo molle tra Talamone e la foce dell'Ombrone (GR). Quad Mus Civ Stor Nat Grosseto 9/10: 45-55

Lenzi M, Salvatori R (1986) Eutrofizzazione e distrofie e produzione ittica della laguna di Orbetello. Atti Mus Civ Stor Nat Livorno 10: 8-17

Lenzi M, Canese S, Alvisi, Micarelli M, Cianchi F (1989) La vegetazione marina tra Talamore e Porto, Stefano S (GR) Quad Mus Stor Nat Livorno 10: 19-38

Maldura C (1929) Le variazioni stagionali dei caratteri chimici e fisici delle aque della laguna di Orbetello. Boll Sicil Idrobiol 5(6) : 962-981

Maldura C (1935) Ricerche chimiche sulla laguna di Orbetello in rapporto alla biologia. Nota I, Rend R Acc Naz Lincei 6(22) (1/2): 65-68; Nota II, Rend R Acc naz Lincei 6(22): 140-145

Naviglio L, Uccelli R, Falchi G, Lenzi M (1988) Risanamento della laguna di Orbetello: indagine preliminare sulla distribuzione e l'abbondanza della vegetazione macrofitica. ENEA RT/PAS/88/11 Casaccia Roma, Regione Toscana

Salvatori R, Angelini M, Lenzi M, Fommei F (1985) Attivita' di acquacoltura nella laguna di Orbetello. Regione Toscana, Firenze

Sartoni GF, Sonni C (1991) *Tribonema marinum* J. Feldmann e *Acinetospora crinita* (Carmichael) Sauvageau nelle formazioni mucillaginose bentoniche observate sulle coste toscane nell'estate 1991. Inf Bot Ital 23: 23-30

T. E. I. Ingegneria per l'Ambiente (1989) Intervento globale di risanamento delle lagune di Orbetello dal fenomeno di eutrofizzazione del corpo idrico. Rapporto finale. Ministero dell'Ambiente, Regione Toscana, Provincia di Grosseto

15 Italy–The Lagoon of Venice

A. Sfriso and A. Marcomini

15.1 Introduction

Although the Italian rock and shore coastline extends for about 3000 km into the Mediterranean Sea and includes several estuarine and lagoon flat areas, knowledge of the vegetation and especially the changes in relation to eutrophication is still very poor. Some papers report on the flora vegetation along the sea coast, especially for the Adriatic Sea (Pignatti and de Cristini 1966; Solazzi 1966, 1968; Giaccone and Pignatti 1967; Giaccone 1974) and the Tyrrhenian Sea (Cinelli et al. 1979; Buia et al. 1985a,b; Buia and Mazzella 1991; Buia et al. 1992). Many of these works take into account seagrasses only and are poor in reporting the changes of vegetation over time. The situation is quite different for lagoons such as Orbetello and Venice, or Sacca di Goro in the Po delta. During the last decades, all these lagoons were found to be exposed to hypertrophic and dystrophic conditions which were considered to be responsible for the observed changes. In the case of Sacca di Goro, comparative data with the past are lacking and only recent vegetation maps are available (Piccoli et al. 1991). In contrast, the Orbetello lagoon is well known for both its plant populations and changes induced by increasing eutrophication (Lenzi 1984; Caprioli et al. 1988; Izzo 1988; Mittempergher et al. 1988; Naviglio et al. 1988; Bucci et al. 1990). The more comprehensive available data on macrophytes refer to the lagoon of Venice.

The taxonomy of the macrophytes living in the lagoon of Venice has been extensively studied since the second half of the nineteenth century. Several reports with descriptions and drawings of macroalgae were published (Zanardini 1876; de Toni and Levi 1885, 1886, 1888; Sighel 1938;

Department of Environmental Science, University of Venice, Calle Larga S. Marta 2137, 30123 Venice, Italy

Ecological Studies, Vol. 123
Schramm/Nienhuis (eds) Marine Benthic Vegetation
© Springer-Verlag Berlin Heidelberg 1996

Schiffner and Vatova 1938). Unfortunately, taxonomic identification was
a difficult task in the past because of both inadequate technical means
and lack of reliable classification keys. Different authors often identified
macroalgae differently, and the same species were frequently classified
under different names. This situation led in the past to the description of
hundreds of different species, subspecies, varieties and forms. Only the
most recent classifications (Pignatti 1962; Sfriso 1987), performed according to well-established taxonomic revisions of some genera (Bliding
1963, 1968; Van Den Hoek 1963; Feldmann 1981) allow comparison of
the changes in the Venice macroalgal typology. Pignatti (1962) studied
macroalgae in a phytosociological context also and characterized the
main associations of benthic algae, determining their vertical distribution as vertical bands or "vegetational zones", which corresponded to the
"etages" of Feldmann (1938, 1959).

As far as our research work is concerned, after comparing the flora
and the vegetational association changes that have occurred in the last
50 years (Sfriso 1987), our attention was addressed to the study of the
dynamics of algal biomass production and decomposition, with particular reference to the lagoon nutrient cycles and balances (Sfriso et al.
1987, 1988b, 1989, 1990). In this chapter, we present a comprehensive
account on the most recently available results.

15.2   The Lagoon Morphology and Hydrography

The Venice lagoon (Fig. 15.1) has a water surface exposed to the tidal
regime of 432 km^2, and a mean depth of about 1 m. Tidal difference is
about ± 31 cm (Pirazzoli 1974). The water exchange between the lagoon
and the sea, through the Lido, Malamocco and Chioggia mouths was
estimated to be $3.1-5.4 \times 10^8$ m^3, during one tidal cycle (12 h), which
approximately corresponds to the total water volume of the lagoon.
Complete water renewal by tidal exchange with the sea occurs in the
areas near the lagoon inlets. Water renewal in the inner parts is much
less efficient (Battiston et al. 1983).

Therefore pollutants which enter the lagoon through the small rivers
and canals at an average flow rate of $30-40$ m^3 s^{-1} (Vazzoler et al. 1987)
may have a quite long residence time (Battiston et al. 1983), and will
probably be enriched in the sediments of the lagoon. From a drainage
basin of approximately 1840 km^2, mostly treated or untreated municipal
effluents from Venice, Mestre, Chioggia, and from the populated islands

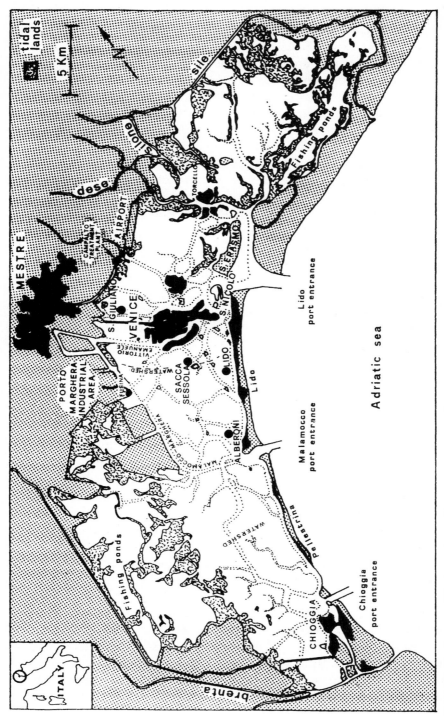

Fig. 15.1. The lagoon of Venice and sampling areas

(accounting for nearly 300 000 inhabitants) enter the lagoon. In addition, treated industrial waste waters are discharged from the chemical and steel plants of the Marghera industrial district (equivalent to 40 000 inhabitants).

Eutrophicating substances (phosphorus and nitrogen compounds) as well as organic pollutants such as PCBs, DDTs or PAHs, together with heavy metals (Hg, Fe, Mn, Zn, Ni, Cd, Cu, Pb, Cr) accumulated in the surface sediments, mainly in the central basin, which receives both industrial and municipal effluents (Marcomini et al. 1986; Pavoni et al. 1986; Orio and Donazzolo 1987).

The spread of chemical pollution in the central lagoon was significantly influenced by the changes of tidal movements as a result of dredging two important canals (Vittorio Emanuele 1919–1930; Malamocco-Marghera 1961–1969; both about 100 m wide and 13–20 m depth). These canals were intended to improve the waterway access to the industrial district of Porto Marghera from the Lido and Malamocco inlets. Strong tidal flows were conveyed toward the industrial district, whereas natural channels ("ghebbi") and wide lagoon areas precluded from tidal water renewal, progressively silted up.

From the literature of the last 30 years (Giordani Soika and Perin 1970, 1974 a,b; Zucchetta 1983; Battaglia et al. 1983; Ministry of Public Works et al. 1985; ENEA 1988, 1990; Consorzio Venezia Nuova 1989), it is apparent that many efforts have been made to reduce the input of eutrophic substances into the lagoon. In particular, two large treatment plants have been built, one in Fusina to process mainly industrial effluents, and one in Campalto mainly for urban sewage treatment.

The main effect of these interventions was a notable reduction in water ammonium concentration which decreased in the area near the Porto Marghera industrial district (Ministry of Public Works et al. 1985; ENEA 1988, 1990) by two orders of magnitude (from $1000-2500 \, \mu mol \, dm^{-1}$ in 1962–1964 to $5-50 \, \mu mol \, dm^{-1}$ in 1985–1990). Unfortunately, changes were not observed for the nitrogen oxidized species (nitrate + nitrate, NO_x) which were recorded at concentrations of $50-70 \, \mu mol \, dm^{-1}$ (Ministry of Public Works et al. 1985). The contribution from agricultural drainage is thought to be the main reason for the constant input of NO_x. Similarly, only minor variations of orthophosphate concentrations were observed from the 1970s on (Consorzio Venezia Nuova 1989), in spite of both the sewage treatment plants which came into operation and the progressive reduction in detergent phosphorus (initially by 8%, then 5% from 1984, 2.5% from 1986, 1% from 1988 and not at all from 1990). The surface sediment concentrations of phosphorus and nitrogen increased by up to 30 and 2.4 times respectively, compared with the levels recorded in the 1960s.

15.3   Temporal Changes of Algal Typology and Associations

Wastewater discharges and the severe hydrodynamic alteration induced by dredging new canals in the central lagoon caused strong selection among the animal and plant species. In particular, a vertical compression of the vegetational zones (Table 15.1) and a general decrease in the number of species (Table 15.2) was observed in the Venice littoral zone (Sfriso 1987). The change in the characteristic species of the Venice macroalgal associations up to the end of the 1980s can be inferred from Tables 15.3 and 15.4.

In more detail, supralittoral and upper mediolittoral zones remained more or less unchanged at 1.4–0.3 and 0.3– − 0.2 m depth ranges, respectively, whereas the lower mediolittoral zone has risen from

Table 15.1. Vertical limits (in metres) of the vegetational zones recorded from Pignatti (1962) and Sfriso (1987)

Vegetational zones	Pignatti's zones	Sfriso's zones
	Mean	Mean SD
	+ 2.0– + 1.5	+ 1.41 ± 0.57
Supralittoral zone[a]		
	+ 0.3	+ 0.32 ± 0.15
Upper Mediolittoral zone[b]		
	− 0.3	− 0.21 ± 0.14
Lower Mediolittoral zone[b]		
	− 1.2	− 0.77 ± 0.18
Infralittoral zone[c]		
	− 5– (− 8)	− 5– (− 10)

[a]Supralittoral zone extends between the mean tide level and the highest level influenced by the sea.
[b]Mediolittoral zone extends between the mean high and the lowest tide level. Lower and upper Mediolittoral zones are separated from the mean low tide level.
[c]Infralittoral zone extends from the lowest limit of the previous vegetational zone down to the bottoms without macroalgae. In the Venice littoral this vegetational zone extends down to − 5– − 10 m depth, depending on the sampling station.

Table 15.2. Number of species and macroalgae percent ratios in the littorals of the Venice lagoon

Authors	Total species	Chlorophyceae No	(%)	Phaeophyceae No	(%)	Rhodophyceae No	(%)
Schiffner and Vatova (1938)	141	35	(25)	18	(13)	88	(62)
Pignatti (1962)	104	21	(20)	15	(15)	68	(65)
Sfriso (1987)	95	32	(34)	14	(15)	49	(51)

No, Number of species.

Table 15.3. Characteristic species of the macroalgal associations found by Pignatti (1962)

Vegetational zones	Characteristic species	
— (1.5–2.0 m)		
Supralittoral zone — (0.3 m)	*Enteromorpha compressa*[a] (L.) Grev.	*Ulotrix pseudoflacca* Wille
Mediolittoral zone	*Bangia fuscopurpurea*[a] (Dillw.) J.Ag. *Antithamnion cruciatum*[a] (C.Ag.) Näg. *Fucus virsoides*[b] J.Ag. *Ceramium barbatum*[b] (Kütz.) J.Ag. *Gelidium pulvinatum*[c] Kütz. *Gelidium spathulatum* (Kütz.) Born. *Ceramium ciliatum* (Ell.) Ducluz. *Corallina elongata* Johnston	*Ceramium diaphanum* (Roth.) Harvey *Gelidium crinale* (Turn.) J. Lam. *Grateloupia filicina* (Wulf.) C.Ag. *Gymnogongrus griffithsiae* (Turn.) Mart. *Cladophora penicillata* Kütz. *Enteromorpha linza* (L.) J.Ag. *Enteromorpha compressa* (L.) Grev.
— (–0.3 m)		
Infralittoral zone	*Gracilaria bursa pastoris*[b] (Gmel.) Silva *Hypnea musciformis*[b] (Wulf.) Lamour. *Ceramium strictum*[b] Kütz. *Cystoseira barbata*[c] J.Ag. *Cystoseira fimbriata*[d] (Desf.) Bory *Dictyopteris membranacea*[d] (Stackh.) Batt. *Bonnesmaisonia asparagoides*[c] (Woodw.)Ag.	*Cladostephus verticillatus*[c] (Lightf.) Lyngb. *Nemastoma dichotoma*[c] Ag. *Gigartina acicularis* (Wulf.) Lamour. *Rhodymenia ardissonei* J. Feldm. *Bryopsis plumosa* (Huds.) C.Ag. *Halymenia floresia* (Clem.) C.Ag.
— (–5 – –8 m)		

[a]Species that actually are characteristic of different vegetational zones.
[b]Rare species scarcely representative of the vegetational zones.
[c]Species disappeared or never found after 1960s.
[d]Species disappeared after 1960s and reappeared from 1990.

$-0.3 - -1.2$ to $-0.2 - -0.8\,m$ (Table 15.1). A similar trend was found in the Trieste Gulf, where a general increase in water turbidity remarkably reduced the beds of rhizophytes of the sea infralittoral zone (Bressan 1991).

From the 1960s to the 1980s, a change in the flora toward species more tolerant to hypertrophic/dystrophic conditions was recorded. The number of Chlorophyceae species increased from 20 to 34% at the expense of the Rhodophyceae, which decreased from 65 to 51% of the total number of algal species (Table 15.2). Nitrophilic species such as *Ulva rigida* C. Ag., *Enteromorpha* spp. and *Cladophora* spp. progressively replaced both the species more sensitive to human activities and rhizophytes, which previously had covered great bottom areas around the city of Venice. Many species previously very abundant, such as *Cystoseira fimbriata, Cystoseira barbata, Dictyopteris membranacea, Cladostephus verticillatus,* or *Taonia atomaria*, disappeared rapidly during the 1960s and 1970s, and the seagrasses *Zostera marina, Cymodocea nodosa,* and *Zostera noltii*, excellent indicators of areas never affected by anoxia (Sfriso 1987; Sfriso et al. 1988a,b), were confined to the northern and sourthern basins. Since 1990, rhizophytes (mostly *Zostera marina*) have reappeared in some areas of the central lagoon, because of the decline of *Ulva* overgrowth and anoxic situations. Moreover, new macroalgal species such as *Grateloupia doryphora* are spreading over large areas of the lagoon. A re-population of *Dictyopteris membranacea* and *Cystoseira fimbriata* was recorded in the southern lagoon near Chioggia, together with abundant *Sargassum muticum*, which has spread from Japan to Pacific North America and the British southern coasts since the 1970s (Chapman and Chapman 1980). Even some specimens of *Undaria pinnatifida* were found in the Giudecca Canal near the historical centre of Venice. The presence of *Sargassum* and *Undaria* is ascribed mainly to the development of commercial exchanges with the Japanese and Chinese sea areas, where these Phaeophyceae are largely cultivated. This is confirmed by the diffusion of these species into the Mediterranean Sea, particularly in the Languedoc region.

The presence of eutrophic species and the poor water exchange in the wide shallow areas of the lagoon split the macroalgal associations of the soft lagoon bottoms into two main typologies: one characterized by the succession of different species, the other by the prevalance of almost monospecific *Ulva rigida* populations (over 95% of the total biomass) responsible for frequent distrophic conditions. From the 1970s, *Ulva* spread in the areas little flushed by tides and caused the disappearance of the other species, which were shaded out by floating, 1–2 m wide *Ulva* plants. They easily reached standing crops of $15–20\,kg\,m^{-2}$ (wet weight)

Table 15.4. Vegetational associations monitored in the Venice littorals. (Sfriso 1987)

Vegetational zones		Sea associations	Port entrance associations	Lagoon associations
Supralittoral zone				
a) Northern exp.	Perennating spp.	–*Blidingia minima*	–*Blidingia minima*	–*Blidingia minima*
	Seasonal spp.	–*Bangia fuscopurpurea* (w-spr)	....	–*Ulotrix pseudoflacca* (w-spr)
b) Southern exp.	Perennating spp.	....	....	....
	Seasonal spp.	–*Bangia fuscopurpurea* (w-spr)	....	–*Ulotrix pseudoflacca* (w-spr)
Upper mediolittoral zone				
a) Northern exp.	Perennating spp.	–*Enteromorpha compressa* –*Gelidium spathulatum*	–*Enteromorpha compressa* –*Gelidium spathulatum*	–*Enteromorpha compressa* –*Enteromorpha prolifera*
	Seasonal spp.	–*Scytosiphon lomentaria* (spr) –*Petalonia fascia* (w)	–*Scytosiphon lomentaria* (spr) –*Petalonia fascia* (w)	–*Petalonia fascia* (w)
b) Southern exp.	Perennating spp.	–*Enteromorpha compressa* –*Gymnogongrus griffithsiae*	–*Enteromorpha compressa* –*Gymnogongrus griffithsiae*	–*Enteromorpha compressa* –*Enteromorpha prolifera*
	Seasonal spp.	–*Petalonia fascia* (w) –*Porphyra leucosticta* (w)	–*Petalonia fascia* (w) –*Ulotrix implexa* (w) –*Navicula* spp. (sum)	–*Petalonia fascia* (w) –*Navicula* spp. (sum)
Lower mediolittoral zone No differences have been observed for the two exposures	Perennating spp.	–*Ceramium ciliatum* –*Corallina elongata* –*Grateloupia filicina*	–*Ceramium ciliatum* –*Polysiphonia breviarticolata* –*Grateloupia filicina*	–*Bryopsis hypnoides* –*Polysiphonia senguinea*
	Seasonal spp.	–*Bryopsis disticha* (spr) –*Ulva fasciata* (sum-a) –*Lomentaria clavellosa* (w)	–*Bryopsis disticha* (spr) –*Ulva fasciata* (sum-a) –*Lomentaria clavellosa* (w)	–*Enteromorpha intestinalis* (spr) –*Ulva fasciata* (sum-a) –*Porphyra leucosticta* (w) –*Antithamnion cruciatum* (w)

Infralittoral zone
No differences have
been observed for
the two exposures

Perennating spp.	-Ulva rigida -Bryopsis plumosa -Polysiphonia sanguinea	-Ulva rigida -Codium fragile -Dictyota dichotoma -Rhodymenia ardissonei	-Ulva rigida -Gracilaria verrucosa
Seasonal spp.	-Enteromorpha linza (sum) -Dasya pedicellata (sum-a) -Antithamnion cruciatum (w)	-Enteromorpha linza (sum) -Halymenia floresia (sum-a) -Antithamnion cruciatum (w)	-Punctaria latifolia (spr) -Cladophora vagabunda (sum-a) -Vaucheria dichotoma (w) -Ectocarpus siliculosus (w-spr)

Exp., Exposure; spp., species; spr, spring; sum, summer; a, autumn; w, winter.

and filled the entire water column up to the surface. At this growth stage, biomass was highly stratified with its productive layer approaching the water surface. Below the canopy, light was considerably reduced (PAR $= 0$–$100\,\mu E\,m^{-2}\,s^{-1}$) and decomposition processes prevailed near the bottom layers. The daily fluctuation of *Ulva* biomass within the water column was caused both by tidal variation and oxygen production during photosynthetic activity. The upper algal layer could remain exposed to air and water loss for many hours or days during neap tides. Beer and Eshel (1983) showed that *Ulva* is neither exceptionally resistant to desiccation, nor does it benefit photosynthetically from emersion. *Ulva* exhibited a negative carbon balance, upon exposure to morning or noon light intensity, after only 30–90 min. Because of the poor water renewal by tides due to macroalga masses filling up the entire water column of the lagoon for many square kilometres, the amount of dissolved oxygen became insufficient to balance the oxygen consumed by the total biomass, and this led the macroalgae to collapse very rapidly. The subsequent release of high amounts of nutrients triggered sudden phytoplankton blooms (especially *Skeletonema costatum* and *Nitzschia closterium*) further contributing to macroalgal decay by shadowing parts of the seaweed. Water and sediment became completely anoxic, and fish and benthic animals died. Only *Chironomus* larvae (*Chironomus salinarius* Kieff.), which have an anaerobic metabolim (Ceretti et al. 1985), took advantage of this situation and grew uncontrolledly. As a result, in August 1985, swarms of such Diptera invaded Venice hindering railway and airport activities (Sfriso et al. 1987). Algal decomposition lasted for 2–3 weeks and phytoplankton blooms for 1–2 months, after which the water became clearer and macroalgae recovered. At first, seasonal species such as *Cladophora* spp. and *Enteromorpha* spp. occurred. Later, *Ulva* again grew reaching a biomass of many kilograms per square metres; it survived the winter and again exhibited a fast growth in the succeeding spring. The annual biomass trend of typical *Ulva* associations is shown for the three lagoonal stations Lido, S. Giuliano and Sacca Sessola (Fig. 15.2a).

In contrast to the areas of the lagoon described above, high species diversity was found near the port entrance (such as Alberoni station), where tidal currents assure a fast water renewal (Fig. 15.2b). Here, *Ulva* represented on average only 15–20% of the total fresh biomass produced yearly. Other species important for biological production, i.e. each contributing more than 1% of the total annual biomass, and replacing each other during the seasons were: *Enteromorpha compressa*, *Enteromorpha intestinalis*, *Gracilaria verrucosa*, *Porphyra leucosticta*, *Ceramium rubrum*, *Dictyota dichotoma*, *Petalonia fascia*, *Ectocarpus siliculosus* and

Macroalgal biomass

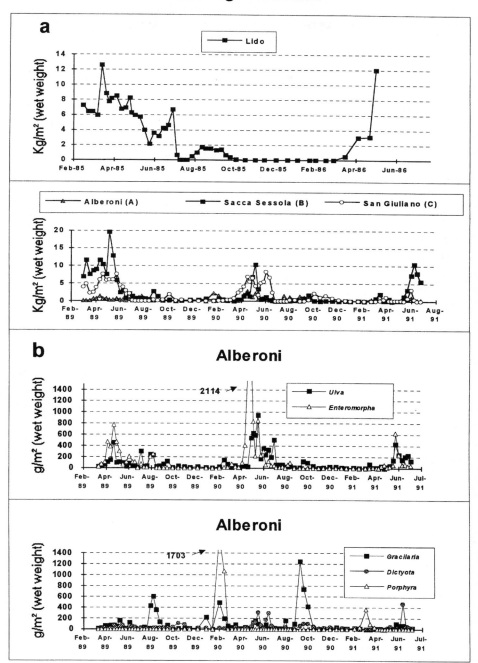

Fig. 15.2. a Annual biomass trends at some stations of the lagoon. **b** Annual biomass trends of the main species (contributing more than 1% of the total production) sampled at Alberoni station

Polysiphonia sanguinea. These species, except for *Gracilaria,* are small and tube or ribbon-like in strucutre, and rarely reach biomasses higher than 2–3 kg m^{-2} (wet weight). The areas inhibited by these species were always flushed by tidal currents and never experienced anoxic conditions typical of the areas covered by *Ulva.*

Spatial distribution of these main macroalgal associations are continuously changing over time: the *Ulva* association expanded from the 1970s until 1990. During the last 5 years, however, this association regressed.

The main cause was meteorological conditions unfavourable for *Ulva* growth during the April–May period that enhanced water turbidity by sediment resuspension and phytoplankton blooms. Nowadays, the coverage and growth of *Ulva* appear primarily governed by natural factors such as the increased sedimentation rates due to the decreased biomass coverage and the grazer pressure enhanced by the disappearance of anoxic crisis (Sfriso and Marcomini 1995). Moreover, biomass harvesting may play a key role in areas affected by quick biomass increases.

15.4 Macroalgal Production and Nutrient Cycling

Until 1986, phytoplankton primarly production was measured only in some restricted areas of the lagoon (Battaglia et al. 1983; Degobbis et al. 1986; Table 15.5). Mass development of macroalgae was a relatively new phenomenon, and the role of the macroalgae as primary producers in the

Table 15.5. Net primary production in the Lagoon of Venice

	Photosynthetic rates	Measurement period	Sampling area	References
Phytopholankton	0.6–16 mg C m^{-3} h^{-1}	Winter	Ind. zone	Battaglia et al.
	17–580 mg C m^{-3} h^{-1}	Summer	Ind. zone	(1993)
	0.4–186 mg C m^{-3} h^{-1}	All year	Ind. zone	Degobbis et al.
	1.0–42 mg C m^{-3} h^{-1}	All year	Sacca Sessola	(1986)
	0.8–20 mg C m^{-3} h^{-1}	All year	Lido inlet	
Macroalgae	up to 30.5 ± 1.6 g C m^{-2} day^{-1}	May 1985	Lido divide	Sfriso et al.
	646 ± 32 g C m^{-2} year^{-1}	(1985–86)	Lido divide	(1988b)
	up to 32.7 ± 1.6 g C m^{-2} day^{-1}	Apr. 1989	Sacca Sessola	Sfriso et al.
	350 ± 17 g C m^{-2} year^{-1}	(1989–90)	Sacca Sessola	(1993)
	up to 12.3 ± 0.6 g C m^{-2} day^{-1}	June 1990	San Giuliano	Sfriso et al.
	357 ± 18 g C m^{-2} year^{-1}	(1989–90)	San Giuliano	(1993)
	up to 3.7 ± 0.2 g C m^{-2} day^{-1}	Apr. 1990	Alberoni	Sfriso et al.
	196 ± 10 g C m^{-2} year^{-1}	(1989–90)	Alberoni	(1993)

lagoon was underestimated until the end of the 1970s. The standardized O_2 and ^{14}C methods, which were then used to evaluate phytoplankton production, were not adequate, because of the huge macroalgal biomass (up to 20 kg m^{-2}, wet weight) and its stratification. In fact, both O_2 and ^{14}C measurements require low biomasses (0.5–5 g wet weight) to be reliable and the results of these methods are difficult to extrapolate to the total biomass. In addition, the oxygen method cannot be applied, since the lagoon waters were mostly air-oversaturated (up to 360%; Sfriso et al. 1988a). Under these conditions, O_2 is very often directly released in the form of gas bubbles. Bidwell and McLachlan (1985) reported that at high O_2 and low CO_2 concentrations in macroalgae photorespiration may occur producing CO_2 and consuming O_2. These gases compete for the active site of ribulose biphosphate carboxylase/oxygenase and produce glycolate by oxygenase activity during photosynthesis (Bidwell 1983). Under photorespiration, the photosynthetic quotient is very different from that normally used in the production calculations (1–1.2) changing from 0.1 to 1, depending on CO_2 and O_2 concentrations. In view of the limitation of the O_2 and ^{14}C methods, measurement of biomass variations with a frequency depending on the seasonal macroalgal growth rate (3–20 days) appeared the most suitable method for the evaluation of macroalgal primary production (Sfriso et al. 1988b, 1991).

For this purpose, an accurate and precise procedure was developed to correctly measure the standing crop of shallow sampling sites populated by macroalgae with different densities and varying distributions (Sfriso et al. 1991). An accuracy within 10% was obtained after only three measurements of the different macroalgal distributions (Fig. 15.3). To reach a higher accuracy within 5%, 4–14 subsamples are needed depending on the different macroalgal dispersions.

Using this procedure, we have sampled over several years, weekly during spring/summer and twice a month during autumn/winter, three

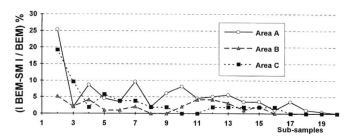

Fig. 15.3. Number of sub samples needed to obtain an SM within a sampling accuracy of 5 and 10%. *BEM* Best estimated mean; *SM* sampling mean

areas of the lagoon characterized by different species and biomass densities. Significant biomass variations between successive sampling periods were determined by applying variance analysis at the 0.01 probability level to each series of annual mean biomass values. This allowed the estimation of daily and yearly net biomass production/degradation rates (Table 15.6). Taking into account the percent dry to wet weight ratio and the mean carbon (C), nitrogen (N) and phosphorus (P) contents of the biomass involved (Table 15.7), the assimilation rates and the amount of P and N uptake or release by macroalgae were also calculated (Table 15.6, Fig. 15.4). Net production rates up to $1506 \, \mathrm{g \, m^{-2} \, day^{-1}}$, wet weight, equivalent to $157 \, \mathrm{g \, m^{-2} \, day^{-1}}$, dry weight and $33 \, \mathrm{g \, C \, m^{-2} \, d^{-1}}$ were recorded, about 1–2 orders of magnitude higher than those reported for phytoplankton in the literature (Table 15.6). Simultaneously, biomass losses down to $995 \, \mathrm{g \, m^{-2} \, day^{-1}}$ wet weight, were found during an anoxic macroalgal collapse.

From the changes in biomass, the percentage of daily biomass variation (%DBV) from the initial biomass was calculated by this equation:

$$\%DBV = [(B_t - B_0)/B_0]/t \times 100 \tag{1}$$

(B_0 = initial biomass; B_t = biomass at day t).

The percentage of progressive daily biomass variation (%PDBV) was computed using the following equation (Lignell et al. 1987):

$$\%PDBV = [(B_t - B_0)^{1/t} - 1]/t \times 100. \tag{2}$$

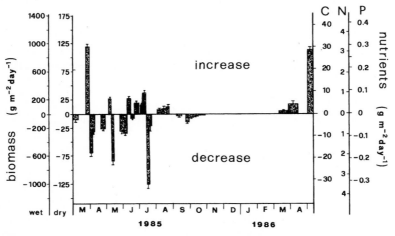

Fig. 15.4. Daily trends of biomass gain and loss per square meter at the Lido station in 1985–1986. From the mean nutrient contents in macroalgae, the amount of carbon and phosphorus assimilated daily, or released by biomass degradation, was also calculated

Table 15.6. Estimated daily and yearly net biomass production and degradation rates

Period (22 March 1989–3 July 1991) (84 sampling campaigns)

		Alberoni			Sacca Sessola			S. Giuliano		
		Total g/m²	Range g/m²/day	Mean	Total g/m²	Range g/m²/day	Mean	Total g/m²	Range g/m²/day	Mean
Standing crop		–	0–2855	359	–	0–1963	1945	–	0–8228	1428
Biomass	Production	16687	0–169	20	52418	0–1506	63	31237	0–566	38
	Loss	16388	0–240	20	54165	0–995	65	35435	0–540	43
Carbon	Uptake	363	0–3.7	0.43	1139	0–32.7	1.4	679	0–12.3	0.8
	Loss	356	0–5.2	0.43	1177	0–21.6	1.5	770	0–11.7	0.9
Nitrogen	Uptake	38	0–0.38	0.05	118	0–3.40	0.14	71	0–1.28	0.09
	Loss	37	0–0.54	0.05	122	0–2.25	0.16	80	0–1.22	0.10
Phosphorus	Uptake	3.3	0–0.033	0.004	10.3	0–0.299	0.013	6.2	0–0.112	0.008
	Loss	3.2	0–0.048	0.004	10.7	0–0.198	0.014	7.0	0–0.107	0.009
		2 May–22 May			26 May–22 May			13 June–18 June		

Period (1989–1990)

		Alberoni			Sacca Sessola			S. Giuliano		
		Total g/m²	Range g/m²/day	Mean	Total g/m²	Range g/m²/day	Mean	Total g/m²	Range g/m²/day	Mean
Standing crop		–	0–2205	385	–	0–12850	1465	–	0–8228	1452
Biomass	Production	9047	0–168	23	16089	0–791	45	16455	0–566	45
	Loss	8181	0–232	21	25480	0–781	71	15887	0–396	43
Carbon	Uptake	196	0–3.6	0.50	350	0–17.2	0.98	358	0–12.3	0.98
	Loss	178	0–5.0	0.46	554	0–17.0	1.54	345	0–8.6	0.93
Nitrogen	Uptake	20	0–0.38	0.05	36	0–1.79	0.10	37	0–1.28	0.10
	Loss	18	0–0.52	0.05	58	0–1.76	0.17	36	0–0.89	0.10
Phosphorus	Uptake	1.8	0–0.033	0.005	3.2	0–0.157	0.009	3.3	0–0.112	0.009
	Loss	1.6	0–0.046	0.004	5.0	0–0.155	0.015	3.1	0–0.078	0.009

Table 15.6. (*Contd.*)

Period (1990–1991)		22 May–23 May			22 May–18 June			18 June–10 June		
		Alberoni			Sacca Sessola			S. Giuliano		
		Total g/m²	Range g/m²/day	Mean	Total g/m²	Range g/m²/day	Mean	Total g/m²	Range g/m²/day	Mean
Standing crop		17–1436		269	–	0–10652	736	–	0–8228	589
Biomass	Production	6074	0–157	17	15507	0–686	40	6833	0–185	19
	Loss	6918	0–240	19	15611	0–995	40	11792	0–540	32
Carbon	Uptake	132	0–3.4	0.37	337	0–14.9	0.9	149	0–4.0	0.41
	Loss	150	0–5.2	0.41	339	0–21.6	0.9	256	0–11.7	0.70
Nitrogen	Uptake	14	0–0.35	0.04	35	0–1.55	0.09	15	0–0.42	0.04
	Loss	16	0–0.54	0.04	35	0–2.25	0.10	27	0–1.22	0.07
Phosphorus	Uptake	1.2	0–0.031	0.003	3.1	0–0.136	0.008	1.4	0–0.04	0.004
	Loss	1.4	0–0.048	0.004	3.1	0–0.198	0.008	2.3	0–0.11	0.006

Table 15.7. Nutrient concentrations found in *Ulva rigida* in different seasons and with different frond ages

Season	Fronds	Dry weight %	Phosphorus	Nitrogen	Carbon	N:P atomic ratio
			g kg^{-1} (dry wt.)			
Winter	New	10.5	0.38	10.2	200.4	59
	Old	11.7	2.91	29.3	195.7	22
Spring	New	9.6	1.40	17.4	230.9	27
	Old	10.2	1.98	22.7	190.0	25
Summer	New	8.7	1.70	22.6	194.0	29
	Old	9.7	2.06	28.3	180.1	30
Autumn	New	10.6	2.41	17.4	227.2	16
	Old	12.4	2.41	25.9	250.0	24
Mean	New	9.8	1.47	16.9	213.1	25
	Old	11.0	2.34	26.5	204.0	25
	Total	10.4	1.91	21.7	208.5	25

These parameters indicate the macroalgal potential to produce biomass starting from a given standing crop level. The yield, normalized to unit weight was very high, when the initial standing crop level was low (Fig. 15.5). For a biomass lower than $100 \, g \, m^{-2}$ wet weight, the recorded DBV [Eq. (1)] was higher than 180% of the initial biomass and very low when the initial biomass exceeded $10 \, kg \, m^{-2}$ wet weight. These results are not surprising, considering that low initial biomasses benefit from conditions (light, nutrients, etc.) favouring a rapid growth, and new fronds can add to the initial ones. The optimal starting biomass to obtain the highest production yield was about $4–6 \, kg \, m^{-2}$ wet weight (Fig. 15.5). According to these results, the area A (Alberoni), characterized by high species diversity and low biomass density, exhibited a higher yield per weight unit, but a low yield per unit area compared to areas B (Sacca Sessola) and C (San Giuliano), where *Ulva rigida* prevailed over the other species and displayed very high standing crops.

Similar results can be inferred from the ratio annual biomass production to highest measured biomass (ABP/HMB) calculated for the same areas. Between 1989 and 1991, this ratio was about 1.51 ± 0.06 in areas such as Sacca Sessola with standing crops of up to $20 \, kg \, m^{-2}$ wet weight. This ratio increased with decreasing biomass densities: it was 2.04 ± 0.06 at San Giuliano area with biomass densities up to $9 \, kg \, m^{-2}$ wet weight, and increased up to 4.5 ± 0.6 at Alberoni station, where the highest biomass density was always no higher than $3 \, kg \, m^{-2}$ wet weight. Using the APB/HMB ratios and the highest standing crops recorded in 178

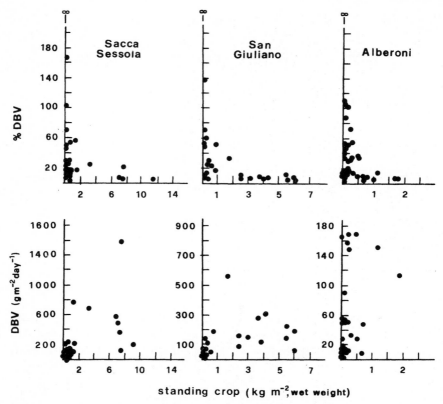

Fig. 15.5. Percent and wet weight daily biomass variation (*DBV*) starting from a predetermined initial biomass. Field measurements are from 1989 to 1991

stations during the highest biomass growth period in 1987 (Fig. 15.6), the estimated macroalgal net primary production (NPP) in the central lagoon was about 1.5 million t wet weight, upto 1990 (Table 15.8). Primary production of seagrasses and macroalgal species, such as *Chaetomorpha linum*, *Cladophora* spp. and *Valonia aegagrophyla*, which prevailed over *Ulva* in the southern and northern parts of the lagoon, were not measured, however, we estimate that the overall NPP in the Venice lagoon would account for a minimum of 3 million t on a wet weight basis.

The balance of phosphorus (which may be considered as a "conservative" element in contrast to nitrogen which may be lost in atmosphere as N_2O and N_2 between macroalgal biomass, surface sediments and settled particulate material (SPM) of an area populated mostly be macroalgae (>95% cover), and without significant import or export processes,

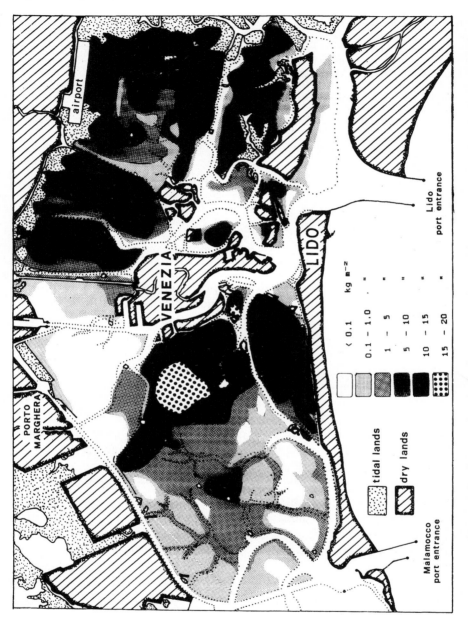

Fig. 15.6. Weight range distribution of macroalgal biomass in the central Venice lagoon during the highest production period in 1987

Table 15.8. Estimation of macroalgal standing crop and biomass production in the central Venice lagoon by the biomass measured during the highest growth period

Biomass Range Mean kg (wet wt.) m^{-2}		Highest	Lagoon surface[a] (km²)	(%)	Standing crop Mean (Tons)	Highest	ABP/HMB[b]	Annual Production (Tons)
15–20	17.5	20	2.3	2.7	40 250	46 000	1.6	73 600
10–15	12.5	15	23.8	27.9	297 500	357 000	1.6	571 200
5–10	7.5	10	19.9	23.3	149 250	199 000	2.0	398 000
1–5	3.0	5	20.1	23.5	60 300	100 500	4.5	452 250
0.1–1	0.6	1	19.3	22.6	11 580	19 300	4.5	86 850
		Total	85.4	100	558 800	721 800		1 581 900

The factors 1.6, 2.0 and 4.5 are the APB/HMB values found in the sampling stations at biomass ranges of 10–20, 5–10 and < 0.5 kg m^{-2}, wet wt.
[a]Central lagoon = 132 km² of water surface.
[b]ABP/HMB = annual biomass production/highest measured biomass.

enabled us to estimate the total gross primary production (GPP) of this area (Sfriso and Marcomini 1994). In more detail, the GPP was inferred from the fraction of macroalgae in the SPM collected during the year by traps settled on the sediment by means of the two equation system:

$$Ax + By = C \tag{3}$$

$$x + y = 100 \tag{4}$$

(A, C = mean concentrations of phosphorus found in *Ulva* and SPM; B = winter background level of phosphorus in the surface sediment; x = macroalgal fraction of phosphorus in SPM; y = sediment fraction of phosphorus in the surface sediment).

The macroalgal fraction multiplied by the annual amount of SPM provides the amount of macroalgae produced in this period. In areas with mean macroalgal densities above > 5 kg m^{-2}, the GPP was six to seven times higher than the macroalgal NPP as determined by the biomass change method (Sfriso and Marcomini 1994). Knowing the ABP/HMB and GPP/HPP ratios, which correspond to different biomass densities and typologies, the macroalgal GPP of stagnant areas can quickly be inferred by measuring the HMB during the period of highest growth of the dominant species, which usually lasts about 1 month. The estimate is obtained from the equation:

$$GPP = HMB \times (ABP/HMB) \times (GPP/NPP) \tag{5}$$

(HMB = highest measured biomass; ABP = annual biomass production). At Sacca Sessola, for macroalgal standing crops up to 20 kg m^{-2} on a wet weight basis, the total GPP between the 1970s and the 1990s was estimated at ~ 200 (i.e. $20 \, \text{kg m}^{-2} \times 1.5 \times 6.7$) kg m^{-2} wet weight (Sfriso et al. 1993).

Macroalgae strongly interfere with the lagoon's nutrient cycles, especially in the areas covered by massive *Ulva rigida* populations. The seasonal variations of phosphorus and nitrogen in the water and surface sediments (0–5 cm) were strictly related to the biomass densities.

During the spring–summer algal growth, total inorganic nitrogen (TIN = sum of NH_4^+-N, NO_2^{-1}-N and NO_3^{-1}-N) in water rapidly reduced from 30–90 to 2–$7 \, \mu\text{mol dm}^{-3}$. In contrast, reactive phosphorus (RP) exhibited during this period the highest concentrations detected over the year (from <0.1 to $16 \, \mu\text{mol dm}^{-3}$; Fig. 15.7). Phosphorus, similarly to nitrogen, was rapidly taken up by macroalgae during their luxuriant growth. The increasing layer of decaying seaweeds close to the sediment surface caused reducing redox potentials (-100 to $-250 \, \text{mV}$), which favoured the release of phosphorus together with reduced forms from the sediments (Golterman 1973; Nixon et al. 1980; Cossu et al. 1983). Concurrently, active bacterial denitrifying processes took place (Nixon et al. 1979; Henriksen et al. 1981; Jenkins and Kemp 1984; Rödnner 1985).

In the sediments, nitrogen and phosphorus as well as organic carbon concentrations exhibited similar behaviour (Fig. 15.8). Macroalgae take up nutrients from waters while growing, and release them onto the surface sediments when biomass collapses. In 1985–86, in a station near the Lido watershed covered by *Ulva*, surface sediment concentrations of phosphorus, nitrogen and organic carbon increased by 75, 160 and 70%, respectively (Fig. 15.8). Sediment nutrient concentrations decreased to pre-spring values during winter because nutrients were slowly released into the overlying water as a result of diagenetic processes.

The opposite behaviour of nutrients that occurs at the water–sediment interface under oxic or anoxic conditions also explains the different trends of water and sediment N:P atomic ratios (Fig. 15.9). The nitrogen to phosphorus ratio in water was high in the cold season (100–300), when macroalgal growth was negligible and decreased nearly to zero in spring-summer, during the highest macroalgal changes. In the upper 5 cm of the surface sediment, the N:P ratio increased temporarily from 7–8 to 11–12, because of the nutrients released from the continuously decaying biomass (N:P atomic ratio in *Ulva* was on average 25 ± 5; Table 15.7).

The total amount of nutrients yearly recycled by macroalgae was estimated from the nutrient concentrations in *Ulva* (accounting for about 90% of the macroalgal biomass produced in the central lagoon

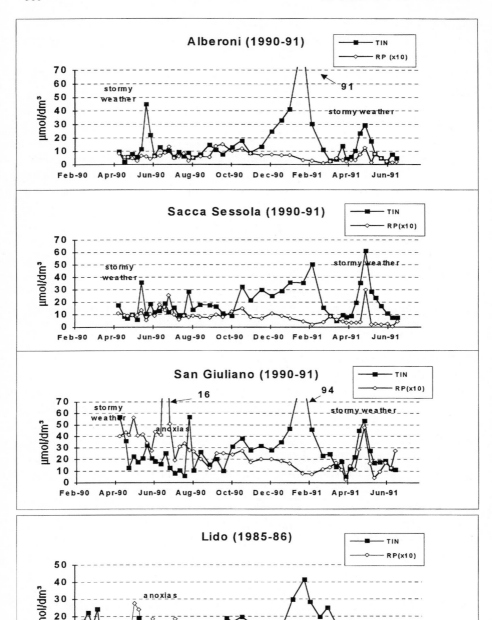

Fig. 15.7. Annual trends of total inorganic nitrogen (TIN: NH_4^+-$N + NO_x$ *Filled square*) and reactive phosphorus (*RP, open diamond*) in seawater from the Alberoni, Sacca Sessola, San Giuliano (1990–1991) and Lido (1985–1986) stations

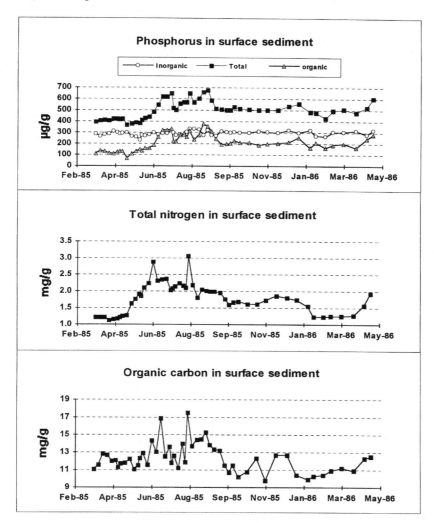

Fig. 15.8. Annual trends of nutrients and organic carbon in the upper 5-cm surface layer of the sediment at the Lido station

between 1989 and 1991) and the GPP in the central lagoon as estimated for 1987 (Table 15.9). Nutrients recycled by macroalgae in 1987 were approximately 23900 t of nitrogen and 2100 t of phosphate. Taking into account the most recent estimations of nutrient loadings in this basin (2578 and 351 t of N and P, respectively; Andreottola et al. 1990), the seaweeds recycled about nine times the total N and six times the P entering the central lagoon annually (Table 15.9). Since the amount of

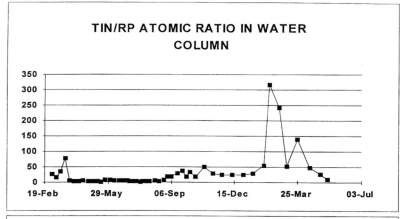

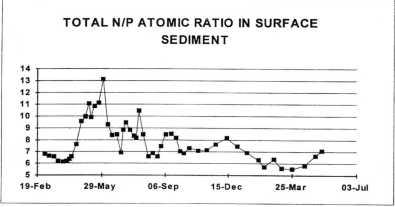

Fig. 15.9. Seasonal variation of N:P atomic ratios (total inorganic nitrogen: reactive phosphorus) in the seawater column and in the sediment at the Lido station

nitrogen and phosphorus temporarily retained by the macroalgal standing crop and surface sediments in spring/summer was about 2.4 times higher than the nutrients entering the same area, the sea was thought to be a nutrient feeder supporting the lagoon gross primary production significantly (Sfriso and Marcomini 1994).

By comparing the behaviour of phosphorus on one side, and nitrogen and carbon on the other side, we tried to evaluate the extent of N and C net losses from the water to the atmosphere (Sfriso and Marcomini 1994). Approximately 12 000 tons of nitrogen and 145 000 tons of carbon, accounting for 51 and 63% of N- and C-GPP, respectively, were transferred into the atmosphere as N_2-N_2O and CO_2-CH_4 (Table 15.10). The loss of N by denitrification was about one order to magnitude higher than

Table 15.9. Total nutrient loads and nutrients recycled by GPP in the central Venice lagoon

Inputs (tons)[a]		Nitrogen	Phosphorus
Total lagoon (432 km²)			
	(direct)	4175	491
	(indirect)	4262	657
	Total	8437	1148
Central lagoon (132 km²)	Total	2578	351
Nutrients recycled by GPP (tons)[b]		23939	2111
Nutrients recycling/input, %		929	601

[a]From Andreottola et al. (1990).
[b]Estimated net primary production (NPP) = 1581900 tons (wet wt.), mean biomass dry/wet weight ratio = 10.41%, mean biomass nitrogen conc. = 21.7 mg/g (dry wt.), mean biomass phosphorus conc. = 1.9 mg/g (dry wt.), gross primary production (GPP) = 6.7 × NPP.

Table 15.10. Atmospheric loss of nutrients in the central Venice lagoon. (Sfriso and Marcomini 1994)

	GPP nutrient recycling	SPM algal fraction	Atmospheric losses	
	($g\,m^{-2}\,year^{-1}$)		(%)	(tons)
Phosphorus	29	29	0	0
Nitrogen	330	162	51	∼ 12000
Carbon	3177	1175	63	∼ 145000

that proposed by ENEA (1990) extrapolating denitrification rates obtained in the northern Adriatic sea.

15.5 Summary and Conclusions

The lagoon of Venice is a good example of an ecosystem with profound changes as a result of increasing human activities, after the World War II. The hydrographic alterations of the lagoon favoured the silting of wide areas, whereas the high nutrient inputs due to increasing industry, agriculture and tourism since the 1970s, accounted for the accumulation of nitrogen and phosphorus in the surface sediments, supporting remarkable changes in benthic associations.

The increasing eutrophication observed over the 1950–1980s, in particular, moved the lower limits of the vegetation upwards, decreased the total number of species living in the Venice littorals and changed the species characteristic of the main associations sampled in the past.

Some species, such as *Ulva rigida*, prevailing in the soft bottoms with weak tidal currents, spread over most of the central lagoon. This species easily reaches biomasses over 20 kg m^{-2} wet weight, forming thick layers in which respiration processes may exceed photosynthetic oxygen production, resulting in anoxic conditions.

Until 1990, net macroalgal production was 1 to 2 orders of magnitude higher than that of phytoplankton, accounting for about 1.5 million t wet weight, in the central lagoon alone.

The estimated total gross primary production which would balance the phosphorus in SPM, surface sediment and algal biomass was six to seven times higher than net macroalgal production.

Under natural conditions, daily macroalgal production yield per metre square strictly depends on the initial biomass: low initial biomass (0.1–0.5 kg m^{-2}, wet weight) can increase as much as 100–200%, while high initial biomass (10–15 kg m^{-2}, wet weight) usually shows low increase (0–5%). The optimal starting biomass to obtain the highest yields was 4–7 kg m^{-2}.

Macroalgae strongly interfere with nutrient cycles. During spring and summer, nitrogen and phosphorus concentrations in the water and sediments are strictly related to biomass changes and, after total decay of the macroalgae, can temporarily increase two to three times in surface sediments. Due to the different behaviour of these elements (release of phosphorus into the water and atmospheric loss of nitrogen from reduced sediments), nitrogen can temporarily become the limiting factor for algal growth.

The amout of nutrients recycled by macroalgae exceeded considerably the nutrient loads entering the lagoon annually. By evaluating the amout of P and N, temporarily stored in standing crop and surface sediments during the warm season, and estimating the loss of N due to denitrification occuring extensively in the anoxic sediments, it follows that the sea must be a significant source of nutrients for algal growth.

References

Andreottola G, Cossu R, Ragazzi M (1990) Nutrient inputs from the Venice watershed: comparison between direct and indirect estimations (in Italian). Ing Ambient 19: 176–185

Battaglia B, Datei C, Dejak C, Gambaretto G, Guarise GB, Perin G, Vianello E, Zingales F (1983) Hydrothermodynamic and biological investigations to determine the environmental consequences of the functionality at full working-order of the ENEL thermoelectric plant in Porto Marghera, vol 4 Regione Veneto, Venice (in Italian)

Battiston L, Giommoni A, Pilan L, Vincenzi S (1983) Salinity exchange induced by tide in the Venice lagoon. Conf Proc on Five centuries of water management in the Venice Territory, Venice, pp 1–7 (in Italian)

Beer S, Eshel A (1983) Photosynthesis of *Ulva* sp. I. Effects of desiccation when exposed to air. J Exp Mar Biol Ecol 70: 91–97

Bidwell RGS (1983) Carbon nutrition of plants: photosynthesis and photorespiration. In: Steward FC, Bidwell RGS (eds) Plant physiology, a treatise, vol VII. Energy and carbon metabolism. Academic Press, New York, pp 287–457

Bidwell RGS, McLachlan (1985) Carbon nutrition of seaweeds: photosynthesis, photorespiration and respiration. J Exp Mar Biol Ecol 86: 15–46

Bliding C (1963) A critical survey of European taxa in Ulvales, part I. *Capsosiphon, Percursaria, Blidingia, Enteromorpha.* Bot Not 8: 1–159

Bliding C (1968) A critical survey of European taxa in Ulvales, part II. *Ulva, Ulvaria, Monostroma, Kornmannia.* Bot Not 129: 535–629

Bressan G (1991) Some space and time changes on the flora and vegetational zones of the northern Adriatic sea. In: Galli F (ed) II Rassegna del Mare "Formazione professionale per la gestione e la protezione dell' ambiente marino". Italprogram Viareggio, Trieste (in Italian)

Bucci M, Ghiara E, Gorelli V, Gragnani R, Naviglio L, Uccelli R (1990) Eutrophication of the Orbetello lagoon (Grosseto, Italy) and suggested actions for its restoration. Proc of Environmental contamination, Barcelona, pp 293–297

Buia MC, Mazzella L (1991) Reproductive phenology of the Mediterranean seagrasses *Posidonea oceanica* (L.) Delile, *Cymodocea nodosa* (Ucria) Aschers and *Zostera noltii* Hornem. Aquat Bot 40: 343–362

Buia MC, Mazzella L, Pirc H, Russo GF (1985a) Flowering of *Zostera noltii* Hornem. at Ischia (Neaples gulf). Obelia 11: 861–862 (in Italian)

Buia MC, Russo GF, Mazzellla L (1985b) Relationship between *Cymodocea nodosus* (Ucria) Aschers and *Zostera noltii* Hornem in a mixed surface bed of the Ischia island. Nova Thalassia 7: 406–408 (in Italian)

Buia MC, Lupo V, Mazzella L (1992) Primary production and growth dynamics in *Posidonia oceanica*. Mar Ecol 13: 2–16

Caprioli R, Ghiara E, Mignuzzi C, Orlandi C (1988) Environmental Restoration of the Orbetello lagoon: seasonal nutrient cycle (March–December 1987) and geochemical characteristics of waters. ENEA, Arti grafiche S. Marcello, Roma, 34 pp (in Italian)

Ceretti G, Ferrarese U, Scattolin (1985) Chironomidae in the Venice Lagoon. Arsenale, Venezia, 59 pp (in Italian)

Chapmam VJ, Chapman DJ (1980) Seaweeds and their uses. Chapman and Hall, New York

Cinelli F, Fresi E, Mazzella L, Ponticelli MP (1979) Deep algal vegetation of the Western Mediterranean. G Bot Ital 113: 173–188

Consorzio Venezia Nuova (1989) The agricultural pollution in the Venice lagoon. Societa' Cooperative Tipografica Padova, Padova, 144 pp (in Italian)

Cossu R, Degobbis D, Donazzolo R, Maslowska E, Orio AA, Pavoni B (1983) Nutrient release from the sediments of the Venice lagoon. Ing Sanit 5/6: 16–23

Degobbis D, Gilmartin M, Orio AA (1986) The relation of nutrient regeneration in the sediments of the northern Adriatic to eutrophication, with special reference to the lagoon of Venice. Sci Total Environ 56: 201–210

De Toni GB, Levi MD (1885) Flora algologica della Venezia. I Floridee. Atti Ist Veneto Sci Lett Arti 3: 1917–2096 (in Italian)

De Toni GB, Levi MD (1886) Flora algologica della Venezia. II Melanoficee. Atti Ist Veneto Sci Lett Arti 4: 1615–1721 (in Italian)

De Toni GB, Levi MD (1888) Flora algologica della Venezia. III Cloroficee. Atti Ist Veneto Sci Lett Arti 5/6: 1511–1593; 95–155; 289–350 (in Italian)

ENEA (1988) Integrative elements on the polluting inputs in the lagoon; final report. Rep No. 1.3.3, Laboratorio Tecnografico ENEA, Venice (in Italian)

ENEA (1990) Final report on the actual conditions of the lagoon ecosystem. Rep No 1.3.9, Laboratorio Tecnografico ENEA, Venice (in Italian)

Feldmann J (1938) Recherches sur la végétation marine de la Méditerranée. Rev Algol 10: 1–340

Feldmann J (1959) Les problemes de l'étagement des peuplements d'algues marines. Ecologie des algues marines. Coll Int Cent Natl Rech Sci 81: 37–44

Feldmann J (1981) Clé des Polysiphonia des Côtes Francaises. Cryptogam Algol 2: 71–77

Giaccone G (1968) New and interestine species of Rhodophyceae collected in the Eastern basin of the Mediterranean. Boll Mus Civ Stor Nat Ven 26: 87–98 (in Italian)

Giaccone G (1974) Features of the lagoon vegetation of the northern Adriatic and changes due to pollution. G Bot Ital 102: 397–414 (in Italian)

Giaccone G, Pignatti S (1967) The vegetation of the Trieste Gulf. Nova Thalassia 3: 1–28 (in Italian)

Giordani Soika A, Perin G (1970) Changes of chemical characteristic and population variation in the lagoon sediments in the last twenty years. I–Changes of animal populations. XI Conf Proc ANLSB Venezia, pp 135–139 (in Italian)

Giordani Soika A, Perin G (1974a) The pollution of the Venice lagoon: study of the chemical changes and population variations in the lagoon sediments in the last twenty years. Boll Mus Civ Stor Nat Ven 26: 25–68 (in Italian)

Giordani Soika A, Perin G (1974b) The pollution of the Venice lagoon: study of the chemical changes and population variations in the lagoon sediments in the last twenty years. Boll Mus Civ Stor Nat Ven 26 (29 charts of the Venice lagoon) (in Italian)

Golterman AL (1973) Vertical movement of phosphate in freshwater. In: Griffith EJ, Beeton A, Spencer JM, Mitchell DT (eds) Environmental phosphorus handbook. Wiley-Interscience, New York, pp 509–538

Henriksen K, Hansen JI, Blackburn TH (1981) Rates of nitrification, distribution of nitrifying bacteria, and nitrate fluxes in different types of sediment from Danish waters. Mar Biol 61: 299–304

Izzo G (1988) Environmental restoration of the Orbetello lagoon: the role of sediment microbic activity in the distrophic conditions of the orbetello lagoon. ENEA, Arti grafiche S. Marcello, Roma, 14 pp (in Italian)

Jenkins MC, Kemp WM (1984) The coupling of nitrification and denitrification in two estuarine sediments. Limnol Oceanogr 29: 609–619

Lenzi M (1984) Investigation of the macrophyte distribution in the Orbetello lagoon. Quad Mus Stor Nat Livorno 5: 7–55 (in Italian)

Lignell A, Ekman P, Peders M (1987) Cultivation technique for marine seaweeds allowing controlled and optimized conditions in the laboratory and on a pilotscale. Bot Mar 30: 417–424

Marcomini A, Sfriso A, Pavoni B (1986) Variable wavelength absorption in detecting environmentally relevant PAHs by high-performance liquid chromatography. Mar Chem 21: 15–23

Ministry of Public Works, Magistrato alle Acque, Consorzio Venezia Nuova (1985) Knowledge of the pollution of the lagoon of Venice lagoon. Consorzio Venezia Nuova (Servizio Informativo) Venice, 4 vols (in Italian)

Mittempergher M, Metalli P, Bucci M, Gragnani R, Izzo G, Naviglio L (1988) Results of integrated multidisciplinary study aimed at environmental restoration of the Orbetello lagoon (Grosseto, Italy). Proc of Environmental contamination, Venice pp 330–335

Naviglio L, Uccelli R, Falchi G, Lenzi M (1988) Environmental restoration of the Orbetello lagoon: preliminary research on the space distribution and density of macrophyte vegetation. ENEA, Arti grafiche S. Marcello, Roma, 67 pp (in Italian)

Nixon SW, Kelly JR, Furnas BN, Oviatt CA, Hale SS (1980) Phosphorus regeneration and the metabolism of castal marine bottom communities. In: Tenore RK, Coull BC (eds) Marine benthic dynamics. South Carolina press, Columbia, pp 219–241

Orio AA, Donazzolo R (1987) Toxic and eutrophicating substances in the lagoon and the gulf of Venice. Ist Veneto Sci Lett Arti 11: 149–215 (in Italian)

Pavoni B, Sfriso A, Marcomini B (1986) Concentration and flux profiles of PCBs, DDTs and PAHs in a dated sediment core from the lagoon of Venice. Mar Chem 21: 25–35

Perin G (1975) The pollution of the Venice lagoon. Summary of seven research years. Atti Tavola Rotonda CDA (Admini stration Council), Venice, 47 pp (in Italian)

Piccoli F, Merloni N, Godini E (1991) Vegetation map of Sacca di Goro. Integrated study on the ecology of Sacca di Goro. Franco Angeli, Milano, pp 173–204 (in Italian)

Pignatti S (1962) Associations of marine algae on the Venice littorals. Mem Ist Veneto Sci Lett Arti 32: 1–134 (in Italian)

Pignatti S, di Cristini P (1966) Associations of seaweeds as pollution indices of the Vallone di Muggia waters near Trieste. Arch Oceanogr Limnol 15: 185–190 (in Italian)

Pirazzoli P (1974) Dati storici sul medio mare a Venezia. Atti Accad Sci Inst Bologna 13: 125–148

Rönner U (1985) Nitrogen transformations in the Baltic proper: denitrification counteracts eutrophication. Ambio 14: 134–138

Schiffner V, Vatova A (1938) The algae of the lagoon: *Chlorophyceae, Phaeophyceae, Rhodophyceae, Myxophyceae*. In: Minio M (ed) The lagoon of Venice, Vol 3. Officine Grafiche C Ferrari, Venice, 250 pp (in Italian)

Sfriso A (1987) Flora and vertical distribution of macroalgae in the lagoon of Venice: a comparison with previous studies. G Bot Ital 121: 69–85

Sfriso A, Marcomini A (1994) Gross primary production and nutrient behaviour in a shallow coastal environment. Bioresource Technol 45: 59–66

Sfriso A, Marcomini A (1995) Recent development of eutrophication in the lagoon of Venice. MEDITERRANEANCHEM, Int Conf on Chemistry and Mediterranean Sea, Taranto, p 85

Sfriso A, Marcomini A, Pavoni B (1987) Relationship between macroalgal biomass and nutrient concentrations in a hypertrophic area of the Venice lagoon. Mar Environ Res 22: 287–312

Sfriso A, Pavoni B, Marcomini A, Orio AA (1988a) Annual variation of nutrients in a lagoon hypertrophic area. Mar Pollut Bull 19: 54–60

Sfriso A, Marcomini A, Pavoni B, Orio AA (1988b) Macroalgal production and nutrient recycling in the lagoon of Venice. Ing Sanit 5: 255–266

Sfriso A, Pavoni B, Marcomini A (1989) Macroalgae and phytoplankton standing crops in the central Venice lagoon: primary production and nutrient balance. Sci Total Environ 80: 139–159

Sfriso A, Marcomini A, Pavoni B, Orio AA (1990) Eutrophication and macroalgae: the lagoon of Venice as study case. Inquinamento 4: 63–78 (in Italian)

Sfriso A, Raccanelli S, Pavoni B, Marcomini A (1991) Sampling strategies for measuring macroalgal biomass in the shallow waters of the Venice lagoon. Environ Technol 12: 263–269

Sfriso A, Marcomini A, Pavoni B, Orio AA (1993) Species composition, biomass and net primary production in shallow coastal waters: the Venice lagoon. Bioresource Technol 44: 235–250

Sighel A (1938) Macroalgal space and time distribution in the lagoon of Venice. Comit Talass Ital Mem CCL, Venice (in Italian)

Solazzi A (1966) Ecological study on the benthic algal vegetation (macrophyte) of the Monte Conero coast 7: 159–192 (in Italian)

Solazzi A (1968) Flora and macroscopic benthic vegetation of Neretina coast (Leece). Atti Relaz Accad Pugl Sci 26: 1–33 (in Italian)

Van den Hoek C (1963) Revision of the European species of *Cladophora*. In: Brill EJ (ed) Brill Publ, Leiden, 248 pp

Vazzoler S, Costa F, Bernardi S (1987) Lagoon of Venice and transfer of freshwaters and pollutants. Ist Veneto Sci Lett Arti 11: 81–124 (in Italian)

Zanardini GG (1876) Iconographia phycologica mediterraneo adriatica, vols 1–3. Venezia

Zucchetta (1983) The pollution of the Venice lagoon, Ateneo Veneto 21: 1–44 (in Italian)

16   The Northern Adriatic Sea

I. M. MUNDA

16.1   Introduction

The Adriatic Sea is a remote branch of the Mediterranean which provides conditions for the development of a heterogeneous flora and vegetation. Due to adverse paleoclimatic effects, tropical, Indo-Pacific, Atlantic, cosmopolitan and endemic species occur side by side, the last being especially well represented in the Adriatic Sea (Feldmann 1958).

The Mediterranean as a whole is a gigantic inland, sea, connected to the Atlantic through the Strait of Gibralter. During geological history, the Mediterranean and its flora and vegetation underwent several stages, from being part of the circumglobal tropical Thetys sea, to a complete closure of the strait of Gibraltar in the Pliocene, when it became an inland lake, to subsequent intrusions of Atlantic water and biota (Gierloff-Emden 1980). During glacial periods, the Mediterranean lost its tropical character. Several Atlantic species as well as glacial relics are still present in colder regions, such as the northern Adriatic.

Besides being an integral part of the Mediterranean flora and vegetation, the Adriatic flora exhibits several particular features and also reflects the recent ecological conditions in this area, such as reduced tidal movements, relatively high water temperatures and salinities with a wide range of seasonal fluctuations (Buljan and Zore-Armanda 1979) and high transparency. The Adriatic vegetation is dominated by fucoids, viz. the endemic *Fucus virsoides* in the eulittoral (Zavodnik 1967; Munda 1972), and by *Cystoseira* and *Sargassum* species in the sublittoral (Ercegović 1952; Španik 1972). Red algae are numerically prevalent in the Adriatic flora and they increase from the northern towards the southern Adriatic (Špan 1980; Špan and Antolić 1983).

Centre for Scientific Research, Slovene Academy of Science and Arts, 6100 Ljubljana, Slovenia

Ecological Studies, Vol. 123
Schramm/Nienhuis (eds) Marine Benthic Vegetation
© Springer-Verlag Berlin Heidelberg 1996

During the last few years, the benthic algal vegetation of the Adriatic Sea has changed most probably because of severe pollution impact. Especially in the shallow shelf area of the northern Adriatic, conditions become locally critical, leading towards eutrophication and anoxia (Degobbis et al. 1979; Degobbis 1989; Justič 1987; Stefanon and Boldrin 1982; Faganeli et al. 1985). This part of the Adriatic Sea receives a heavy load of domestic wastes and industrial discharges from the north Italian rivers (Po, Adige, Brenta, Piave, Tagliamento, Isonzo) as well as from direct urban effluents.

In this chapter attention will be paid to the flora of the northern Adriatic, where deleterious effects of sewage and industrial wastes on marine ecosystems are the most drastic.

Local degradation of benthic communities was found on the one hand (Fedra et al. 1976; Zavodnik 1976, 1977; Ott and Fedra 1977 and Stachowitsch 1984; Stachowitsch and Avčin 1988), and frequent and increased plankton blooms on the other (Revelante and Gilmartin 1976; Gilmartin and Revelante 1983; Fanuko 1989; Malej 1983). Large inputs of suspended and colloidally dispersed organic matter into the northern Adriatic, together with excessive phytoplankton blooms occasionally create amorphous mucose aggregations (Herndl 1988; Herndl and Peduzzi 1988; Stachowitsch et al. 1990), which have an additional impact on the benthic communities.

The increased eutrophication of the Adriatic Sea and in particular of its northern shelf area has profoundly changed the benthic algal vegetation in terms of zonation patterns, leading algal association, floristic diversity and biomass (Pignatti 1962; Golubič 1968, 1970; Munda 1974a, 1980a,b, 1982a, 1993a,b; Pignatti and de Christini 1976). The physico-chemical basis for these changes is also: the increased amounts of suspended and colloidally dispersed particles, resulting in increased turbidity; the input of fecal coliform bacteria; detritus; detergents and heavy metals.

The level of pollution in the Gulf of Trieste was estimated by Ghirardelli et al. (1973) and recently by Tušnik et al. (1989). During the last few years, the northern Adriatic was repeatedly subjected to oxygen deficiencies in the near-bottom water (Fedra et al. 1976; Faganeli et al. 1985). Critical values, where the presence of H_2S could be expected, were found locally. Simultaneously, a hypersaturation with oxygen at the surface became increasingly more apparent.

Marine benthic algae are useful as indicator organisms in monitoring various effects of pollution, as they are sessile and integrate the deleterious effects of diverse organic and inorganic pollutants over a longer time span. As such they were studied in different areas (Borowitzka 1972; Edwards 1972, 1975; Katzmann 1972, 1975; Rizzi-Longo 1972; Rueness

1973; Giaccone and Rizzi-Longo 1974; Littler and Murray 1975, 1978; Murray and Littler 1976, 1978, 1984; Bokn and Lein 1978; Fonselius 1978; Klavestad 1978; Giaccone 1981, 1991; May 1985; Marchetti 1987; Sfriso et al. 1987, 1988, 1989, 1993; Fredriksen 1987; Cormacci and Furnari 1991; Rueness and Fredriksen 1991).

In the present chapter, a descriptive account on pollution- and eutrophication-induced changes of the macrophytobenthos in the northern Adriatic is presented, while the microphytobenthos and bacteria were recently studied by Herndl et al. (1989).

16.2   Study Area and Ecological Conditions

The northern Adriatic is a shallow shelf with a maximum depth of 50 m offshore and 25 m nearshore in the study area (Fig.16.1). In the Gulf of

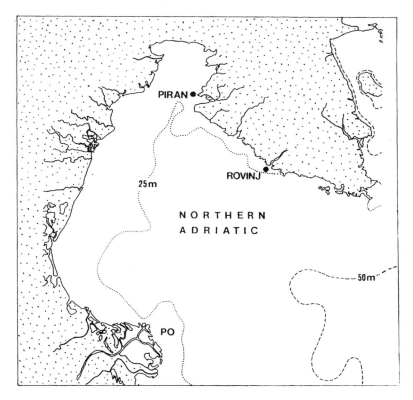

Fig. 16.1. Map of the northern Adriatic

Trieste there are only two depressions at the capes of Strunjan and Punta Madonna near Piran (Orožen-Adamič 1981).

The bottom topography of the northern Adriatic was determined by several pre- and postglacial events. The sea bottom in the Gulf of Trieste is mostly Paleocene flysch, while Cretaceous limestone prevails farther south along the Istrain coast. Alluvial Holocene sediments are usual below about 10 m depth in both areas mentioned above (Meischner 1973).

The bottom topography is responsible for the course of the main currents in the Adriatic as a whole (Zore-Armanda 1968). In the northern Adriatic, a regular circular convection current transfers river-borne and urban pollutants in the surface water layers. Local circuits of a changing direction were noticed in the Gulf of Trieste (Rajer 1990), propably responsible for the transport of pollutants during calm weather. In addition, there is, however, a cyclonic circulation over the Adriatic as a whole. In the northernmost shelf area the surface water is under the impact of freshwater runoff from the north Italian rivers, while the open northern Adriatic is influenced by advection from the south. Long-term collected hydrographic data have revealed the existence of a frontal zone between colder, low-salinity water from the north and the saltier, warm water from the south (Zore-Armanda et al. 1983). The position of this semi-permanent frontal zone is influenced by the "bora" winds, which blow in an offshore direction. In the northernmost part of the Adriatic the Po river affects the formation of a peculiar circulation. A line between the Po river mouth and Rovinj separates two gyres, an anticyclonic south and a cyclonic north of this line (Zore-Armanda and Vucak 1984). The response of this current field to the "bora" wind has been examined by current data analyses (Mosetti 1972; Zore-Armanda and Gačić 1987). The influence of the Po river, which carries a heavy load of pollutants, is, however, most pronounced during the "bora" wind and during summer stratification (Marchetti et al. 1989).

Temperature and salinity data were collected by us in the investigated area at Piran in the Gulf of Trieste (Table 16.1). Measurements were carried out at the surface and at 1,3, and 7 m depths.

In February/March the water temperature was at its minimum and increased with depth. The opposite trend was noticed in April while homeothermic conditions were indicated in May and October. During summer stratification, the water temperature decreased with depth, whereas in November the opposite trend was observed. Seasonal variations of salinity were opposite to those of the water temperature, i.e. minima in June/July and maxima in February/March. Except for August, there was a tendency of salinity increase with depth.

Table 16.1. Seasonal distribution of temperatures and salinities at different depths near Piran

Depth (m)	F	M	A	M	J	J	A	O	N
					Temperature (°C)				
0	7.8	7.7	11.5	12.5	19.9	22.5	25.5	18.5	15.0
1	8.2	7.9	10.2	12.5	19.6	21.8	25.5	18.5	15.1
3	8.4	8.5	10.1	12.2	19.4	21.5	25.0	18.3	15.1
7	8.6	9.9	10.0	12.1	18.5	20.7	24.0	18.3	15.6
Depth (m)					Salinity (‰)				
0	37.3	37.8	37.6	37.9	33.9	34.6	35.2	35.2	36.2
1	37.5	38.2	37.6	37.7	33.8	34.2	35.2	35.5	36.7
3	37.6	38.3	37.8	38.0	33.7	34.6	35.3	35.3	36.8
7	38.4	38.3	38.0	38.1	34.0	35.1	35.3	36.6	37.1

Concentrations of primary nutrients at different depths in the area around Piran are presented in Fig.16.2. The data are based on short-term measurements carried out at the marine biological station at Piran. While concentrations of inorganic nitrogen decreased with depth, the vertical distribution of phosphorus did not show significant trends. Further available data for the same area are given in Tables 16.2 and 16.3. A decrease in inorganic N and P was observed during summer, when phytoplankton blooms deplete the water of nutrients.

According to Tušnik et al. (1989), the inorganic N:P ratio of 13.3 (atomic weight) is rather close to the Redfield ratio. The largest quantities of nutrients are obviously contributed by the sewage outflow from the town of Koper (204 and 42 t year^{-1} for nitrogen and phosphorus, respectively) into this part of the northern Adriatic.

Table 16.2. Average values of primary nutrients in the surface water at Piran

	PO_4-P	NO_3-N	NO_2-N
January	0.24	4.13	0.99
February	0.27	3.08	0.28
March	0.25	2.19	0.15
April	0.12	1.88	0.09
May	0.06	1.86	0.01
June	0.03	0.53	0.06
July	0.01	0.31	0.01
August	0.02	0.21	0.01
October	0.28	2.30	0.04
November	0.31	4.39	0.04

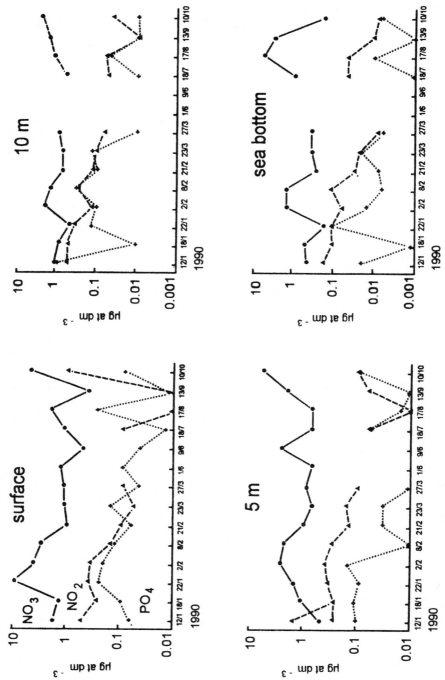

Fig. 16.2. Seasonal variations in the concentration of primary nutrients at Piran at 1 and 5 m (*left*); at 10 m and the bottom (*right*)

Table 16.3. Main algal associations in the northern Adriatic

	Relatively undisturbed site	Moderately polluted site	Heavily polluted site
Eulittoral	*Fucus virsoides* *Cystoseira compressa*	*Enteromorpha* spp. *Cladophora* spp. and diverse seasonal annuals *Gelidium pusillum*	*Enteromorpha* spp. *Blidingia minima*
Eulittoral/sublittoral junction	*Ceramium* spp. and *Polysiphonia* spp. *in spring* *Laurencia obtusa* *in summer*	*Ulva rigida* and *Scytosiphon lomentaria*	*Ulva rigida* and *Scytosiphon lomentaria* or *Mytillus galloprovincialis*
Upper sublittoral	*Cystoseira compressa* *C. barbata* *C. adriatica*	*Halopteris scoparia* *Dictyota dichotoma* *Padina pavonica* or *Halopythis incurvus*	Bare slopes with sea urchins
Lower sublittoral	*C. spicata* *C. crinita* *C. corniculata* *Sargassum* species Crustose floristic elements as undergrowth	Crustose floristic elements (*Peyssonelia* species, *Zanardinia prototypus*, crustose corallines), *Codium vermilara*	Bare slopes

Further, extensive data on temperature and salinity conditions and the distribution of primary nutrients in the northern Adriatic were collected during the cruises with the research vessel *Vila Velebita* (Škrivanič and Barič 1979).

At Piran in the Gulf of Trieste, a heavy load of domestic wastes is the main source of pollution. Pollutants from sewage outlets, fish-canning, ship-building and chemical industries are transferred to this area also from the neighbouring harbours of Isola, Koper and Trieste. The Istrian area around Rovinj receives pollutants from a fish canning factory ("Mirna") and from domestic sewage, which increased drastically during the last few years due to tourism and extended urbanization of the area. The tidal range in the investigated areas of the northern Adriatic is about 25 cm around Rovinj and 95 cm farther north, in the Gulf of Trieste.

16.3   The Benthic Algal Vegetation of Undisturbed Sites

A high degree of endemism, high florisitic diversity and predominance of red algae are characteristic features of the Adriatic flora and vegetation. Before the severe pollution impact the Adriatic benthic algal vegetation was dominated by fucoids. This is still the case in undisturbed sites, distant from sources of pollution, where representatives of the polymorphic genus *Cystoseira* colonize the grater part of the sublittoral slopes and occupy a similar level (niche) as Laminarians from the Atlantic coasts. In the lower sublittoral, *Sargassum* species were rather common (Ercegović 1952; Špan 1972). In the eulittoral, the endemic fucoid *Fucus virsoides* is common in the northern Adriatic (Zavodnik 1967; Munda 1972). It extends, however, as far south as to Dubrovnik, but is most abundant in the northern regions. The northern Adriatic flora and vegetation were investigated by us in the Gulf of Trieste, around Piran, along the Istrian coast and on some islands in the Gulf of Quarnero.

Earlier, extensive floristic and vegetational data for the area around Rovinj refer to unpublished diaries of P.Kuckuck and are based on his field studies in the area between 1884 and 1901. Older published data for the Adriatic are available in the classical works of Hauck (1885) and Lorenz (1863), in the extensive compilatory contribution of Vatova (1928) as well as in the works of Techet (1906), Schiller (1914), Vouk (1914), Schiffner (1916), Schiffner and Vatova (1938), and some others.

More recent data dealing with a rather undisturbed flora and vegetation refer to the surroudings of Rovinj (Munda 1972, 1973, 1975, 1979), the Quarner region (Rizzi-Longo 1972) and the coast below Velebit (Zalokar 1942). Algal associations at Piran were studied by Vukovič (1980).

A complete floristic list for the Adriatic as a whole was published by Giaccone (1978) and is based on different records through the last decades.

In undisturbed sites of the northern Adriatic, the level of the littoral fringe is usually populated by *Catenella caespitosa*, which occurs in patches. The most characteristic feature is, however, the *Fucus virsoides* association (Zavodnik 1967; Munda 1972) which is prolific in moderately exposed habitats. The *Fucus* settlements are less dense in exposed sites. Their biomass varied between approximately 2000 g m^{-2} under optimum conditions, and 200 g m^{-2} under conditions of strong exposure (Munda 1972, 1973, 1990).

In some highly exposed habitats *Fucus* settlements are replaced by either *Cystoseira compressa* or *C. adriatica*. The floristic diversity within these high-level Fucacean associations decreases with increasing exposure of the habitats, but is in all cases highest in spring, due to the admixture of several seasonal floristic elements belonging to the genera *Ceramium, Callithamnion, Griffithsia, Champia, Chylocladia, Antithamnion, Polysiphonia,* and *Ectocarpus.*

Cladophora dalmatica is common in spring, in both the epiphytic cover of *Fucus* and in small eulittoral depressions. In the understorey of the high-level Fucacean associations, several crustose floristic elements are common (*Phymatolithon lenormandii, P. Polymorphum,*

Hildenbrandia rubra, Lithoderma sp., *Ralfsia verrucosa, Lithophyllum incrustans)* and are shared with the upper sublittoral. Several high-level spring annuals can likewise join the lowermost stratum of the associations, such as *Bangia atropurpurea, Ulothrix* species, *Porphyra leucosticta, Blidingia minima,* dwarf *Ulva rigida* and various *Enteromorpha* species.

The level of eulittoral/sublittoral junction, i.e. the intermediate zone between *Fucus virsoides* and *Cystoseira* belts, is usually occupied by various spring annuals, first of all by *Ceramium rubrum, C. diaphanum, Callithamnion corymbosum, Polysiphonia furcellata, P. elongata, Chylocladia verticillata, Champia parvula, Lomentaria clavellosa, Polysiphonia sertularioides, Gastroclonium clavatum, Nitophyllum punctatum, Antithamnion, Aglaothamnion* and *Griffithsia* species. During summer, this level is usually colonized by *Laurencia obtusa, Acetabularia acetabulum* and *Nemalion helminthoides.*

The sublittoral of the northern Adriatic is dominated by associations of various *Cystoseira* species, which occur on hard substrata of undisturbed sites. They were even more prolific before the impact of severe pollution (Munda 1979). *Cystoseira barbata* (Pignatti 1962) is dominant among the *Cystoseira* species of the northern Adriatic.It prefers moderately exposed sites and its association normally occupies the greater part of the sublittoral slopes, i.e. between belts of *Fucus virsoides* and *Sargassum* species or other low-level *Cystoseira* stands. The biomass within this association was high, with pronounced summer maxima and ranged from approximately 2800 to 9800 g m^{-2} on a fresh weight basis (Munda 1973, 1979).

The highest floristic diversity within this as well as other *Cystoseira* associations was obvious in the strata of the epiphytes and companion species, with peaks during spring (Munda 1979). The biomass maxima during summer usually coincide with minima in species diversity in most *Cystoseira* associations from the northern Adriatic. Next in frequency to the *Cystoseira barbata* association is that of *Cystoseira spicata* (*C. stricta* var. *spicata*) which is a vicariate to the latter in exposed habitats. Its zonal position is usually between the high-level *Cystoseira compressa* association and that of *Sargassum* species on rocks or of *Cymodocea nodosa* on sandy substrata. A high biomass, upto 7000 g m^{-2} fresh weight was found in summer. Where the exposure is not too extreme, a gradual transition, viz. mixed stands of *Cystoseira barbata* and *C. spicata* was found on the mainland coasts. On the islands an improverishment of floristic diversity was obvious and the understorey greatly influenced the physiognomy of the settlements.

The vicariate species to the Adriatic *Cystoseira spicata* are *C. stricta* and *C. mediterranea* in the western Mediterranean and *C. amentacea* in the eastern Mediterranean. They form similar associations under comparable ecological conditions.

The *Cystoseira compressa* association appeared in two different variants. Whereas the high-level variant is characteristic of exposed sites, the sublittoral one is common in sheltered habitats and is situated higher up the sublittoral than *Cystoseira barbata*. Nitrophilous codominants (*Dictyota dichotoma, Halopteris scoparia, Ulva rigida*) were already found within this association before the impact of severe pollution.

Further Fucacean associations, characteristic of undisturbed habitats, are those of *Cystoseira adriatica*, and *C. crinita*, which can tolerate a wide scale of exposure conditions and are found at different levels throughout the sublittoral. *C. corniculata* usually grows in deeper water layers. The low sublittoral *Cystoseira* species, viz. *C. ercegovichii* and *C. fucoides*, appear usually in patches or as single specimens in between *Sargassum* stands and do not deserve an association rank in the northern Adriatic.

Characterisitc of deeper water layers of undisturbed habitats, and apparently most sensitive, are settlements of *Sargassum* species, viz. *Sargassum acinarium* and *S. hornschouchii*. Their associations were described by Munda (1979). It is obvious that seasonal variations in biomass and floristic composition of the low-level Fucacean associations are less pronounced than in those occupying the upper sublittoral slopes.

It is noteworthy, however, that several tropical floristic elements were found in the lower sublittoral of the northern Adriatic, such as *Halimeda tuna, Udotea petiolata, Anadyomene stellata* along with several *Codium* species and the prostrate green alga *Palmophyllum crassum*.

A scheme of the zonal distribution of main algal belts in undisturbed sites is presented in Fig. 16.3. A clear zonal distribution of the main fucoids is obvious along with an intermediate red algal belt at the level of the eulittoral/sublittoral junction, and the dominance of crustose floristic elements in deeper water layers (crustose corallines, *Peyssonnelia polymorpha P. rubra, P. squamaria, Zanardinia prototypus*). *Corallina officinalis* was characteristic for undisturbed sites under conditions of extreme surf, mainly on steep rocky surfaces around the eulittoral/sublittoral junction and in the upper sublittoral.

16.4   Changes of the Benthic Algal Vegetation Under the Influence of Eutrophication

By the late 1970s, profound changes in the benthic algal vegetation had become obvious. They comprised a disappearance of the main Fucacean

Undisturbed site

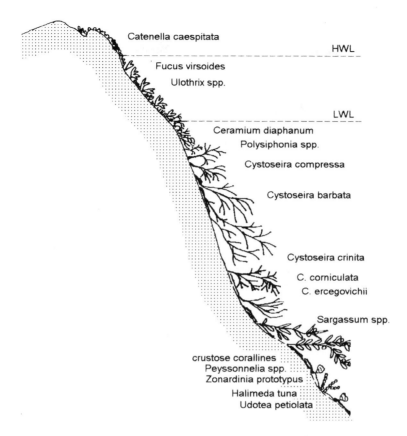

Catenella caespitata

HWL

Fucus virsoides

Ulothrix spp.

LWL

Ceramium diaphanum

Polysiphonia spp.

Cystoseira compressa

Cystoseira barbata

Cystoseira crinita

C. corniculata

C. ercegovichii

Sargassum spp.

crustose corallines

Peyssonnelia spp.

Zonardinia prototypus

Halimeda tuna

Udotea petiolata

Fig. 16.3. Scheme of zonation pattern at an undisturbed site; *HWL* high water level; *LWL* low water level

association, decreased biomass, changed zonation patterns, uplift of the lower limit of the vegetation and decreased floristic diversity. Sensitive species, first of all several red algae, disappeared and some declined in quantity. Species with a wide ecological valence are able to prosper in eutrophicated biotopes and replace the previous Fucacean stands and red algal belts.

Even in relatively undisturbed sites, distant from the outfalls of sewage (Fig.16.4) several changes became obvious, first of all the admixture of nitrophilous species into the Fucacean populations. The *Fucus virsoides* stands were less dense and considerable amounts of *Ulva rigida* and *Enteromorpha* species were obvious in both, the eulittoral and upper sublittoral. The intermediate red algal belt was reduced or absent. Populations of *Dictyota dichotoma* and *Halopteris scoparia (Stypocaulon*

Relatively undisturbed site

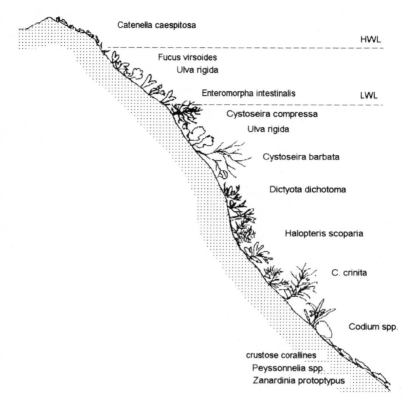

Catenella caespitosa

HWL

Fucus virsoides
Ulva rigida

Enteromorpha intestinalis   LWL

Cystoseira compressa
Ulva rigida

Cystoseira barbata

Dictyota dichotoma

Halopteris scoparia

C. crinita

Codium spp.

crustose corallines
Peyssonnelia spp.
Zanardinia protoptypus

Fig. 16.4. Scheme of zonation pattern at a relatively undisturbed site, distant from the outfalls; *HWL* high water level; *LWL* low water level

scoparium) occurred in patches, interrupting the *Cystoseira barbata* and *C. crinita* associations. In the lower sublittoral, i.e. between 7 and 10 m depth, deep-water *Cystoseira* species *(C. ercegovichii, C. fucoides)* and *Sargassum* species were absent and a rise of the lower limit of macro-algae was noted. Crustose floristic elements, mentioned above, were dominant here, whereas they formed the undervegetation of fucoid stands in undisturbed sites. *Codium* species *(C. bursa, C. vermilara)* became outstanding in the lower sublittoral, due to lack of competition with fucoids.

Changes were even more pronounced in moderately polluted sites, close to outfalls in the towns and harbours (Fig. 16.5). Here, the level of the littoral fringe was frequently occupied by Cyanobacteria *(Rivularia*

Moderately polluted site

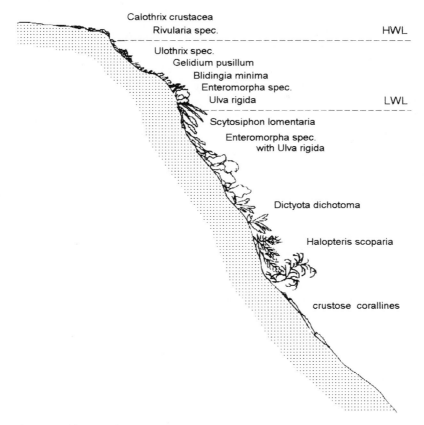

Calothrix crustacea
Rivularia spec.   HWL

Ulothrix spec.
Gelidium pusillum
Blidingia minima
Enteromorpha spec.
Ulva rigida   LWL

Scytosiphon lomentaria
Enteromorpha spec.
with Ulva rigida

Dictyota dichotoma

Halopteris scoparia

crustose corallines

Fig. 16.5. Scheme of zonation pattern at a moderately polluted site; *HWL* high water level; *LWL* low water level

spp., *Microcoleus chtonoplastes, Calothrix crustacea, Lithoderma adriaticum* and *Oscillatoria* spp.). In the eulittoral, the *Fucus virsoides* and/or *Cystoseira compressa* associations deteriorated. They were replaced by seasonal annuals, mainly by green algae belonging to the genera *Blidingia, Enteromorpha, Cladophora* and *Ulva.* Perennial turf-like mats of *Gelidium pusillum* with interwoven *Gelidiella* species and *Laurencia papillosa* occupied rocky surfaces in the lower eulittoral. During spring, seasonal species, which could form the understorey of *Fucus virsoides* stands in undisturbed sites, occupied eulittoral slopes in distinct zones: *Bangia atropurpurea, Ulothrix* species, *Blidingia minima, Porphyra leucosticta* (in patches), *Enteromorpha intestinalis,* dwarf *Ulva rigida* and lowermost *Scytosiphon lomentaria.*

The algal belts in the eulittoral zone shift with the seasons. A complete zonation, as outlined above, was found in spring and was followed by dense mats of *Blidingia minima* and *Enteromorpha* species, whereas during the summer the eulittoral slopes used to be bare of algal growth. Due to lack of competition with other macrophytes, Cyanobacteria locally occupied the eulittoral slopes in dense mats during summer, for example, *Oscillatoria nigroviridis* and other *Oscillatoria* species.

The level of the eulittoral/sublittoral junction was devoid of the prolific red algae belts, viz. *Ceramium, Polysiphonia, Chylocladia, Champia, Griffithsia, Antithamnion* and *Callithamnion* species during spring and *Laurencia obtusa* with *Wrangelia penicillata* during summer. The latter species extended, however, to the upper sublittoral. In moderately polluted sites, the vegetation was depleted of this characteristic intermediate belt. This level was usually bare during summer, whereas in spring prolific stands of *Scytosiphon lomentaria* and *Ulva rigida* together with some *Enteromorpha* species were common. This vernal association, characteristic of polluted habitats, extends locally into the upper sublittoral. *Scytosiphon lomentaria* was likely to increase in size and density of the settlements with increasing eutrophication, reaching its most prolific growth in the harbour waters. *Ulva rigida* was found in patches throughout the sublittoral slopes also in moderately polluted habitats.

The main characteristic feature of moderately polluted sites in the northern Adriatic are associations of *Halopteris scoparia* (*Stypocaulon scoparium*) and of *Dictyota dichotoma*. They cover the sublittoral slopes in a mosaic-like pattern, interrupting *Cystoseira* stands. In Rovinj, however, the sublittoral was devoid of fucoids during the period of the most severe pollution impact, in the late 1970s and early 1980s. Later, a partial restitution of both *Cystoseira* and *Fucus virsoides* populations took place, although they were subordinate in the vegetation. In the Gulf of Trieste, *Cystoseira* species, above all *C. compressa* and *C. barbata*, were still relatively abundant. There, *Halopythis incurvus* was outstanding in the sublittoral vegetation, forming dense stands. Erect corallines, such as *Corallina officinalis* and *Jania rubens*, were likewise prolific in some spots, while they were rare around Rovinj. Further floristic elements, which occurred in notable amounts in moderately polluted habitats of the northern Adriatic were *Dictyopteris membranacea, Padina pavonica, Ulva rigida, Pterocladia capillacea, Codium fragile* var. *tomentosoides, Hypnea musciformis*. The last species was locally belt-forming during summer at the level of the eulittoral/sublittoral junction.

The spring annual *Nitophyllum punctatum* was abundant at lower water levels also during summer, while it deteriorated near the surface. Also *Ceramium ciliatum*, which was found partly embedded in sand, showed a relatively great tolerance to eutrophication. In the lower sublittoral, *Colpomenia sinuosa* occupied considerable surfaces, particularly in the area around Rovinj.

It was obvious that the deterioration of fucoid stands was more severe around Rovinj than in the Gulf of Trieste. An extinction of deep-water fucoids was, however, observed in both areas. The main remnants at these levels were crustose floristic elements (*Lithophyllum incrustans, Pseudolithophyllum expansum, Lithophyllum fasciculatum, Peyssonnelia* species, *Zanardinia prototypus*), which had previously formed their understorey.

In heavily polluted sites, changes in vegetation patterns were even more extreme (Fig. 16.6). The eulittoral zone became colonized by opportunistic annual algae and was devoid of perennial ones, except for occasional crusts of red and brown algae. *Blidingia minima, Enteromorpha intestinalis, Cladophora dalmatica, C. ruchingeri* and dwarf *Ulva rigida* were usually found at this level during spring. In the

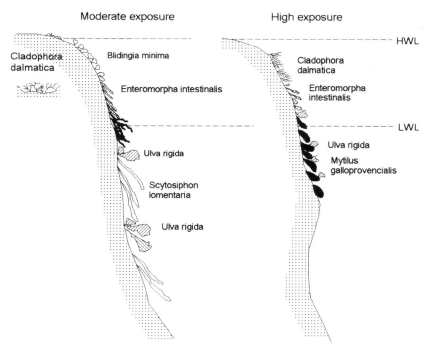

Fig. 16.6. Scheme of zonation pattern at a heavily polluted site; *HWL* high water level; *LWL* low water level

Rovinj area *Cladophora* species occupied wide eulittoral surfaces in heavily polluted habitats.

A characteristic feature of heavily polluted and rather exposed sites was the dominance of macro-invertebrates, above all *Mytilus galloprovincialis* Lam. A monopolization of eulittoral and upper sublittoral slopes by *Mytilus* displaced most of the macroalgae. *Ulva rigida* and some *Enteromorpha* species were found attached to the shells. Further down the sublittoral, bare slopes were encountered, grazed by sea urchins, with only some remnants of crustose corallines.

On shores devoid of *Mytilus*, the above named *Scytosiphon lomentaria-Ulva rigida* association was prolific and extended into the sublittoral zone. *Enteromorpha* species were its only companions. In several heavily polluted sites with increased turbidity, *Enteromorpha* species formed dense stands, replacing *Cystoseira* species.

Looking at the schematic profiles of the vegetation in differently polluted habitats (Figs. 16.3. to 16.6), a rise of the lower vegetation limit with increasing pollution is evident. Thus, increasing eutrophication is likely to exclude perennial algae from the vegetation in favour of short-lived species with simpler thalli. Under extreme conditions an intensive settling of macro-invertebrates exerts an additional deteriorating effect on benthic macroalgae. Intermediate conditions, however, favour the development of dense settlements of nitrophilous species, which replace or interrupt *Cystoseira* stands. As mentioned above, associations of *Dictyota dichotoma* and *Halopteris scoparia* and locally also of *Halopythis incurvus* are the main characteristic features of moderately polluted habitats in the northern Adriatic.

16.5   Number of Species

The effect of eutrophication on the total number of species from a moderately polluted site near Piran is presented in Fig. 16.7A,B. The number of species shifted with the seasons and differences between the littoral levels were observed (Munda 1993a,b).

The pollution impact has eliminated several species from the vegetation and given preference to opportunistic species with a short life span and a wide ecological valence. Sensitive species, which disappeared from the vegetation, were deep-water fucoids (*Sargassum acinarium, S. hornschouchii, Cystoseira fucoides, C. ercegovichii*) along with *Liebmannia leveillei, Phyllophora palmetoides, Rytiphlaea tinctoria, Acrosymphytum*

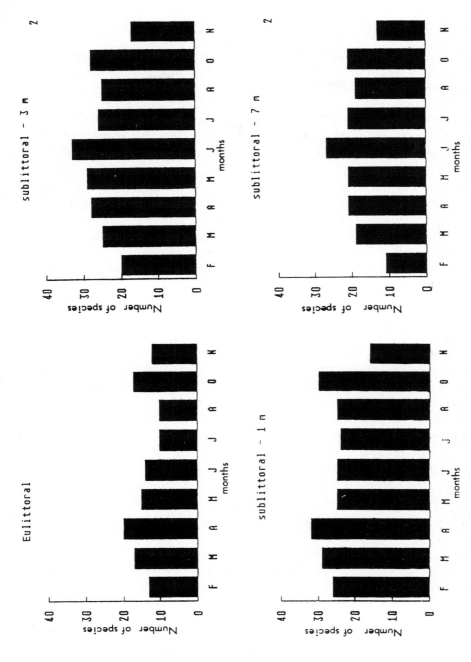

Fig. 16.7. Seasonal variations in the number of species at different depths. A at 0 and 1 m; B at 3 and 7 m. (Munda 1993a)

purpuriferum, Dudresnaya verticillata and several representatives of the genera *Ceramium, Polysiphonia, Dasya, Antithamnion, Callithamnion, Aglaothamnion, Bryopsis* and some others. Some of the species declined in quantity and were only found in single specimens in undisturbed or relatively undisturbed localities, distant from the outfalls.

The number of species was lowest in the eulittoral zone, which was devoid of *Fucus virsoides* (Fig. 16.7A). Seasonal maxima were found in April and October. The number of species was notably higher at 1 and 3 m depths and decreased again in the lower sublittoral, at 7 m depth. In the eulittoral and at 1 m depth, seasonal maxima were found during the same seasons, i.e. in April and October, due to the appearance of spring annuals and several green algae in autumn. At 3 and 7 m depths, the seasonal maxima in species number occurred in June.

The decrease in the species number in the lower sublittoral, at 7 m, is due to the disappearance of deep-water fucoids along with their epiphytic cover. Crustose floristic elements predominate at this level and thus seasonal variations were less pronounced than at 3 m depth.

At the intermediate depth of 1 and 3 m the number of recorded species also varied due to changes in the epiphytic cover of the still present *Cystoseira* species and *Halopythis incurvus*.

Looking at the heavily polluted sites, the number of species notably decreased due to the eradication of all the fucoids and red algae from the residual vegetation. *Scytosiphon lomentaria* and *Ectocarpus siliculosus* were the only representatives of the Phaeophyta there, with occasional findings of *Pilayella littoralis*. Among the Chlorophyta, representatives of the genera *Ulva, Enteromorpha, Blidingia, Chaetomorpha* and *Cladophora* were common and Cyanobacteria proved also to be successful in some eutrophicated habitats (*Oscillatoria, Lyngbya* and *Spirulina* species).

A significant decline in the number of species towards the sources of pollution was demonstrated also for the harbours of Makarska and Split in the middle Adriatic (Špan and Antolić 1991). In the harbours, a twofold decrease in species number was recorded compared to areas just outside, and again a 2.5- to 3-fold decrease close to the sewage outlets. The decrease in species number was most pronounced for the red algae and less for the brown and green algae. The latter remained numerically constant within the harbour and close to the sources of pollution. In control areas Rhodophyta were numerically dominant, while they were in a minority within the harbour areas, where finally Chlorophyta dominated.

Long-term observations in other areas (Rueness and Fredriksen 1991) have revealed a prominent decrease in species number in the Oslofjord during the last four decades. These changes were particularly conspicuous below 10 m depth.

16.6 Algal Biomass

The northern Adriatic benthic algal vegetation exhibited relatively high biomass values, in particular in the upper sublittoral (Špan 1969; Munda 1972, 1973, 1979).

A further effect of eutrophication is obvious in changes in the algal biomass. A degradation or absence of associations of perennial algae led to a decrease in biomass in most habitats. In heavily polluted sites, a transitent increase in the fresh weight biomass was observed for annual species, which prosper in nutrient-rich habitats (e.g. *Ulva rigida, Enteromorpha intestinalis, E. prolifera, E. clathrata, Blidingia minima,* several *Cladophora* species, *Ectocarpus siliculosus, Pilayella littoralis* and *Scytosiphon lomentaria*; Fig. 16.8).

Some data on annual variations in biomass in relation to exposure for the area around Rovinj are presented in Fig. 16.8. In the late sixties, the eulittoral zone was dominated by *Fucus virsoides*. Year to year variations exhibited a gradual decrease in biomass of the *Fucus* settlements along the mainland coasts, whereas on the islands an increase was observed. In highly exposed habitats, the *Fucus* settlements were replaced by *Ceramium diaphanum* during the years before the severe pollution impact. In the late 1970s when observations in the same area were repeated the *Fucus* settlements were replaced by stands of *Blidingia minima, Cladophora* species and *Enteromorpha intestinalis.* The average fresh weight biomass of the eulittoral algal settlements decreased on the islands and in highly exposed mainland habitats. At other mainland sites the fresh weight biomass of the spring annuals exceeded that of the previous *Fucus* stands or was at the same level. The average yearly biomass values are based on vernal observations, since during summer the eulittoral slopes were bare.

At the eulittoral/sublittoral junction *Ceramium* species along with the other red algae declined gradually during the first years of the field studies. After the severe pollution impact they were replaced by the association of codominant *Scytosiphon lomentaria–Ulva rigida* with accompanying *Enteromorpha* species. Thus the fresh weight biomass was approximately fourfold that of the previous red algae stands during spring (Fig. 16.8).

In the upper sublittoral, *Cystoseira* settlements were dominant, with irregular year to year variations in both exposed and sheltered habitats. *Cystoseira adriatica* and *C. crinita* were dominant in the former, *C. compressa* and *C. barbata* in the latter habitats. These *Cystoseira* stands deteriorated after the severe pollution impact in the Rovinj area and were

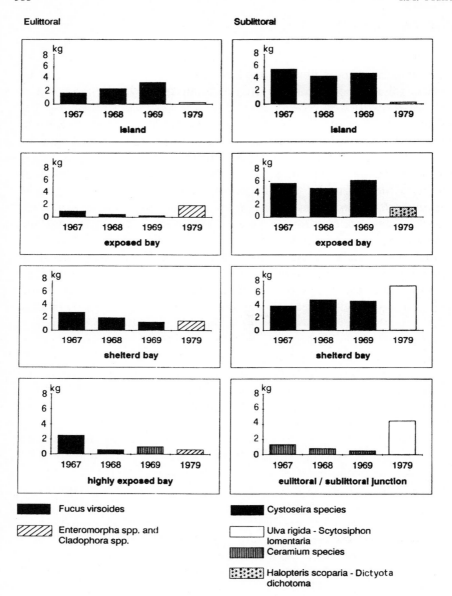

Fig. 16.8. Annual variations in fresh weight algal biomass (kg m^{-2}) in the eulittoral and sublittoral. (Munda 1993b)

absent from the vegetation in the late 1970s and early 1980s. They were replaced by an association of *Halopteris scoparia* (*Stypocaulon scoparium*) and *Dictyota dichotoma*, which exhibited a notably decreased biomass. In some sheltered and heavily polluted sites, the above-mentioned

Scytosiphon lomentaria – Ulva rigida association extended into the upper sublittoral. Its fresh weight biomass was notably higher than that of the previous *Cystoseira* stands. The data collected refer to averages during spring months, from February to June.

Seasonal variations in fresh weight biomass at different levels presented in Fig. 16.9a–c refer to observations in the polluted area around Piran in the Gulf of Trieste (Munda 1993a,b). In the eulittoral, the biomass was low, with a seasonal maximum during spring (April), coinciding with the appearance of spring annuals. The main algal biomass was concentrated at intermediate depths of 1 and 3 m (Fig. 16.9 A,B). *Cystoseira barbata, Halopteris scoparia, Dictyota dichotoma, Padina pavonica* and *Ulva rigida* contributed the major part to the biomass at these levels. Minor contributors were *Dictyopteris membranacea, Pterocladia capillacea, Cladophora rupestris* and *Padina pavonica*. At 1 m depth, the seasonal maximum was found on account of spring annuals in April. At 3 and 7 m, a biomass maximum occurred later, in May/June, possibly due to the fully grown and fruiting *Cystoseira* thalli, which grew

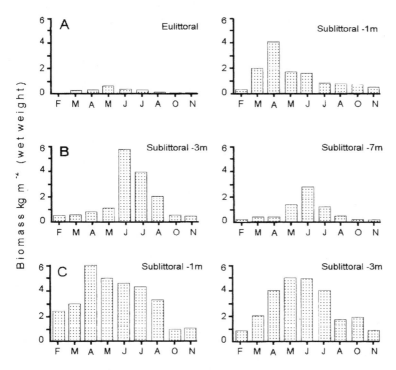

Fig. 16.9. Seasonal variations in fresh weight biomass (g m^{-2}). **A** at 0 and 1 m; **B** at 3 and 7 m; **C** at 1 and 3 m in a relatively undisturbed sites. (Munda 1993a, b)

in patches at these levels. The biomass was notably lower at 7 m, where crustose floristic elements dominate.

In a moderately polluted habitat from the same area (Fig. 16.9C), the biomass was higher at the two intermediate depths of 1 and 3 m. *Halopythis incurvus* contributed a great proportion to the total biomass. *Cystoseira species* were better represented here than in the heavily eutrophicated locality. In the eulittoral and at 7 m depth, differences between the transects were less conspicuous.

A scheme of the average percentage species composition of the fresh weight biomass is given in Fig. 16.10 (Munda 1993b). In the eulittoral, *Fucus virsoides* dominated in undisturbed sites, with a small proportion of the green algal component. With increasing pollution the proportion of the green algae within the *Fucus* settlements increased and turf-like mats of *Gelidium pusillum* became notable. In moderately polluted sites, these *Gelidium* mats may contribute the greater part to the total biomass (60%), while the rest are green algae (*Blidingia minima, Ulva rigida, Enteromorpha* spp. and *Cladophora* species) Green algae could be predominant in the biomass of some sites. In heavily polluted sites the eulittoral slopes were covered by diverse spring annuals, but the bulk of the biomass consisted of nearly 100% of either *Blidingia minima* or *Cladophora* species.

In undisturbed sites, the level of the eulittoral/sublittoral junction was covered by either *Ceramium* or *Polysiphonia* species, each of them contributing 100% to the total biomass of the transects investigated.

In moderately polluted sites, the algal biomass was dominated by *Scytosiphon lomentaria* and green algae (*Ulva rigida, Enteromorpha* spp.) occurring in different proportions. In heavily polluted sites, this level was usually occupied by *Mytilus galloprovincialis. Ulva rigida*, attached to it, formed only a negligible proportion of the total biomass.

In the upper sublittoral, *Cystoseira* species dominated (*C. compressa, C. barbata*), with a small proportion of *Corallina officinalis* and locally also *Dictyota dichotoma* in the fresh weight biomass. In moderately polluted sites, this level was occupied by mixed populations, dominated either by *Halopteris scoparia* or *Dictyota dichotoma*. Other components contributing to the biomass could be *Halopythis incurvus*, green algae and *Cystoseira* species, which were, however, in the minority. Locally, *Ulva rigida* also contributed to the total biomass.

It is noteworthy that *Pterocladia capillacea* and *Padina pavonica* occurred, in terms of fresh weight biomass, in negligible quantities and were not included in the measurements.

In heavily polluted sites, the biomass was negligible due to grazing by sea urchins, crustose corallines being the only remnants of the previous

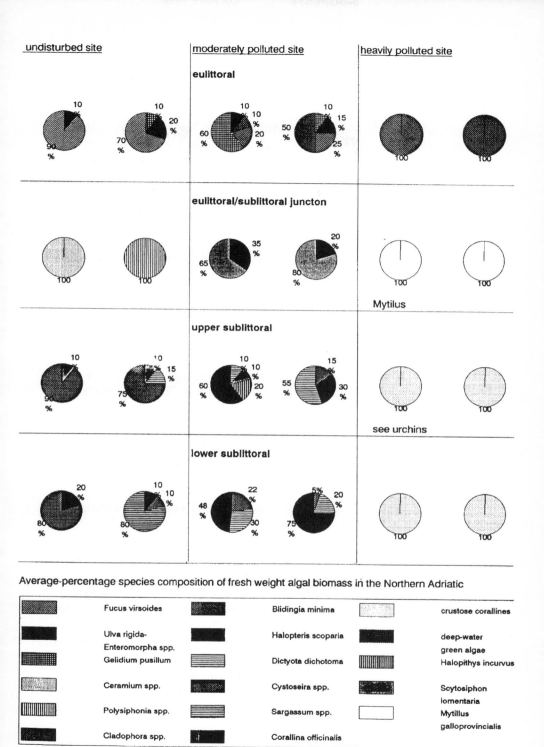

Fig. 16.10. Average percentage species composition of algal biomass (fresh weight) in the northern Adriatic. (Munda 1993b)

macroalgal settlements. The same holds true for the lower sublittoral. There, *Cystoseira* species were dominant in undisturbed sites (*C. corniculata, C. crinita*); and *Dictyota dichotoma* under the influence of slight pollution. Further contributors to the fresh weight biomass at this level were *Halopteris scoparia* and deep water green algae (*Halimeda tuna, Anadyomene stellata, Udotea petiolata, Codium* species). In moderately polluted sites either *Halopteris scoparia* or *Dictyota dichotoma* dominated in the total biomass with an admixture of the above-mentioned green algae. In the schemes, only components which were conspicuous in the total fresh weight biomass along the studied transects were considered.

16.7   Possible Causes for the Extinction of Fucoids in the Eutrophicated Habitats

A combination of environmental stresses, caused by the discharge of sewage, agricultural runoff or industrial effluents has profoundly changed the benthic algal vegetation of the north Adriatic shelf area. An analysis of the causality of these changes is a difficult and complicated task. Field studies revealed that fucoids disappeared from areas affected by sewage in accordance with their possibly genetically conditioned susceptibility. On the basis of the distribution pattern of individual species, a sensitivity scale was worked out (Munda 1982a). *Fucus virsoides* is probably the most resistant among the fucoids from the Rovinj area and is able to tolerate a certain degree of pollution. This might be in accordance with its relatively wide ecological valence, as it tolerates lowered salinities as well. Next to it were *Cystoseira compressa* and *C. barbata*, followed by *C. spicata, C. adriatica* and *C. crinita*. The highest susceptibility towards organic pollution was found in *Sargassum* species as well as in the deep water *Cystoseira* species (*C. fucoides, C. ercegovichii*). It seems likely, however, that sensitivity towards organic pollution and eutrophication increases with the depth penetration of the fucoids, which means that eulittoral fucoids are more tolerant compared to the sublittoral ones.

Chemical analyses of diverse benthic algae from polluted habitats have revealed certain differences from those growing in unpolluted areas. Changes in chemical composition under the influence of eutrophication proved to be consistent with those induced by lowered salinities. A pronounced increase in protein content was found in both brown and green algae from polluted habitats (Munda 1988, 1990). The mannitol content in brown algae was usually reduced compared to the controls, indicating a decreased assimilatory activity. In *Fucus virsoides* plants protruding into low-salinity areas, similar changes in the chemical composition were noted, along with a pronounced reduction in the thallus size, related to the salinity gradient. A similar reduction in size was found also in plants from moderately polluted habitats.

Autecological tests carried out with fucoids in vitro (Munda 1982b) showed that excess nutrient salts as well as untreated harbour water severely depresses growth performance in *Fucus virsoides* and can be lethal for *Cystoseira* species, the latter deteriorating in nutrient-enriched media after 1 to 2 weeks. *Fucus virsoides*, on the other hand, was able to survive, growing slightly, when nitrate was supplied separately. In the experiments, growth was reduced at the higher temperature, i.e. at 10°C. The separate addition of phosphate revealed another pattern: a considerable temperature–enhanced growth performance at its lower concentration and severe growth depressions at the higher concentration at 5°C, or lethal effects at 10°C. A combination of both salts severely decreased growth rates and induced fruiting at 5°C, but was lethal at 10°C.

In general, susceptibility towards organic pollutants (untreated harbour water) or excess nutrients increases with temperature. The higher experimental temperature of 10°C, however, corresponds to the winter temperature in the northern Adriatic. We may expect that even higher temperatures would induce a lethal effect of the pollutants mentioned, without previous growth.

From this, we assume that an excess of nutrients in combination with organic substances excludes fucoids from sewage-influenced areas because of negative effects on their growth performance and metabolic activities.

Previous experiments with Atlantic fucoids also showed that untreated harbour water severely depressed the assimilation rate and at the same time enhanced respiration (Munda 1974b).

A further cause for reduced settlement of benthic marine algae is the occurrence of occasional extensive mucus aggregations, which occurred in the northern Adriatic waters during summer months of the last few years. These mucus aggregations. deriving from plankton blooms, are mixed with terrigenous humic substances and contain assemblages of phytoplankton, bacteria and protozoa, embedded into an amorphous matrix. They contribute significantly to the metabolism of the pelagic system (Herndl 1988; Herndl and Peduzzi 1988; Herndl et al. 1989). The aggregations pass through different stages, from floccula over "stringers" to large coagulations, which can cover macroalgae and benthic animals in a cobweb-like form. Such phenomena were common in the Gulf of Trieste during summer months of the last few years and have caused modifications as well as mortality within benthic communities (Stachowitsch 1984; Stachowitsch and Avcin 1988; Stachowitsch et al. 1990).

Further deteriorations of the sublittoral vegetation during summer months of the years, when extensive mucus aggregations were formed, were observed locally in the Gulf of Trieste, around Piran (M. Richter, pers. comm.).

Eulittoral surfaces covered solely by mats of Cyanobacteria, mainly *Oscillatoria* species, were followed downwards to the depth of approximately 1 m by *Licmophora gracilis* mats, in sites where the coast slopes moderately. Below the *Licmophora* mats, stones were covered with patches of a low-growing form of *Cladophora* sp. with partly decayed thalli. It seems that Cyanobacteria and diatoms thrive in sites affected by mucus aggregations.

In polluted sites, a monopolization of the substrata by macroinvertebrates in the eulittoral and upper sublittoral had excluded most of the macroalgae (cf. Figs. 16.6 and 16.8), with only *Ulva rigida* and *Enteromorpha* species growing attached to the mussels. Lower sublittoral slopes, grazed by sea urchins, were bare of macroalgae. Surfaces covered by patches of *Spirulina* species were frequently observed there during the last few years, indicating polysaprobic conditions and also anoxia. Remnants of the previous undergrowth of crustose corallines were still observed. Mats of *Mytilus galloprovincialis* were common especially in polluted sites of the Rovinj area, and also on the adjacent islands. Comparable conditions have been reported for other areas, such as the Baltic and the Oslofjord (e.g. Lundälv and Christie 1986; Rueness and Fredriksen 1991).

16.8 Conclusions

Increased eutrophication in the marine environment leads to excessive algal growth in the pelagic system, including unusually large phytoplankton blooms, whereas only a few opportunistic benthic algal species are capable of mass development under such conditions(e.g. *Ulva rigida* in the lagoon of Venice).

Perennial, more sensitive species with rather complicated thalli, such as fucoids and most of the red algae, either die away or decline in quantity. Main algal associations, found before the severe pollution impact in the northern Adriatic, were eradicated from the vegetation close to the sources of pollution. They are, however, still found in relatively unpolluted sites, though with some modifications in their floristic composition and degree of cover.

There is, however, a consistency of algal associations with environmental conditions. In particular Fucacean association rather clearly reflect the pollution gradients and are able to record environmental conditions or stress over time, such as the nutrient input by sewage and the nitrate/phosphate cycles, which are again conditioned by pelagic events (cf. Schramm et al. 1988).

The *Fucus virsoides* association disappeared from polluted sites and was succeeded by diverse spring annuals and algal turfs, containing *Gelidium pusillum*, interwoven with *Gelidiella* and *Laurencia* species. Under extreme conditions, the eulittoral level was occupied by mats of *Blidingia minima, Enteromorpha intestinalis* or *Cladophora* species. The eulittoral slopes were usually bare of macroalgae during mid-summer (Munda 1972, 1980a).

Cystoseira associations were either reduced or excluded from the vegetation. Their distribution pattern in the investigated areas of the northern Adriatic revealed a scale of susceptibility towards pollution by sewage (Munda 1982a). *Cystoseira compressa* and *C. barbata* are the least susceptible among other representatives of this genus. As such, they can at first replace as vicariates other *Cystoseira* associations, when pollution is not yet too severe. Similar changes of the benthic algal associations were observed in the middle and southern Adriatic (Špan et al. 1989; Cormaci and Furnari 1991). Also in these areas of the Adriatic sea, a decrease or disappearance of *Cystoseira* associations, reduction in species number and admixture of nitrophilous species were found.

The succession in relation to pollution gradients was demonstrated for benthic communities by Pearson and Rosenberg (1978). For benthic marine algae, some characteristic associations were found in moderately polluted habitats, representing intermediate conditions between extreme pollution-induced changes and an undisturbed vegetation pattern.

Associations of the nitrophilous species *Dictyota dichotoma* and *Halopteris scoparia*, which either interrupt or replace *Cystoseira* stands, are the most conspicuous characteristic of moderately polluted habitats all over the northern Adriatic (Munda 1980a, 1993a,b). Further characteristic species in these associations were *Dictyopteris membranacea, Padina pavonica, Cladophora rupestris, Halopythis incurvus, Pterocladia capillacea, Ulva rigida* and *Hypnea musciformis*. It is noteworthy that erect corallines, for example *Jania rubens* and *Corallina officinalis*, were still relatively abundant or even belt-forming in the upper sublittoral. Furthermore, an upwards migration of deep-water species, such as *Halimeda tuna, Pterocladia capillacea, Gigartina acicularis, Gracilaria verrucosa, Codium effusum, Codium vermilara*, was recently observed. The upwards migration of these sciaphilic species might be related to the increased turbidity of the northern Adriatic water.

In heavily polluted sites, a series of disturbances is even more pronounced. Among such disturbances are excess of nutrients, input of suspended and colloidally dispersed materials, bacteria, humic substances, detergents and heavy metals, along with increased turbidity. A consequence of eutrophication is also a decreased oxygen concentration, leading locally to anoxic conditions. The interacting effects of such extremely adverse conditions have excluded fucoids and most red algae from the vegetation and totally changed the zonation patterns and the dominant algal associations (cf. Figs. 16.3 to 16.6).

Only species with a wide ecological valence are able to exist and prosper in eutrophicated habitats. We are therefore concerned mainly with seasonal floristic elements with simple thalli and a short life span, mostly vernal species. These algae are often found in estuaries and other brackish habitats, since they are able to tolerate wide ranges and rapid fluctuations of temperature, salinity, dissolved nutrients and other substances drained in by the rivers (e.g. green algae of the genera *Blidingia, Enteromorpha* and *Cladophora; Ulva rigida, Scytosiphon lomentaria, Ectocarpus siliculosus, Pilayella littoralis*). Only a few red algae can tolerate such extreme conditions. It is noteworthy that *Ceramium ciliatum*, in contrast to other species of this genus, was found partly buried in sand in highly polluted habitats.

A comparison of the benthic algal vegetation of differently polluted sites from the northern Adriatic (Munda 1993a,b) revealed different zonation patterns (Figs. 16.3 to 16.6) along with a decreased depth penetration of macroalgae. This phenomenon, caused by increased turbidity in eutrophicated habitats as well as grazing by sea urchins, was also observed in other areas (Kautsky et al. 1986; Breuer and Schramm 1988; Rueness and Fredriksen 1991; Vogt and Schramm 1991).

Pronounced year to year variations within algal associations were followed in the investigated areas of the northern Adriatic. Transient restitution of *Cystoseira* and *Fucus virsoides* stands with subsequent degradations were observed, as well as the disappearance of some species with simultaneous reappearance of others (e.g. *Nemastoma dichotoma, Petalonia fascia, Cladostephus spongiosus* f. *verticillatus, Acrosorium venulosum, Dasya* spp.). In the eulittoral, during spring and summer months, a dense girdle of *Codium fragile* var. *tomentosoides* appeared in the little harbour of Piran in 1990 to 1992 and totally disappeared later. Further examples of year to year variations are the prolific growth *Chylocladia verticillata* and *Lomentaria clavellosa* during spring of the last 2 years, the appearance of *Acinetospora crinita* in the lower sublittoral, where it covers other macro-algae with dense mats in the spring, and finally the appearance of Diatoms and Cyanobacteria, which interrupt macroalgal stands (e.g. *Licmophora gracilis* mats in the eulittoral, patches of *Oscillatoria* and *Lyngbya* in the eulittoral and upper sublittoral and *Spirulina* mats in the lower sublittoral). An interruption of the macrophytobenthos by microphytobenthos is thus a further characteristic feature of pollution- and eutrophication-induced changes during the last few years. Although it is not possible to relate these changes directly to certain environmental parameters, we might safely assume that the benthic algal vegetation of this area reflects the history of nutrient input by sewage as well as other biotic and abiotic factors in the unstable environment of the northern Adriatic shelf.

Acknowledgments. The author is grateful to Dr. Milan Orožen-Adamič, Mr. Marjan Richter and members of the diving club, Piran, for their help during the fieldwork.

References

Bokn T, Lein TE (1978) Long-term changes in fucoid associations of the inner Oslofjord, Norway. Norw J Bot 15: 9–14

Borowitzka MA (1972) Intertidal algal species diversity and the effects of pollution. Aust J Mar Freshw alter Res 23: 73–74

Breuer G, Schramm W (1988) Changes in macroalgal vegetation in the Kiel Bight (western Baltic Sea) during the past 20 years. Kiel Meeresf. Orsolv Sonderheft 6: 241–255

Buljan M, Zore-Armanda M (1979) Hydrographic properties of the Adriatic Sea in the period from 1965 through 1970. Acta Adriat 20: 1–368

Chiaudani G, Vighi M (1982) Multistep approach to identification of limiting nutrients in the northern Adriatic eutrophied coastal waters. Water Res 16: 1161–1166

Cormaci M, Furnari G (1991) Phytobenthic communities as monitor of the environmental conditions of the Brindisi coastline. Oebalia 17 Suppl: 223–235

Degobbis D (1989) Increased eutrophication of the northern Adriatic Sea – 2nd act. Mar Poll Bull 20: 452–457

Degobbis D, Smodlaka N, Pojed I, Škrivanič A, Precali R (1979) Increased eutrophication of the northern Adriatic Sea. Mar Poll Bull 10: 298–301

Edwards P (1972) Benthic algae in polluted estuaries. Mar Poll Bull 3: 55–60

Edwards P (1975) An assessment of possible pollution effects over a century on the benthic marine algae of Co. Durham, England. J Linn Soc Bot 109: 269–305

Ercegovic A (1952) Jadranske cistozire. Split. Fauna Flora Jadrana 2: 1–212

Faganeli J, Avčin A, Fanuko N, Malej A, Tušnik P, Vrisěr B, Vukovic A (1985) Bottom layer anoxia in the central part of the Gulf of Trieste in the late summer of 1983. Mar Poll Bull 16: 75–78

Fanuko N (1989) Possible relation between a bloom of Distephanus speculum (Silicoflagellata) and anoxia in bottom waters of the northern Adriatic, 1983. J Plankton Res 11: 75–84

Fedra K, Ölscher EM, Scherübel M, Stachowitsch M, Wurzian RS (1976) The ecology of North Adriatic benthic communities: distribution, standing crop and composition of macrophytobenthos. Mar Biol 38: 129–145

Feldmann J (1958) Origine et affinités du péuplement végétal benthique de la Mediterranée. Rapp Int mer N S 14: 515–518

Fonselius H (1978) On nutrients and their role as production-limiting factors in the Baltic. Acta Hydrochim Hydrobiol 6: 329–339

Fredriksen OT (1987) The fight against eutrophication in the Odense Fjord by reaping sea lettuce (Ulva lactuca). Water Sci Technol 19: 81–87

Fuks D (1979) Cruises of the research vessel "Vila Velebita" in the Kvarner region of the Adriatic Sea. VII. Total concentrations of coliforms and biochemical oxygen demand. Thalassia Jugosl 15: 143–148

Ghirardelli E, Orel G, Giaccone G (1973) L'inquinamento del Golfo di Trieste Atti Mus civ Stor Nat Trieste 28: 431–450

Giaccone G (1978) Revisione della flora del Mare Adriatico. Supplemento dell annuario 1977. In: Marine Park of Miramare Monitoring Station, vol 6, No 19 WWF Italia Trieste, 118 pp

Giaccone G (1981) Il fitobentos: indicatori biologici e mecanismi di ricciclagio nei fenomeni eutrofici. Quardorni Lab Techn. Pesca, Ancona. 3 Suppl: 385–407

Giaccone G (1991) The algae in the management and treatment of wastewater disposal. Oebalia 27 (Suppl): 121–130

Giaccone G, Rizzi-Longo L (1974) Structure et évaluation de la végétation marine dans les environments pollués. Rev Int Oceanogr Med 349: 67–72

Gierloff-Emden HG (1980) Geographie des Meeres. Ozeane and Küsten. Teil 1 and 2. Ed. W. de Gruyter, Berlin, 310 pp

Gilmartin M, Revelante N (1983) The phytoplancton of the Adriatic: standing crop and primary production. Thalassia Jugosl 19: 173–188

Golubić S (1968) Die Verteilung der Algevegetation in der Umgebung von Rovinj (Istrien) unter den Einfluss häuslicher und industrieller Abwässer. Wasser Abwasserforsch 3: 87–95

Golubić S (1970) Effect of organic pollution of benthic communities. Mar Pollut Bull 1: 56–57

Hauck F (1885) Die Meeresalgen Deutschlands and Österreichs. In: Rabenhorst L Kyptogamen – Flora von Deutschland, Oesterreich und der Schweiz. (ed) Edward Kummer, Leipzig 575 pp

Herndl GJ (1988) Diel and spatial variations in bacterial density in a stratified water column in the Gulf of Trieste. Progr Oceanogr 21: 119–129

Herndl GJ, Peduzzi P (1988) The ecology of amorphous aggregations (marine snow) in the northern Adriatic Sea. I. General considerations. PSZNI Mar Ecol 9: 79–90

Herndl GJ, Peduzzi P, Fanuko N (1989) Benthic community metabolism and microbial dynamics in the Gulf of Trieste (northern Adriatic Sea). Mar Ecol Progr Ser 53: 169–178

Justić D (1987) Long-term eutrophication in the northern Adriatic sea. Mar Poll Bull 18: 281–284

Katzmann W (1972) Regression von Braunalgenbeständen unter den Einfluss von Abwässern. Naturwiss Rundsch 5: 182–186

Kautsky N, Kaustsky H, Kautsky U, Waern M (1986) Decreased depth penetration of *Fucus vesiculosus* since the 1940s indicates eutrophication of the Baltic Sea. Mar Ecol Progr Ser 28: 1–8

Klavestad N (1978) The marine algae in the polluted inner part of the Oslo fjord. Bot Mar 21: 71–97

Littler MM, Murray NS (1975) Impact of sewage on the distribution, abundance and community structure of rocky intertidal macro-organisms. Mar Biol 30: 277–291

Littler MM, Murray NA (1978) Influence of domestic wastes on energetic pathways in rocky intertidal communities. J Appl Ecol 15: 583–595

Lorenz JR (1863) Physikalische Verhältnisse and Verteilung der Organismen im Quarnerischen Golfe. Königl Akad Wiss Wien, 379 pp

Lundälv T, Christie H (1986) Comparative trends and ecological patterns of rocky subtidal communities in the Swedish and Norwegian Skagerrak area. Hydrobiologia 142: 71–80

Malej A (1983) Observations on *Noctiluca miliaris* Suiray red tide in the Gulf of Trieste during 1980. Thalassia Jugosl 19: 261–269

Marchetti R (1987) L'Eutroficazione. F Angeli, Milano, 315 pp

Marchetti RA, Provini A, Crosa C (1989) Nutrial load carried by the river Po into the northern Adriatic Sea, 1968–1987. Mar Pollut Bull 20: 168–172

May V (1985) Observations on algal flora close to two sewage outlets. Cunninghamia 1: 385–394

Meischner D (1973) Formation process and dispersal patterns of the sediments along the Istrian coast of the Adriatic. Rapp Comm Int Mer Médit 21: 843–486

Mosetti F (1972) Alcune richerche sulle correnti del Golfo di Trieste. Riv Ital Geofis 21: 33–38

Munda IM (1972) Seasonal and ecologically conditioned variations in the *Fucus virsoides* association from the Istrian coast (northern Adriatic). Dissertationes SAZU (Slovene Academy of Science and Arts) Ljubljana 15: 1–33

Munda IM (1973) The production of biomass in the settlements of benthic marine algae in the northern Adriatic. Bot Mar 15: 37–52

Munda IM (1974a) Changes and succession of benthic algal associations of slightly polluted habitats. Rev Int Oceanogr Med 34: 37–52

Munda IM (1974b) The effect of polluted water on the assimilation rate of the brown algae *Ascophyllum nodosum* (L.) Le Jol. and *Fucus vesiculosus* L. Rev Int Oceanogr Med 34: 53–66

Munda IM (1975) Some seasonal associations of benthic marine algae from the northern Adriatic. Rapp Comm Int Mer Médit 23: 65–66

Munda IM (1979) Some Fucacean associations from the vicinity of Rovinj, Istrian coast, North Adriatic. Nova Hedwigia 31: 607–666

Munda IM (1980a) Changes of the benthic algal associations of the vicinity of Rovinj (Istrian coast, North Adriatic) caused by organic wastes. Acta Adriat 21: 299–332

Munda IM (1980b) Survey of the algal biomass in the polluted area around Rovinj (Istrian coast, North Adriatic). Acta Adri at 21: 333–354

Munda IM (1982a) The effect of organic pollution on the distribution of fucoid algae from the Istrian coast (vicinity of Rovinj). Acta Adriat 23: 329–337

Munda IM (1982b) The effect of different pollutants on benthic marine algae. Rapp Com. Int Mer Médit VI J Poll Cannes: 721–726

Munda IM (1988) A note on the chemical composition of some common benthic algae from a polluted area in the northern Adriatic (Rovinj). Rapp Comm Int Mer Médit 31: 2

Munda IM (1990) Resources and possibilities of exploitation on North Adriatic seaweeds. Hydrobiologia 204: 309–315

Munda IM (1993a) Impact of pollution on benthic marine algae from the northern Adriatic. Int J Environ Stud 43: 185–199

Munda IM (1993b) Changes and degradation of seaweed stands in the northern Adriatic. Hydrobiologia 260/261: 239–253

Murray SN, Littler MM (1976) An experimental analysis of sewage impact on macrophyte dominated rocky intertidal community. J Phycol 12: 15–16

Murray SN, Littler MM (1978) Patterns of algal succession in a perturbated marine intertidal community. J Phycol 14: 506–512

Murray SN, Littler MM (1984) Analysis of seaweed communities in a disturbed rocky intertidal environment near Whites Point, Los Angeles Calif. USA. Hydrobiologia 116/117: 374–382

Orožen–Adamič M (1981) Prispevek K poznavanju izoblikovanosti podvodnega reliefa slovenske obale. Geogr Vestnik Ljubljana 52: 39–46

Ott J, Fedra K (1977) Stabilizing properties of a high biomass benthic community in a fluctuating ecosystem. Helgol wiss Meeresunters 30: 485–494

Pearson TH, Rosenberg R (1978) Macrobenthic succession and relation to organic enrichment and pollution in the marine environment. Oceanogr Mar Biol Ann Rev 16: 228–231

Pignatti S (1962) Associazioni di alghe marine sulla costa veneziana. Mem Ist Sci Lett Atti 32: 1–234

Pignatti S, de Christini P (1976) Associazioni di alghe marine come indicatori d' inquinamento delle aque nel Vallone di Muggia presso Trieste. Arch Oceanogr Limnol 15 (Suppl): 185–191

Rajer R (1990) Mathematical modelling of currents and dispersion in the Northern Adriatic. Rapp Comm Int mer Médit 32: 179

Revelante N, Gilmartin M (1976) The effects of Po river discharge on phytoplancton dynamics in the northern Adriatic Sea. Mar Biol 34: 259–271

Rizzi–Longo L (1972) Compionamenti di alghe benthonice nel Quarnero. Atti Mus Civ Stor Nat Trieste 28: 147–165

Rueness J (1973) Pollution effects on littoral algal communities in the inner Oslofjord, with special reference to *Ascophyllum nodosum*. Helgol wiss Meeresunters 249: 466–454

Rueness J, Fredriksen S (1991) An assessment of possible pollution effects on the benthic algae of the outer Oslofjord, Norway. Oebalia 17 (Suppl): 223–235

Schiffner V (1916) Studien Über die Algen des Adriatischen Meeres. Helgol wiss Meeresunters NF 11: 127–198

Schiffner V, Vatova A (1938) Le alghe della Laguna di Venezia. Monogr del La laguna de Venezia 3,5,9: 87–250

Schiller J (1914) Österreichische Adriaforschung. Bericht Über die allgemeinen biologischen Verhältnisse der Flora des Adriatischen Meeres. Int Rev gesamten Hydrobiol Hydrogr Biol Suppl 6: 1–35 Wien

Schramm W, Abele D, Breuer G (1988) Nitrogen and phosphorus nutrition and productivity of two community forming seaweeds (*Fucus vesiculosus, Phycodrys rubens*) from the western Baltic (Kiel Bight) in the light of eutrophication processes. Kiel Meeresforsch Sonderh 6: 221–241

Sfriso A, Marcomini A, Pavoni B (1987) Relationship between macroalgal biomass and nutrient concentrations in a hypertrophic area of the Venice lagoon. Mar Environ Res 22: 297–312

Sfriso A, Marcomini A, Pavoni B, Orio AA (1988) Macroalgal production and nutrient recycling in the lagoon of Venice. Int Sanit 5: 256–266

Sfriso A, Pavoni B, Marcomini A (1989) Macroalgae and phytoplankton standing crops in the central Venice lagoon: primary production and nutrient balance. Sci Total Environ 80: 139–159

Sfriso A, Marcomini A, Pavoni B, Orio AA (1993) Species composition, biomass and net primary production in shallow coastal waters: the Venice lagoon. Bioresource Technol 44: 235–250

Škrivanič A, Barič A (1979) Cruises of the research vessel "Vila Velebita" in the Quarner region of the Adriatic Sea. III. Hydrographic conditions. IV. Distribution of primary nutrients. Thalassia Jugosl 15: 55–60, 61–88

Špan A (1969) Quantities of the most frequent *Cystoseira* species and their distribution in the central and northern Adriatic. Proc 6th Int Seaweed Symp, Sept. 1968, Santiago de Compostela 383–387 pp

Špan A (1972) Rod *Sargassum* u Jadranu. Morfolosko–sistematska i ekoloska obrada. Thesis, University of Zagreb, Zagreb, 110 pp

Špan A (1980) Composition et zonation de la flore et végétation benthique de l'ile de Hvar (Adriatique moyenne). Acta Adriat 22: 169–194

Špan A, Antolič B (1983) Prilog K poznavanju fitobentosa crnogorskog primorja (južni Jadran). Stud 13, 14: 87–110

Špan A, Antolič B (1991) Procjena utjecaja na okolja acy–marine u Makarskoj. Bentoske zivotne zajednice. Studije i elaborati. Oceanografski institut, Split 114: 32–42

Špan A, Simunovic A, Antolič B, Grubelič I (1989) Kontrola kvalitete mora (Vir–Konavle) 1988. životne zajednice morskog dna. Studije i elaborati Oceanografski institut, Split 98: 83–110

Stachowitsch M (1984) Mass mortality in the Gulf of Trieste. The course of community destruction. PSZNI Mar Ecol 5: 243–264

Stachowitsch M, Avčin A (1988) Eutrophication–induced modifications of benthic communities. UNESCO Rep Mar Sci 49: 67–81

Stachowitsch M, Fanuko N, Richter M (1990) Mucus aggregates in the Adriatic Sea: An overview of stages and occurrence. Mar Ecol 11: 327–350

Stefanon A, Boldrin A (1982) The oxygen crisis of the North Adriatic in 1977 and its effects on benthic communities. Proc 6th Symp CMAS, 167–175 pp

Techet K (1906) Über die marine Vegetation des Triester Golfes. Abh kgl Zool Bot Ges Wien: 31–52

Tewari A, Joshi HV (1988) Effect of domestic sewage and industrial effluents on biomass and species diversity of seaweeds. Bot Mar 31: 389–397

Tušnik P, Turk V, Planinc R (1989) Assessment of the level of pollution in the coastal area in the eastern part of the Gulf of Trieste. Biol Vestnik Ljubljana 37: 47–64

Vatova A (1928) Compendio della flora e fauna del Mare Adriatico press Rovigno. R. Comm Thalassiogr Ital Mem 143: 1–553

Vogt H, Schramm W (1991) Conspicuous decline of *Fucus vesiculosus* in Kiel Bay (western Baltic): what are the causes ? Mar Ecol Prog Ser 69: 189–194

Vouk V (1914) O istraživanju fitobentosa u Kvarnerskom zalijevu. Prirodosl Istražjrojz Hrvatske i Slavonije 2: 20–30 ; 5: 21–30

Vukovič A (1980) Asocijacije morskih bentoškich alg v Piranskem zalivu. Biol Vestnik Ljubljana 28: 102–124

Zalokar M (1942) Les associations sous-marines de la côte Adriatique au dessus de Velebit. Bull Soc Bot Genève 33: 1–24

Zavodnik D (1967) The community of *Fucus virsoides* (Don. J. Ag. on a rocky shore near Rovinj (northern Adriatic). Thalassia Jugosl 3: 105–113

Zavodnik D (1976) Investigation of pollution-stressed littoral communities in the northern Adriatic.In: Meyers SP (ed) Proc Int Symp Mar Poll Res, 80–88 pp

Zavodnik D (1977) Benthic communities in the Adriatic Sea: reflects of pollution. Thalassia Jugosl 13: 413–422

Zore-Armanda M (1963) Les masses d'eau de la mer Adriatique. Acta Adriat 10: 1–94

Zore-Armanda M (1968) The system of currents in the Adriatic Sea. Stud Rev Gen Fish Cons Médit 34: 1–48

Zore-Armanda M, Gacič M (1987) Effects of bora on the circulation in the North Adriatic. Ann Geophys 58: 93–102

Zore-Armanda M, Vučak Z (1984) Some properties of the residual circulation in the Northern Adriatic. Acta Adriat 25: 101–117

Zore-Armanda M, Dadič V, Gačič M, Morovič M, Vučetič T (1983) MEDALPEX of the North Adriatic. Preliminary Report. Notes 508: 1–8

17 Greece

S. Haritonidis

17.1 Introduction

Attention has been paid to the marine vegetation of the Greek coasts only in the last 30 years (Haritonidis 1978; Nikolaidis 1985). Recently, we have come to witness the great qualitative and quantitative changes in these communities which probably are the outcome of eutrophication.

During the past decades, conditions in the eutrophicated and polluted areas along the Greek coasts, particularly in enclosed gulfs and estuaries have changed, and degradation of the macrobenthos together with increasing phytoplankton blooms occurred. There was an analogous decrease in the marine phanerogam meadows in the gulfs.

A reduction of the number or absence of some genera or species (e.g. *Cystoseira*) was the first sign of eutrophication, whereas the biomass of the eutrophilic species increased (by > 20%). Macroalgae, such as *Ulva rigida*, *Enteromorpha compressa* or *Gracilaria verrucosa*, which were recognized as biological indicators for eutrophicaton, proved to be very tolerant of other types of pollution, for example heavy metal bioaccumulation.

17.2 Marine Vegetation and Eutrophication in Greece

During the past three decades, the taxonomy and chorology of marine macroalgae and phanerogams as well as of unicellular algae were mainly studied. About 550 different species of macrophytes are encountered along the Greek coasts nowadays. The greatest variation and diversity are observed in the Aegean Sea (Fig. 17.1; Haritonidis et al. 1994).

Botanical Institute, Aristotle University, 54006 Thessaloniki, Greece

Ecological Studies, Vol. 123
Schramm/Nienhuis (eds) Marine Benthic Vegetation
© Springer-Verlag Berlin Heidelberg 1996

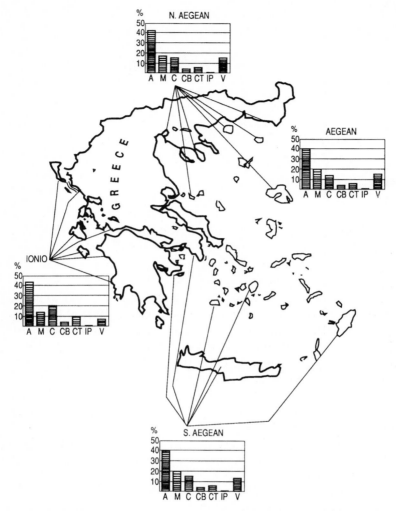

Fig. 17.1. Map of Greece with the Marine elements of macrophytes. *A* Atlantic; *M* Mediterranean; *C* Cosmopolitan; *CB* Circa-Boreal; *CT* Circa-Tropical; *IP* Indo-Pacific; *V* various. (Haritonidis 1978)

Analyses of part of the marine elements indicate that Atlantic and Mediterranean species prevail (Feldmann 1938; Haritonidis 1978; Haritonidis et al. 1986). Cosmopolitan species, most of which are found in great numbers and show maximal growth during the winter months (Fig. 17.2), are especially interesting because many of them constitute the so-called eutrophication index (Haritonidis et al. 1986).

Cosmopolitan

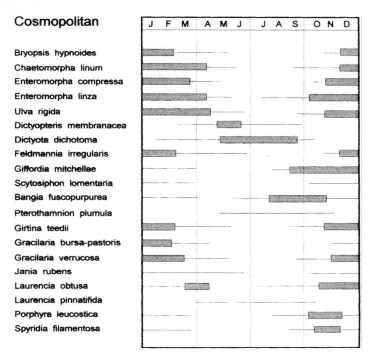

Fig. 17.2. Seasonal occurrence of cosmopolitan algal species in Greek waters. Most of these are "winter" species. (Haritonidis 1978)

Investigations of this type were initiated at various universities, in particular those of the big cities Athens and Thessaloniki. After initial comparative studies of the flora and the vegetational associations and their changes which occurred during the past 30 years (Haritonidis et al. 1986), attention was later focused on the study of the dynamics of algal biomass production with particular reference to the nutrient cycles of the Gulf of Thermaikos (Nikolaidis and Moustaka-Gouni 1992). Nowadays, we have come to witness the great qualitative and quantitative changes in these communities which probably is the outcome of eutrophication. Eutrophication is more evident in enclosed bays and near large populated areas or cities, with the Gulfs of Thermaikos (Thessaloniki; Fig. 17.3), Saronikos (Athens) and Pagasitikos (Volos) as the most conspicuous examples.

Five hundred and fifty different species of macrophytes have been identified in the Greek waters so far, indicating that diversity is still quite satisfactory. However, if we look separately at enclosed and shallow basins, we shall see that the number of macrophytic species has

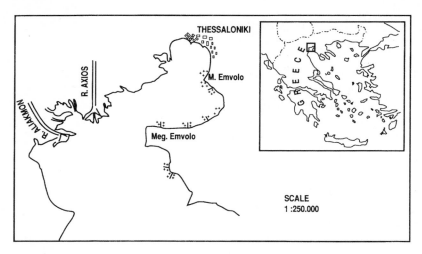

Fig. 17.3. Thermaikos Gulf. The *crosses* indicate the location for sampling of algae. (Haritonidis 1978)

considerably decreased (Haritonidis et al. 1994; Diapoulis 1982; Nikolaidis 1985).

The valuation of an ecosystem should not only be based on biological parameters (organisms, etc.), but also on the knowledge of the physicochemical parameters. Such information is now sufficiently available for the Greek coasts to support a more or less complete hydrobiological study (Haritonidis 1978; Nikolaidis and Moustaka-Gouni 1990).

The coasts of northern Greece, as in Greece in general, display a diverse configuration which is due to the volcanic nature of the substrates. Studies conducted in the bays of Chalkidiki, Thasos and other parts of northern Greece have revealed a large number of seaweeds forming various communities down to depths of more than 40 m. Similar numbers have been observed in association with communities of marine phanerogams, which in clear waters and outside of bays develop meadows down to 40 m depth.

As already pointed out, the density and covered area of the macroalgal and phanerogam communities have diminished dangerously recently (Haritonidis et al. 1994). Probably, this was caused mainly by eutrophication, i.e. by the increase in nutrients as a result of discharge of untreated domestic sewage (Nikolaidis 1985; Nikolaidis and Moustaka-Gouni 1990).

17.3 The Gulf of Thessaloniki

The Gulf of Thessaloniki has some specific features which constitute its peculiar environment: the shallow depth, the narrow opening to the northern Aegean, and the rivers which discharge thousands of tons of wastewaters daily (annual average flow 150 m^3 s^{-1}), enriched with residues from the surrounding area (Table 17.1; Figs. 17.3, 17.4).

With these factors together with the fact that wastes from a city of 1 million inhabitants (200 000 m^3 every day) are flowing into the Gulf, we may very well realize how critical the problem of eutrophication is for this area. In recent years, particularly in spring 1993, the appearance of phytoplankton as well as of macroalgal "blooms" (*Ulva rigida*, *Enteromorpha linza* and a few species of *Cladophora* and *Chaetomorpha*) has been remarkable. Blooms of microalgae occurred on the coasts of northern Greece, especially in the enclosed bays at the beginning of spring and summer (end of June) during the past 10 years.

In addition, "red tides" have occurred in the Gulf of Thessaloniki, especially during afternoons when the sea is rough. However, no increased numbers of the known toxic Dinophyceae have been observed

Table 17.1. Physical and chemical characteristics of the Thermaikos Gulf–Estuary

Area (km^2)	518
Volume ($m^3 \times 10^6$)	560
Maximum depth (m)	28
Average depth (m)	5
Residence time (days)	–
Freshwater load (from rivers)	500
Freshwater (from raw domestic effluents)	150^3 s^{-1}
Volume:load	1:1
Tides	15 cm
Stratification	None
Extinction coefficient (m^{-1})	1.5–4.4
pH	8–8.3
Temperature	10–26 °C
Chlorinity (%)	16–18
Nitrogen load (g N m^{-2} $year^{-1}$)	45–60
Winter conc. of N (mg dm^{-3})	25–30
Phosphorus load (g P m^{-2} $year^{-1}$)	9–27
SiO_2 (μg/l)	21.6–376
Autumn conc. P-PO_4 (mg dm^{-3})	9–14
Summer Chl conc. (mg Chl m^{-3})	10–14

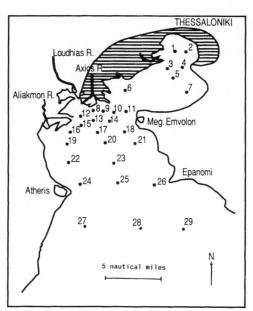

Fig. 17.4. Thermaikos Gulf. Sampling stations where large quantities of *Codium* have been observed since 1980. (Haritonidis 1978)

(Nikolaidis 1985; Nikolaidis and Moustaka-Gouni 1990). Of the Dinophyceae found in Thermaikos Gulf, six *Prorocentrum* species are the dominant forms which, however, do not cause any problems, Fig. 17.5 shows the seasonal variation and growth of *Prorocentrum micans* in relation to depth (Seferlis 1990).

Monthly data for nutrients and other hydrographic parameters in 1988–1989 are compiled in Table 17.2. There are also other data on the hydrography (tides, currents, waves) of this bay (Zoi-Morou 1981). The small depth (28 m in the center of the Gulf and 8 m in the bay), the narrow opening (2–3 km), and the absence of significant tides and currents are the main causes of the high eutrophication observed in this area.

Measurements of the nutrient load (NO_3, NO_2, NH_4, org-N, total N, PO_4-P, total P) were carried out over the past 5 years with monthly and sometimes by-weekly sampling (Nikolaidis 1985).

The major sources of eutrophication are the rivers which carry the nutrients from the drainage basins together with the domestic wastewaters into the estuary. It was only 2 years ago that a treatment plant was constructed to process mainly domestic sewage. The industrial effluents have so far been partially under control. However, the greatest problem, which is almost impossible to cope with, derives from agriculture. The all-over nutrient discharge through the rivers, however, has slightly decreased the last few years, due to the reduced discharge from the rivers

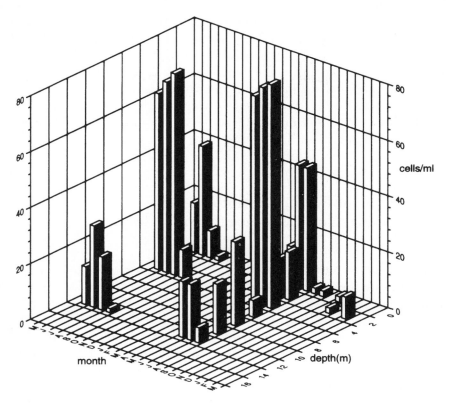

Fig. 17.5. Seasonal fluctuation of *Prorocentrum micans* in relation to depths. (Seferlis 1990)

Table 17.2. Monthly physical and chemical data from Thermaikos Gulf in the period May 1988-April 1989. (Nikolaidis and Moustaka-Gouni 1990)

Year	1988								1989			
Month	M	J	J	A	S	O	N	D	J	F	M	A
Temperature (°C)	21	24	26	27	22	18	14	11	10	11	13	17
Secchi depth (m)	1.5	3.2	4.4	3.8	3.0	3.2	2.6	4.0	3.0	3.2	2.4	2.2
Salinity (PSU)	24.5	28.5	31.6	33.3	31.5	31.4	31.4	34.5	34.0	32.5	32.0	30.5
pH	8.2	8.2	8.2	8.1	8.2	8.4	8.1	8.2	8.1	8.2	8.3	8.3
Inorg.nitrogen (μmol dm^{-3})	3.2	3.0	3.0	4.5	5.0	3.2	7.2	8.2	1.8	3.7	4.4	2.3
PO_4-P (μmol dm^{-3})	0.5	0.9	0.6	0.6	0.3	2.3	0.5	1.1	0.7	2.6	2.8	2.2
SiO_2-Si (μmol dm^{-3})	2.6	6.9	0.8	5.5	13.4	1.1	1.4	7.5	1.0	4.6	0.3	4.4

Axios and Aliakmonas as a result of extensive use of river water for irrigation during the summer months. All the eutrophicating substances flowing into the Gulf and their annual cycling are given in the detailed studies by Nikolaidis and Moustaka-Gouni (1990, 1992).

Based on the information of the past 20 years on the eutrophication system, we gather that variation in the amounts of nutrients in the Gulf throughout the year is attributed to the biological activities taking place there. In addition, sewage of different types tends to augment the problem. We believe that without these additional elements the Gulf could well be characterized as oligotrophic, and probably changes in nutrient levels will be evident within the coming years after the sewage plant has been set in full operation.

17.4 Effects of Eutrophication: Algal Typology and Community Evaluation

Qualitative and quantitative investigations on the phytoplankton composition were conducted every month, always in conjunction with nutrient measurements (Nikolaidis and Moustaka-Gouni 1990). Apart from phytoplankton blooms, red tides sometimes appear in the bay.

Very little information is available on other pelagic organisms. In contrast, there is considerable data on benthic macrofauna, although these are more of a general nature and not directly connected with eutrophication processes (Koukouras 1979; Zarkanellas 1980).

Studies on the marine flora and vegetation of the Greek area have unfortunately been conducted only the last 30 years (Haritonidis 1978). However, it was a fortunate coincidence that in the course of 20 years two comparative studies on the flora and vegetation of Thermaikos Gulf have been produced (Haritonidis 1978; Nikolaidis 1985).

The first investigations on the composition of Greek seaweeds appeared in the 1960s (Anagnostidis 1968), followed by more detailed studies, including seasonal collection of macroalgae and phanerogams, in the 1970s (Haritonidis 1978).

The sandy and muddy bottom of the inner gulf does not provide very clear limits of the supra- and mid-littoral zone. Small tidal fluctuations (on average 25 cm) make the situation more complicated, resulting in the dominance of the sublittoral zone with high coverage of benthic flora (Zoi-Morou 1981).

The supra- and mid-littoral zones are inhabited by only few species, mainly Chlorophyceae (*Ulva rigida* and *Enteromorpha compressa*) with the Phaeophyceae *Cystoseira fimbriata* appearing on the midlittoral fringe. At present, this species grows only in the outer Gulf (Epanomi station; Fig. 17.4). Only the species *Ulva rigida*, *Enteromorpha linza*, *E.*

compressa, *Gracilaria verrucosa* and *Punctaria latifolia* are nowadays encountered in these two zones. *Scytosiphon lomentaria*, which previously grew in this zone during the winter months, is now restricted to the outer gulf. Thus the species number in the above two zones has been slightly reduced, although diversity has never been large in the area.

The most conspicuous changes were observed in the sublittoral zone. Within one decade, the number of macrophyte species was dramatically reduced from 127 to 30 by 1985 (Nikolaidis 1985). The decrease has been both qualitative and quantitative. Total biomass decreased by 1000–1500 g m^{-2} (Haritonidis 1978; Haritonidis et al. 1986). The more sensitive Phaeophyceae, previously forming the major portion of the plant biomass, have nearly disappeared from the inner Thermaikos Gulf. The species which are now dominant belong mainly to Chlorophyceae (Haritonidis 1978; Haritonidis et al. 1986; Fig. 17.6). Before 1975, the dominant species were *Cystoseira fimbriata*, *Cystoseira barbata*, *Dictyopteris membranacea*, *Cladostephus verticillatus*, *Padina pavonica* and *Taonia atomaria*, but in 1985, *Ulva rigida*, *Enteromorpha linza*, *Enteromorpha compressa*, *Punctaria latifolia*, *Scytosiphon lomentaria* and *Codium* spp. were dominant. *Codium* was seldom found in the Gulf up to 1975 (Haritonidis 1978). Since 1980, mass development of this seaweed has occurred in both the inner and outer Thermaikos Gulf during the winter months.

It is known that *Codium* strongly accumulates nitrate (Asare and Harlin 1983; Lüning 1990). Thus, the increase in nitrate in the Gulf may possibly account for the explosive development of this seaweed. Most of the thalli are comprised of *Codium fragile* with a diameter of 15–25 cm.

From 1967 on, since the beginning of seasonal investigations and measurements *Ulva rigida* has been observed to gradually cover great areas in the Gulf. At depths of 1–2 m, its biomass reached 10–12 kg m^{-2} wet material. The biomass of *Ulva* displays seasonal variations with 5–6 kg m^{-2} in spring, 10–12 kg m^{-2} in summer, 6–7 kg m^{-2} in autumn, and 2–3 kg m^{-2} in winter (Haritonidis et al. 1986).

The nitrophilous nature of *Ulva* and *Enteromorpha* was also checked under experimental conditions using a medium of sewage and seawater in which 95% of nitrate and other nutrients were absorbed within 2–3 days (Haritonidis et al. 1991). The nitrophilous characters of the species were also verified by the tremendous increase in the *Ulva* sp. biomass which grows in enriched seawater medium with a 15:1 ratio of N:P. The highest relative growth – up to 95% weight increase per experimental period (10 days) – was observed in culture when nitrogen was mainly supplied in the form of NH_4-N (Fig. 17.7; Kotropoulos et al. 1991).

S. Haritonidis

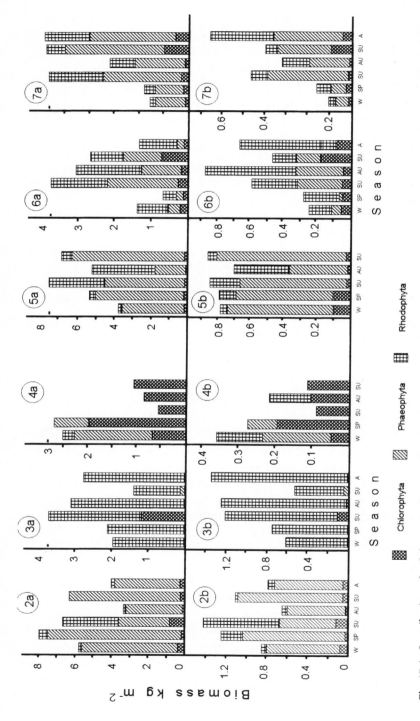

Fig. 17.6. Contribution of red, brown and green algal to biomass (*a* wet weight; *b* dry weight) in marine plant communities of different Greek coasts. 2 Kovios-Chalkidi; 3 Parga-west coasts; 4 Kavala harbour; 5 Kavala-Palio; 6 Pylos-west coasts; 7 Parga village. (Haritonidis et al. 1986)

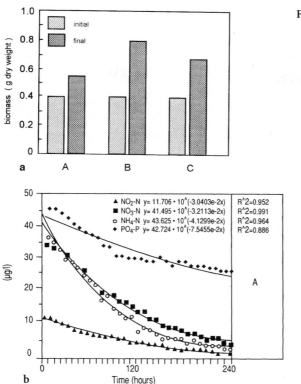

Fig. 17.7a,b.

Generally, macroalgae strongly interfere with the lagoon nutrient cycles mostly in places where *Ulva rigida* and *Enteromorpha linza* populations are dense. The seasonal fluctuation of nutrient concentrations is strongly related to the biomass densities (Kotropoulos et al. 1991; Nikolaidis 1985; Nikolaidis and Moustaka-Gouni 1992). Apart from isolated species, the qualitative and quantitative variations have also been evident in communities of macroalgae.

Table 17.3 shows the typical plant communities of the Thermaikos Gulf and other Greek coasts, based on many years of study (Haritonidis 1978). A comparative survey conducted in various Greek polluted and non-polluted biotopes displayed a slight decrease in the dry biomass of the polluted areas and an increase in the undisturbed ones (Haritonidis et al. 1986). Figure 17.6 shows the dominant community with a clear seasonality. Dry biomass reaches 8 kg m^{-2} in non-polluted areas and 250 g m^{-2} in polluted areas, respectively (Haritonidis et al. 1986). In polluted and non-polluted biotopes high biomass values are noted when the dominant species are large brown algae (e.g. *Cystoseira*; Figs. 17.3,

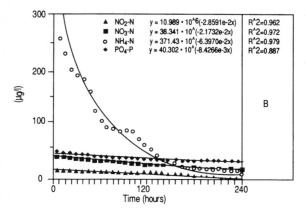

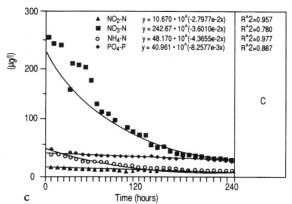

Fig. 17.7. **a** Initial and final biomass and **bc** nutrient uptake of *Ulva* sp. after 10 days of culture with different nutrient supples (wet weight increase and decrease in nutrient concentration in three laboratory experiments **A, B C**). (Kotropulos et al. 1991)

17.2, 17.5). On the contrary, in harbours where the wet biomass is comparatively higher (up to 4 kg m^{-2}), dry weight is sometimes comparatively lower (300 g m^{-2}). We believe that the biomass denoting the wet:dry ratio from 1:8 to 1:14 is characteristic for polluted biotopes, and possibly could serve as an eutrophication indicator.

The community-forming species appearing in Thermaikos Gulf since 1985 are *Porphyra leucosticta, Scytosiphon lomentaria, Enteromorpha compressa* and *Ulva rigida*. These communities show only a very small diversity with only three to four associated species, probably as a result of increasing nitrogen levels (Nikolaidis 1985).

Little information is available on marine phanerogam communities which inhabit the Greek coasts at depths ranging from a few centimeters to 30 m. Some more general research work on phanerogams was done along the Greek coasts (Haritonidis et al. 1986, 1990), and Panayotidis (1979) worked with the *Posidonia oceanica* beds in Thermaikos Gulf

Table 17.3. Plant communities on Greek coasts

Supralittoral zone

Verrucaria sp. (Feldmann 1938)

Midlittoral zone (or Tidal zone)
Higher Midlittoral zone
Neogoniolithon-lithophylletum (Molinier 1959/60)

Lower midlittoral zone

Ass. *Porphyra leucosticta* (Feldmann 1938)
Scytosiphon lomentaria (Ollivier 1929)
Enteromorpha compressa (Boudouresque 1971)

Sublittoral zone
Higher sublittoral zone
Pterocladia-Ulvacetum (Molinier 1959/60)
Cystoseira fimbriata (Coppejans 1974)

Lower sublittoral zone

Cysroseira crinita
Ass. *C. discors* and *C. barbata* (Feldmann 1938)
Padina pavonica (Boudouresque 1970)
Peup. *Dictyotales* (Boudouresque 1970)
Peup. *Corallina of. officinalis* (Bellan-Santini 1962)

(Fig. 17.8). These contributions are the only work done specifically in Greek waters.

Safriel and Lipkin (1976) examined the migration of *Halophila stipulacea* after the opening of the Suez Canal (Fig. 17.9). *Halophila stipulacea* is now found in the Korinthos gulf (Haritonidis and Diapoulis 1990), and seems to be moving on to the north. In Pagasitikos Gulf, for example, a meadow was found at 15 m depth in August 1993 (Amoutzo-poulou and Haritonidis 1994).

Recently, we have noted the decrease in certain marine phanerogam meadows, especially in closed gulfs, which may be a response to seawater pollution. All our observations are based on data of papers published in the last 20 years and on personal research (Haritonidis 1978; Haritonidis et al. 1986).

The communities of marine phanerogams in the Gulf of Thermaikos comprise mainly beds of *Cymodocea nodosa* and *Posidonia oceanica* (Fig. 17.8). *Zostera noltii* communities are found at depths to 0.5 m in the inner Thermaikos Gulf and on its southeastern side. Well-developed beds of *Cymodocea nodosa* occur at depths of 0.5 to 2 m in the outer area of the gulf, and beds of *Posidonia oceanica* at depths to 12 m. Over the past 20 years, considerable losses have been observed, in both surface cover and shoot density. Remarkable changes have also been noted in

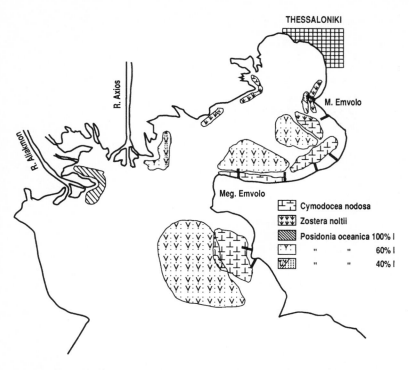

Fig. 17.8. Distribution of marine phanerogams in Thermaikos Gulf. (Haritonidis et al. 1990)

the epiphytic community of the macro- and microphyceae of marine phanerogams.

Considering that beds of *Posidonia oceanica* are more sensitive to natural and human stress compared to *Cymodocea nodosa* and *Zostera noltii*, their decline in the Gulf of Thermaikos, which has lately been heavily polluted with domestic and industrial wastes, is not surprising.

Changes in the distribution of the marine phanerogams are directly linked to the ecological evolution of the gulf. The great sensitivity of the leaves of *Posidonia oceanica* to the increasing pollution of the gulf results in their continuing degradation and reduction. Strong anthropogenic influences in the form of municipal sewage and industrial waste have also played an important role in the decline of other phanerogam communities (Haritonidis 1978; Haritonidis et al. 1990).

Although our knowledge on the present situation of the phanerogams in the Gulf of Thermaikos is limited, we may conclude that during the past 20 years the decrease in *Cymodocea nodosa* is about 10%, which is less than that of *Posidonia oceanica* which suffered the greatest

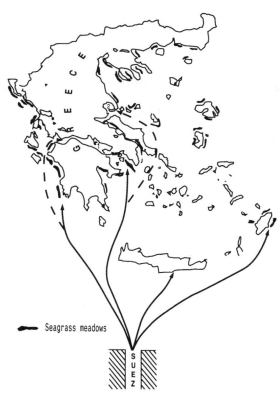

Fig. 17.9. Marine phanerogam distribution (mainly *Posidonia oceanica* meadows) and propagation in Greece. The *arrows* indicate the pathways of *Halophila stipulacea*. (Haritonidis et al. 1990)

reduction. Even the leaf parts cannot be traced in the coasts up to the limit of Megalo Emvolo (Fig. 17.8; Haritonidis et al. 1990). The type of bottom, which is muddy at places, has contributed to this loss. It is estimated that the total coverage of *Posidonia* has been reduced by 30%. It seems that *Zostera noltii* is the most tolerant species of all three phanerogams, and it has managed not only to survive, but even to expand remarkably.

17.5 Conclusions

The gulf estuaries or closed gulfs are very changeable ecosystems because of the increasing human activities. Thermaikos Gulf is an example which apart from the sewage effluents is the recipient of a great amount of industrial and agricultural wastes. The ecology of the Gulf has thus changed dramatically over the past 20 years, and as a result the marine flora is in great danger.

The problems we are currently facing are:

1. Qualitative and quantitative decrease in the marine algal flora and the marine phanerogams.
2. Many species have disappeared from the Gulf, others have prevailed (*Ulva rigida*, even on soft bottoms).
3. Wet biomass of the biological indicators has increased (only Chlorophyceae).
4. Nitrate, phosphorus and other nutrients are now in abundance in the Gulf.
5. It is too early to conclude that the sewage treatment plant of the city Thessaloniki, which has been in operation since the beginning of 1992, has had a positive effect on the ecosystem of the Gulf.

References

Amoutzopoulou H, Haritonidis S (1994) Marine phanerogam communities in Pagassitikos gulf, Magnesia, Greece. Proc 16th Sym Greek Biological Society, Volos, May 1994, pp 2.43–2.44

Anagnostidis K (1968) Investigations on the salt and freshwater thiobiocoenoses (sulphuretum) of Greece. Thesis, University of Thessaloniki. Annu Sci Rep Phys Math Fac Univ Thessaloniki 10: 406–460 (in Greek with German summary)

Asare SO, Harlin MM (1983) Seasonal fluctuations in tissue nitrogen for five species of perennial macroalgae in Rhode Island Sound. J Phycol 19: 254–257

Bellan-Santini D (1962) Contribution à l' étude des peuplements infralittoraux sur substrat rocheux (étude qualitative et quantitative de la fringe supérieure). Rec Trav Stu Mar Endoume 47 (63): 1–294

Boudoureque C-F (1970) Recherches de bionomie analytique, strcutrale et expérimentale sur les peuplements benthiques sciaphiles de Méditerranée occidentale (fraction algale). Thèse Doct Etat, Univ Aix-Marseille Luminy, 624 pp

Boudouresque C-F (1971) Méthodes d'étude qualitative et quantitative du benthos (eu particulier du phytobenthos. Téthys 3(1): 79–104

Coppejans E (1974) A preliminary study of the marine algal communities on the islands of Milos and Sikinos (Cyclades-Greece). Bull Soc R Bot Belg 107: 387–406

Diapoulis A (1982) Quantitative and qualitative investigation of the phythobenthos of Saronikos Gulf. Thesis, University of Thessaloniki, 171 pp. (in Greek with English summary)

Feldmann G (1938) Recherches sur la vegetation marine de la Mediterrranee. La cote des Alberes. Rev Algol 10: 1–139

Haritonidis S (1978) Contribution to the research of marine plant macrophyceae (Chloro-, Phaeo- and Rhodophyceae) of Thermaikos Gulf. Dissertation Thessaloniki, 175 pp (in Greek with English summary)

Haritonidis S, Diapoulis A (1990) Evolution of Greek marine phanerogam meadows over the last 20 years. Posidonia News 3(2): 5-10

Haritonidis S, Nikolaidis G, Tsekos I (1986) Seasonal variation in the biomass of marine macrophyta from Greek coasts. Mar Ecol 7: 359-370

Haritonidis S, Diapoulis A, Nikolaidis G (1990) First results of the localisation of the herbiers of marine phanerogams in the Gulf of Thermaikos. Posidonia News 3(2): 11-18

Haritonidis S, Nikolaidis G, Tryfon H, Gartsonis K (1991) Cultures of macroalgae in wastewater-treatment. I. Biomass. Toxicol Environ. Chem 31-32: 515-520

Haritonidis S, Lazaridou E, Orfanidis S (1994) Checklist of the benthic marine algae of the Greek coasts (Aegean and Ionian Seas). Thalassographica (in press)

Kotropoulos D, Nikolaidis G, Haritonidis S (1991) Biomass response of the macrophyte Ulva sp. to nitrogen enriched seawater. Oebelia 17(1): 65-72

Koukouras A (1979) Bionomic investigation of the macrofauna from the mobile substrate of mediolittoral zone in Strymonikos and Thermaikos Gulf. Dissertation, Thessaloniki, 205 pp (Greek with English summary)

Lüning K (1990) Seaweeds: their environment, biogeography and ecophysiology. John Wiley, London, 527 pp

Molinier R (1959/60) Etude de biocénoses marines due Cap Corse. Vegetatio 9(3-5): 121-292, 217-312

Molinier Moustaka-Gouni M, Nikolaidis G, Alias H (1992) Nutrients, Chlorophyll-a and phytoplankton composition of Axios river, Macedonia, Greece. Fresenius Environ Bull 1: 244-249

Nikolaidis G (1985) Qualitative and quantitative investigation of the marine plant macrophyta (Chlorophyceae, Phaeophyceae, and Rhodophyceae) in different polluted (Thessaloniki gulf) and not polluted areas (Chalkidiki). Thesis, University of Thessaloniki, 163 pp (in Greek with English summary)

Nikolaidis G, Moustaka-Gouni M (1990) The structure and dynamics of phytoplankton assemblages from the inner part of the Thermaikos Gulf, Greece. I. Phytoplankton composition and biomass from May 1988 to April 1989. Helgol Meeresunters 44: 487-501

Nikolaidis G, Moustaks-Gouni M (1992) Nutrient distribution and eutrophication effects in a polluted coastal area of Thermaikos Gulf, Macedonia, Greece. Fresenius Environ Bull 1: 250-255

Ollivier G (1929) Sur les bromuques de diverses Céramiacées. Ebenda, T 186

Panayotidis P (1979) Etude phytosociologique de deux aspects saisonniers de la flore epiphyte des feuilles de Podisonia oceanica (Linnaeus) Delile, dans la Golfe de Thessaloniki (Mer egee, Grece). Thalassographica G 1(3): 93-104

Safriel V, Lipkin J (1975) Pattern of colonization of the eastern Mediterranean intertidal zone by red immigrants. J Ecol 63: 61-63

Seferlis M (1990) Description of the species from the genus Prorocentrum Ehrenberg in Thermaikos Gulf and notes on their abundance during the period May 1988-May 1990. Diploma investigation, Thessaloniki 1990, 27 pp

Zarkanellas A (1980) Ecological investigation of the macrobenthos from Thermaikos Gulf. Dissertation, Thessaloniki, 147 pp (in Greek with English summary)

Zoi-Morou A (1981) Documents of the tide levels from the Greek harbours. Greek Hydrogr Serv Athens 13: 1-77

18   Turkey

H. Güner and V. Aysel

18.1   Introduction

In Turkey most investigations on marine plants have mainly concentrated on the floristic, taxonomic or systematic aspects of the subject. In comparison, the number of vegetational or phytobenthic studies is small. It was only during 1978 that attention was bent towards the benthic flora and fauna in Turkey. So far, approximately 30 vegetational studies exist on algal groups, representing the coastal zones of the Aegean, Marmara and Turkish Mediterranean seas. Studies on the benthic vegetation in the Black Sea are still lacking.

This chapter includes mainly information on our own research activities during the years 1974–1993, as well as a general review of observations collected recently.

18.2   Hydrological and Hydrochemical Characteristics of Turkish Seas

Turkey is surrounded by four different seas: the Black Sea in the north, the Marmara Sea in the northwest, the Aegean Sea in the west, and the Mediterranean Sea in the south (Fig. 18.1). The total length of the shoreline of the country is 8333 km; 6480 km in Asia Minor (Anatolian shore), 786 km in Thrace (Europe) and 1067 km shoreline surrounding islands.

Faculty of Science, Ege University, 35100 Bornova, Izmir, Turkey

Ecological Studies, Vol. 123
Schramm/Nienhuis (eds) Marine Benthic Vegetation
© Springer-Verlag Berlin Heidelberg 1996

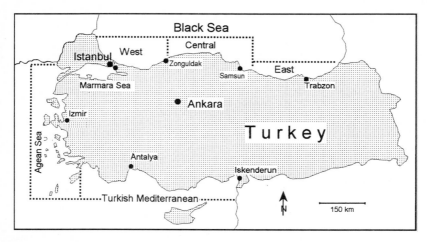

Fig. 18.1. The geographic position of the seas surrounding Turkey

18.2.1   The Black Sea

The Black Sea with a total surface of 423×10^3 km^2, a volume of 537×10^3 km^3, and an average depth of 1271 m (maximum depth 2245 m) is characterized by its pronounced stratification. The average water balance of the Black Sea, taking into account the changes in recent years, is shown in Table 18.1. The inflow of river water into the Black Sea is unevenly distributed over the coastal areas (Table 18.2).

The surface currents in the Black Sea are generated by inflow from the rivers and from the Azov Sea. The direction and velocity of the currents are strongly influenced by the winds and by the configuration of the coast (Fig. 18.2).

The regime of the two currents in the Bosphorus Strait is an important factor. The salinity of the upper Bosphorus current is about 18 PSU, and the lower Bosphorus current about 35 PSU. The water of the Sea of

Table 18.1. The average annual water balance of the Black Sea

Gains	km^3	Losses	km^3
River water	294	Evaporation	301
Rain	254		
Flow from the		Flow into the	
Azov Sea	38	Azov Sea	29
Lower Bosphorus current	229	Upper Bosphorus current	485
Total	815	Total	815

Table 18.2. Average annual inflow of river water into the Black Sea

Area	Inflow of river water	
	(km³)	(%)
Northwest part	234	79.59
Crimean coasts	1	0.34
Caucasian coasts	35	11.90
Turkish coasts	23	7.83
Bulgarian coasts	1	0.34

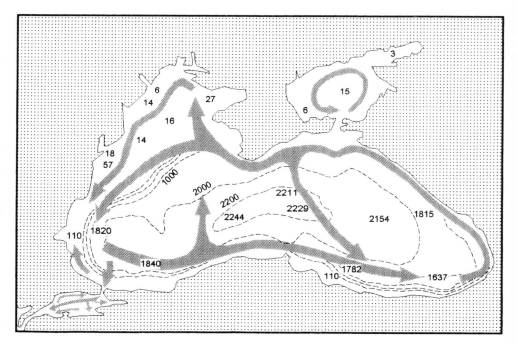

Fig.18.2. Schematic map of the current systems (*arrows*) and depths (in meters) in the Black Sea and the Marmara Sea

Marmara with a salinity of about 24–25 PSU is diluted with Pontic (Black Sea) sea water with a salinity of 18 PSU before these currents penetrate into the deep layers of the Black Sea. Seasonal variations of salinity as well as temperature are most pronounced in the coastal areas. In the open sea, these are clearly noticeable in the upper layer down to 75 m, where the autumn mixing takes place.

Dissolved oxygen is found in the superficial, oxygenated layer in which the main aerobic biological processes take place. In the central parts of the sea, oxygen is detectable down to depths of about 250 m. The distribution limits of oxygen and hydrogen sulphide, which dominates the deeper waters, vary between years and seasons.

Another feature of the Black Sea is the nutrient concentration. Phosphate predominates over nitrate which is virtually absent in the waters between 200 and 300 m depth. In periods of intensive production of phytoplankton, especially in summer, nitrate may completely disappear, whereas phosphate remains available in sufficient quantities.

18.2.2   The Sea of Marmara

The Bosphorus (Istanbul Strait), the Sea of Marmara and Dardanelles (Canakkale Strait) together constitute the system of Turkey Straits. These systems share unique topographical and hydrographical attributes that determine their biological and ecological characteristics. Hydrographical studies on the Marmara Sea and Straits date back to the seventeenth century, but biological research, in particular phycological studies started at the end of the last century. Since the 1970s, studies with the pollution of the Marmara Sea and Straits have gained prominence. The Marmara Sea, which constitutes the centre of the system of Turkish Straits, has the characteristics of an inland sea. It is connected to the Black Sea via the Bosphorus and to the Mediterranean through the Dardanelles. While the surface waters of the Marmara Sea are effected by the Black Sea, its deeper waters remain under Mediterranean influence. Therefore, the Marmara Sea has the typical attributes of an estuary. The two water layers do not mix and a stable density gradient is established. Numerous locations suitable for marine aquaculture exist around the Marmara Sea which, however, is confronted by a constantly growing threat of pollution.

The Marmara Sea is the smallest of the Turkish Seas with respect to both its surface area and volume (total surface 11 500 km², volume 3378 km³, depths average 35 m, maximal 110 m) with a coastline length of 927 km.

The drainage area of the Marmara Sea makes up 3.1% of Turkey's land area (24 100 km²). The annual discharge of the rivers in the drainage area is about 6.6×10^9 m³.

Tides are barely noticeable in the Marmara Sea. The hydrologial properties are influenced by both the neighbouring Black Sea and Mediterranean, and by wind. Sometimes under the effect of very strong winds water levels may fluctuate 1 m along the northern and southern shores.

The temperatures of the surface waters of the Bosphorus are under the influence of the Black Sea and range between 4 and 24°C. The salinity varies between 18 and 10 PSU. Deep waters show pronounced changes in salinity and temperature. The salinity at 20 m rises to 30 PSU and at 40–50 m to 37 PSU.

A pronounced stratification due to these temperature and salinity gradients is typical, which greatly affects the nutrient dynamics in this sea. Little work has been done on the nutrients of the Marmara Sea. The few investigations on nutrients indicate that nutrient-rich water of the Black Sea enters the Marmara from the surface in winter. In the spring and summer, the nutrient flow decreases because of the high level of primary productivity in the Black Sea.

18.2.3 The Aegean Sea

The Aegean Sea, located between Turkey and Greece, is a semi-enclosed sea (total surface 241 000 km^2, volume 7404 km^3), separated from the Mediterranean in the south by the islands of Crete, Rhodes and Carpathos. The average depth of the Aegean Sea is 350 m, however, a significant part of it has depths of from 100 to 500 m. Deep trenches are found, which divide the Aegean into east and west plateaus.

The Aegean Sea, which is a transition area between the Mediterranean and the Black Sea with regard to oceanographic characteristics, may be divided into the northern, middle and southern sections. Of these, the northern section is affected by less saline and colder water originating in the Black Sea and passing through the Marmara Sea and Dardanelles. The southern and middle sections are affected by Mediterranean waters. Thus, the northern section is less saline and colder, while the middle and southern sections are warmer and very saline. Although the Aegean waters are relatively low in the concentration of nutritive salts, they are higher compared to the other areas of the Mediterranean.

Two current systems can be distinguished in the Aegean Sea. One of these is the main current, coming from the Mediterranean and carrying to the area water masses warm and rich in salt, and moving around in a clockwise direction. The second is a current system consisting of waters originating in the Black Sea which gain salinity as they move through the Marmara Sea (Fig. 18.2).

18.2.4 The Turkish Mediterranean

Variation in meteorological conditions is one of the distinguishing characteristics of the Mediterranean Sea bordering Turkey. The region functions as a pathway for extra-tropical cyclones during winter and spring.

During winter, westerly winds prevail, while in summer and early autumn winds generally blow from northwesterly directions.

The Levantine Basin is the second largest basin of the eastern Mediterranean; its volume amounts to 7.5×10 km^3 and the maximum water depth is 4300 m. The water masses in the Levantine basin consist of: (1) Atlantic water; (2) Levantine intermediate water; and (3) cold water from deeper areas. During summer, the Atlantic water mass is overtopped by warmer and more saline surface water down to 40 m water depth. Salinity concentrations during late summer rise to 39.5 PSU. In winter, vertical mixing processes change this distribution pattern of water masses.

The nutrient supply to the eastern Mediterranean is poor. The phosphate concentration of the deep water in the Levantine basin is only one-sixth of the concentration in the Atlantic Ocean, which in turn is rather low comapred to the other oceans. The eastern Mediterranean surface waters are oligotrophic, resulting in occasional limitation of phytoplankton primary production. Land-based sources, however, such as agriculture, industries and tourism-related activities, continuously supply the nearshore waters with new nutrients and pollutants. Further research is needed to quantify the effects of these discharges.

18.3 The Benthic Flora in Turkish Seas and Possible Effects of Eutrophication and Pollution

The seas around Turkey show marked differences in the floristic composition of the marine benthic vegetation (Table 18.3). The taxonomic composition and ecological characteristics of the Black Sea flora are

Table 18.3. The number of taxa in the Turkish seas

Taxonomic group	Black Sea	Black Sea Turkey	Marmara Sea	Aegean Sea	Mediterranean Turkey
Cyanophyta	20	13	11	26	45
Chlorophyta	99	50	87	82	82
Phaeophyta	104	53	94	95	85
Rhodophyta	190	140	229	237	228
Xanthophyta	1	—	—	1	—
Spermatophyta	3	3	—	5	5
Charophyta	—	2	—	2	1
Total	417	261	421	448	446

mainly determined by two factors: temperature and salinity. Because of the low salinity level of the Black Sea, Chlorophyta can be expected to dominate over Rhodophyta and Phaeophyta as for example observed in the coastal areas of Istanbul, Iğneada, Sile, Bartin, Sinop, Trabzon and Rize (C.A. Agardh 1823; J.G. Agardh 1851, 1876; Stockmayer 1909; Öztürk 1980; Altindağ 1990; Aysel et al. 1990; Aysel and Erduğan 1995). The total biomass of the Black Sea macrophytes has been estimated as 16×10^6 t, and the annual production was estimated at 17.6×10^6 t wet weight.

There have only been a few investigations on the algae of the Marmara Sea. The first records from the area, at the species level, came from Fritsch (1899), and further research was done by Öztiğ (1957). In the following years, research focused on macroalgae (Güner 1973; Zeybek and Güner 1973; Aysel and Güner 1979a, 1980; Aysel et al. 1984, 1991, 1993; Güner et al. 1987). In the Marmara Sea, which receives water from both the Black Sea and the Aegean Sea, the numbers of green and brown algal taxa are generally at the same level, although at times variations may occur, depending on the intensity of the currents.

The Turkish strait system lies in the country's most populated and industrialized region. Istanbul, with its surrounding districts, accommodates one-sixth of Turkey's population. Nearly 766 million m^3 domestic and industrial wastewater is discharged into the Marmara Sea annually. In addition, excessive sea-going traffic, illegal discharge of bilge water as well as tanker accidents also cause oil pollution. Organic waste products dumped into the Marmara Sea have led to the decline of certain organisms, including algae. Since 1986, a substantial increase in eutrophication has been observed.

The Aegaen Sea has a rich biological productivity and its hores (the Turkish part) include many areas suitable for algal growth. There have been many investigations on the floristic composition of the Aegean Sea. The first record from the area belonging to Turkish shore at the species level is from Karamanoğlu (1964), the second from Zeybek (1966, 1973), followed by a number of works on macroalgae by Aysel (1979, 1980, 1981, 1984, 1987, 1989), Aysel et al. (1977, 1986), and Aysel and Güner (1977, 1983a, 1985) between 1979 to 1989. In addition, some other workers have studied the macro- and microalgae (Güner and Aysel 1978a; Öztürk 1980; Dural et al. 1989, 1990, 1992; Cirik et al. 1990).

Some economic algae grow in the area, such as *Ulva lactuca, Gracilaria verrucosa, Gelidium capillaceum, Gigartina* sp., *Hypnea musciformis, Cystoseira* sp., *Dictyopteris membranacea,* and *Halopteris* sp. (Aysel and Güner 1978).

In the Aegean Sea, Mediterranean algae occur due to the effects of currents from Mediterranean deep waters, whereas due to the effects of surface currents of Marmara origin, algae from the Black Sea and the Marmara Sea dominate in northern Aegean waters and occur sometimes in the south. The boundary line of the distribution pattern of specific algae of Mediterranean origin is very sharp between the Aegean Sea and the Marmara Sea and is related to the effects of ecological factors. Roughly 90% of the Aegean algae can be found along the Mediterranean coasts, however, some characteristic species grow exclusively on the Aegean coasts.

Like the Mediterranean, the Aegean Sea is under the influence of heavy domestic and industrial pollution: 11% of the total pollution entering the Mediterranean reaches the Aegean Sea.

There have only been a few investigations on the Turkish Mediterranean algal flora. The first record from the area at the species level was made by Karamanoğlu (1964). Further research was performed by Zeybek (1969). In recent years, a few researchers (Cirik 1978; Öztürk 1985; Aysel et al. 1986; Zeybek et al. 1983, 1993; Aysel 1992) have studied the composition of the flora in the region. The flora comprises some algae of economic importance (e.g. *Cystoseira* sp., *Gelidium* sp.).

Surveys of some of the typical communities of benthic vegetation were carried out between 1978 and 1984 (Güner and Aysel 1978b, 1984; Aysel and Guner 1979b, 1982) and in 1992 (Güner et al. 1992; Sukatar et al. 1992). Some of the algal communities are summarized in Table 18.4.

Table 18.4. Littoral benthic communities along the Turkish coasts

A. Supralittoral zone
    1. *Rivularia biasolettiana* community (AS)
    2. *Littorina neritoides* (AS)

B. Mediolittoral zone
    3. *Chthamalus-Gloeocapsa-Entophysalis-Brachytrichia* community (AS)
    4. *Prophyra leucosticta* community (AS, BS)
    5. *Nemalion-Polysiphonia* community (AS)
    6. *Gelidiella pannosa* community (AS)
    7. *Ceramium ciliatum* community (AS, BS, MS)
    8. *C. diaphanum* (BS)
    9. *C. rubrum* (BS)
   10. *Rivularia-Patella* community (AS)

C. Intralittoral zone - upper layer

  (a) Photophilic communities
    11. *Codium dichotomum* (MS, AS)
    12. *Ulva lactuca* (MS, AS, TM)
    13. *Dasycladus - Anadyomene* (AS, TM)

Table 18.4. *(Contd.)*

14. *Padina - Dilophus* (MS, AS, TM)
15. *Dictyopteris membranacea* (MS, AS, TM)
16. *Stypocaulon scoparium* (MS, AS, TM)
17. *Cystoseira compressa* (AS, TM)
18. *C. corniculata* (AS, TM)
19. *C. crinita* (MS, AS, TM)
20. *G. elegans* (AS, TM)
21. *C. ercegovicii* (AS, TM)
22. *Corallina mediterranea* (MS, AS, TM)
23. *Tenarea undulosa* (AS, TM)
24. *Gracilaria verrucosa* (MS, AS)
25. *G. bursa pastoris* (AS, TM)
26. *Hypnea musciformis* (AS, TM)
27. *Gelidium capillaceum* (AS, TM)
28. *Halopithys incurvus* (MS, AS, TM)
29. *Laurencia obtusa* (MS, AS, TM)
30. *L. papillosa* (AS, TM)
(b) Sciaphilic communities
31. *Udotea - Peyssonnelia* (AS)
32. *Rhytiphloea - Spyridia* (AS)

BS, Black Sea; MS, Marmara Sea; AS, Aegean Sea; TM, Turkish Mediterranean

Some remarks can be made regarding the effects of eutrophication and pollution on the benthic flora along the coasts of Turkey. In certain coastal areas, pollution is increasing, accompanied by mass mortality of the macroalgae, mainly during summer after the phytoplankton bloom. This process causes a significant decrease in oxygen concentration and a concomitant production of hydrogen sulphide.

The environmental effects related to eutrophication can be outlined as follows:

1. Increase in numbers of Chlorophyta species and biomass due to an increase in loads of organic wastes, particularly an increase in Ulvales along the Black Sea, Marmara Sea and Aegean Sea.
2. Stimulating effects of chemical residues on mass growth and biomass resulting in an increase in specific green and red algae (Ulvales and *Gracilaria verrucosa*).
3. Complete dominance of green algae, with some occasional brown and red algae, in areas where millions of tons of fresh water are poured into inlets of large rivers.
4. Pollution and eutrophication may affect considerable numbers of macroalgae negatively, such as brown algae and particularly red algae of the Cryptonemiales and Corallinaceae.

5. Intensive collection of some economically important macroalgae may result in exhaustion of the available resources; these negative effects should be avoided.
6. Marine accidents (oil spills, etc.) may also have detrimental effects on the growth and reproduction of macroalgae.

In conclusion, the long coasts of Turkey offer a good habitat for biological life, and are suitable for diverse macroalgal growth, with the exception of some local areas such as the bays of Izmir or Iskenderun.

References

Agardh CA (1823) Species algarum.I. Gryp his, 531 pp
Agardh JG (1851) Species genera et ordines Floridearum II:Lund
Agardh JG (1876) Species genera et ordines algarum.III:Lipsiae
Altindag S (1990) Some species of *Ceramium* in the western Black Sea. EU Fish Coll Fish 6(21–24): 31–49 (in Turkish)
Aysel V (1979) Studies on some of the *Polysiphonia* Grev. species (Rhodomelaceae, Rhodophyta) from the bay of Izmir. EU Fac Sci Ser B 3(1–4): 19–42 (in Turkish)
Aysel V (1980) Some species of the *Polysiphonia* Grev. (Rhodophyta,Rhodomelaceae) from the Aegean coasts, Turkey. STRCT VII Sci Congr Bio Sec Kusadasi, Aydin pp 841–855 (in Turkish)
Aysel V (1981) Taxonomy and ecology of the Rhodomelaceae species in the upper infralittoral zone of the Aegean coasts, Turkey. Diss Thesis, 132 pp (in Turkish).
Aysel V (1984) The *Polysiphonia* Grev. species (Rhodomelaceae, Ceramiales) of the Aegean coast, Turkey. I.Sect: *Oligosiphonia*. Doga A2 8(1): 29–42 (in Turkish)
Aysel V (1987) The flora of the Aegean Sea, Turkey. II. Red algae (= Rhodophyta). Doga Tu Bot 11(1): 1–21 (in Turkish)
Aysel V (1989) The *Polysiphonia* Grev. species (Rhodomelaceae, Ceramiales) of the Aegean coast, Turkey. II. Sect: *Polysiphonia*. Doğa Tu Bot 13(3): 488–501 (in Turkish)
Aysel V (1992) The algal flora of the Turkish Mediterranean. E U Investigation Fund N 1988/019, EU Fac Sci Bio Dep. Bornova, Izmir, 267pp
Aysel V, Erdugan, H (1995) Check–list of Black Sea seaweeds, Turkey (1823–1994). STRTC, Doga TrJ Bot 19: 1–10 (in Turkish)
Aysel V, Güner H (1977) Some *Punctaria* species in Izmir Bay and distributional areas. EU Fac Sci Ser B 1(1): 375–384 (in Turkish)
Aysel V, Güner H (1978) The seasonal ecology of some useful algae in the Aegean coast. EU Fac Sci Ser B 2(1): 73–91 (in Turkish)
Aysel V, Güner H (1979a) Qualitative and quantitative studies of the algae population in the Aegean and Manrmara Seas. 3. *Gracilaria verrucosa* (Huds.) Papenf. population. EU Fac Sci Ser B 3(1–4): 111–118 (in Turkish)
Aysel V, Güner H (1979b) Qualitative and quantitative studies of the algae population in the Aegean and Marmara Seas. 2. *Dictyopteris membranacea* (Stackh.) Batt. population. EU Fac Sci Ser B 3(1–4): 85–93 (in Turkish)

Aysel V, Güner H (1980) Qualitative and quantitative studies of the algae population in the Aegean and Marmara Seas. 4. *Gelidium capillaceum* (Gmel.) Kütz population. EU Fac Sci Ser B 4(1–4): 141–153 (in Turkish)

Aysel V, Güner H (1982) Qualitative and quantitative studies of the algae population in the Aegean and Marmara Seas. 6. *Laurencia obtusa* (Huds.) Lam var. *obtusa* population (Rhodomelaceae, Ceramiales, Rhodophyta). Doga 6(3): 97–103 (in Turkish)

Aysel V, Güner H (1983a). the *Chondria* Agardh species (Rhodophyta, Ceramiales) of the Aegean coast, Turkey. Doga 7: 47–57 (in Turkish)

Aysel V, Güner H (1983b) The *Lophosiphonia* Falkenberg species (Ceramiales, Rhodophyta) of the Aegean coast, Turkey. Doga Tu Bio 10(3): 254–264 (in Turkish)

Aysel V, Güner H (1985) The *Alsidium* Agardh species (Ceramiales, Rhodophyta) of the Aegean coast, Turkey. Doga A₂ 9(3): 493–499 (in Turkish)

Aysel V, Zeybek N, Güner H (1977) New records from Turkish seashores I. *Liebmannia leveillei* J Ag. EU Fac Sci Ser B 1(1);275–280 (in Turkish)

Aysel V, Güner H, Zeybek N (1984) Some of the deep seaweeds of Turkey. II. Phaeophyta (=brown algae). Doga A₂ 8(2): 183–192 (in Turkish)

Aysel V, Güner H, Zyebek N, Sukatar A (1986) Some deep seaweds of Turkey. III. Rhodophyta (=Red algae). Doga Tu Bio 10(1): 8–29 (in Turkish)

Aysel V, Kesercioglu T, Güner H, Akçay H (1990) Trabzon marine algae. X Nat Biol Congr, 18–20 July, Erzurum 2: 183–192 (in Turkish)

Aysel V, Güner H, Dural B (1991) The flora of the Marmara Sea, Turkey. I. Cyanophyta and Chlorophyta. EU Fish Fac Fish Symp, 12–14 Nov 1991, Atatürk, C C Izmir, pp 74–112 (in Turkish)

Aysel V, Güner H, Dural B (1993) The flora of the Marmara Sea, Turkey. II. Phaeophyta and Rhodophyta. EU Fish J 10(37–39): 115–167 (in Turkish)

Cirik S (1978) Recheches sur la Végétation marine des cotes Turques de la mer Egée. Etude particuliere des Peyssonneliacées de Turquie. These Doct, III Cycle Biol Vég, Univ P et M Curie Paris, 172 pp

Cirik S, Zeybek N, Aysel V, Cirik S (1990) Note preliminaire sur la végétation marine de l'ile de Gökçeada (Mer Egeé, Nord, Turquie). Thalassografica 13 (Suppl 1): 33–47

Dural B, Güner H, Aysel V (1989) Taxonomical studies on the Ulvales order in Candarh Bay, Turkey II.Ulvaceae A. *Ulva* L. species. Doga Tu Bot 13(3): 474–487 (in Turkish)

Dural B, Güner H, Aysel V (1990) Algal flora of Yassicaada Island, Izmir Bay, Turkey. X Nat Biol Congr, 18–20 July, Erzurum. Bot Cire 2: 205–219 (in Turkish)

Dural B, Güner H, Aysel V (1992) The comparison of marine flora of Çeşme-Eskifoça with Turkey and Mediterranean. EU Fac Sci Ser B 14(2): 65–77 (in Turkish)

Fritsch K (1899) Flora von Constantinopol I.Kryptogamen. Densk. Mat Nat Cl.LXVIII, pp 219–250

Güner H (1973) Preliminary studies on the algal vegetation of Istanbul islands. STRCT IV Sci Congr, 5–8 Nov 1973, Ankara, pp1–9 (in Turkish)

Güner H, Aysel V (1978a) New records from Turkish seashores. II. *Catenella repens* (Lightf.) Batt. Plant 5(1): 85–90 (in Turkish)

Güner H, Aysel V (1978b) Qualitative and quantitative studies of the algae population in the Aegean and Marmara Seas. I. *Ulva lactuca* L. population (Chlorophyta). EU Fac Sci Ser B 2(1): 55–71 (in Turkish)

Güner H, Aysel V (1984) Qualitative and quantitative studies of the algae population in the Aegean and Marmara Seas. 5. *Hypnea musciformis* (Wulf) Lam. population (Hypneaceae, Gigartinales, Rhodophyta). Doga A₂ 8(3): 343–349 (in Turkish)

Güner H, Aysel V, Sukatar A (1987) Taxonomical studies on the coastal algae of Marmara Sea, Türkiye. VIII Natl Biol Congr, 3–5 Sept 1986 Izmir II: 483–493 (in Turkish)

Güner H, Aysel V, Sukatar A (1992) Preliminary studies on the algae of Gencelli Bay Turkey. XI. Natl Biol Congr, 24–27 June, 1992 Elazig. FU Hidr and Env Bio Sec. pp145–153 (in Turkish)

Karamanoglu K (1964) some marine algae of Mamaris and Güllük coasts. Türk Biol J 14(3): 32–38 (in Turkish)

Öztig F (1957) Marine vegetation of the Erdek coasts. Türk Biol 7(1): 12–13 (in Turkish)

Öztürk M (1980) Distribution and taxonomy of some Dictyotaceae (Phaeophyta) from the bay of Izmir. EU Fac Sci, 48 pp

Öztürk M (1985) Taxonomy and distribution of the members of Phaeophyta (=brown algae) from the Aegean and Mediterranean coastal areas of Turkey. Diss, Thesis EU Fac Sci Bio Dep, 155pp (in Turkish)

Öztürk M (1988) A research on plant organisms at the upper infralittoral of the Akliman and Hamsaros bays. IX. Natl Biol Congr, 21–23 Sept 1988. CU Sci-Arts Fac Biol Dep, Sivas Gen and Sist Bot Sec 3: 329–343 (in Turkish)

Stockmayer S (1909) In: Handel – Mazetti, H.F. Ergebnisse einer botanischen Reise in das Pontische Randgebirge in Sands. Trapezunt Ann K.K.Nat Hofm Wien 23: 6–212

Sukatar A, Güner H, Aysel V (1992) Qualitative and quantitative studies on *Cystoseira ercegovicii* Giaccone population in the southern Aegean, Turkey. XI. Natl Biol Congr, 24–27 June, 1992. Elazig. FU Hidr and Env Bio Sec, pp199–206 (in Turkish)

Zeybek N (1966) Some algae of Aegean coasts. EU Fac Sci Scientific Rep Ser N 27: 1–29 (in Turkish)

Zeybek N (1969) The marine algae of Turkish Mediterranean. I. Bodrum to Finike. EU Fac Sci. TBAG-24 N Proj, 40 p

Zeybek N (1973) Meeresalgen aus der Turkei.I.Die Buchten von Edremit und Saros am Ägäischen Meer. 2. Die Küste von Igneada bis Sile am Schwarzen Meer. Sep Verh Schweiz Nat Forsch Gesellschaft: 95–100

Zeybek N, Güner H (1973) Marine algae of the Dardanelles and Bozcaada. EU Fac Sci, Scientific Rep Ser N 145: 1–19 (in Turkish)

Zeybek N, Güner H, Aysel V (1983) Some deep seaweeds of Turkey. I. Chlorophyta (=green algae). Doga A 7(3): 547–556 (in Turkish)

Zeybek N, Güner H, Aysel V (1993) The marine algae of Turkey. Proc 5th OPTIMA Meet, 8–15 Sep 1986, Istanbul. IU Fac Sci, pp 169–197

Part B.III
The Black Sea

19 The Black Sea

19.1 Hydrological and Chemical Characterization of the Black Sea

In order to understand the changes to the Black Sea ecosystems caused by human impact especially during the past 20–30 years, it is necessary to describe in short the physical and chemical peculiarities for which the Black Sea is also sometimes named the *"Unicum hydrobiologicum"*.

The Black Sea is an intercontinental basin, connected with the Mediterranean and the ocean only by the narrow Bosphorus Straits. Geologically, the Bosphorus is a tectonic ditch, only 250 to 550 m wide and 27.5 to 120 m deep. Two bodies of water flowing in opposite directions are separated from each other by a sharp salinity gradient of approximately 18 PSU in the surface layer to 37–40 PSU in the water below. The boundary layer slopes down from about 10 m depth in the south to 40 m in the north, depending on the volumes of flow and on the metereological conditions.

The geomorphological "threshold" features of the Bosphorus Straits are a barrier for: (1) the influence of Mediterranean tidal fluctuations on the Black Sea; (2) the exchange of water between the Mediterranean and the Black Sea; and (3) the migration of plants and animals.

Another specific feature of the Black Sea is its role as a recipient of a drainage area of about 2.4 million km², covering 13 different countries.

The hydrological balance is shown in Table 19.1. River discharges contribute about 54% of the total input, however, they represent only 0.075% of the 537 000 km³ volume of the Black Sea. The brackish character of the so-called Pontic Basin must also be related to the origin of this sea, which, in fact, is a mixture of three different water bodies:

Complexui Muzeal de Stiinte ale Naturii–Constanta, Bul. Mamaia nr. 255, 8700 Constanta, Romania

Ecological Studies, Vol. 123
Schramm/Nienhuis (eds) Marine Benthic Vegetation
© Springer-Verlag Berlin Heidelberg 1996

Table 19.1. The water balance of the Black Sea

Sources	Inputs (Km³)	(%)	Outputs (Km³)	(%)
Rivers	336	53.16	–	–
Bosphorus	123	19.46	260	41.14
Kerch (Azov Sea)	53	8.39	32	5.06
Precipitation	120	18.99	–	–
Evaporation	–	–	340	53.80
Total	632	100.00	632	100.00

Table 19.2. Dissolved and suspended materials (t year^{-1}) carried by the Danube River into the Black Sea. (After Serbanescu et al. 1978)

Materials	Quantities (t/year)
Total suspended materials	37 912 000
PO_4^{-3}-P	21 850
Si	405 700
NO_3-N	30 044
NO_2-N	4 078
Organic matter (BOD mg O_2/l)	1 236 000
Cl	3 360 000

(1) the brackish waters of the primary Pontic Lake; (2) the fresh water from precipitation and continental flows; and (3) the Mediterranean marine waters.

The vertical stratification with distinct gradients in salinity, temperature, as well as in oxygen and hydrogen sulfide concentrations is another important feature of tha Black Sea (Bacescu et al. 1971).

The stable stratification is the cause for almost no vertical water exchange between the surface layer down to approximately 200 m depth and the deep water below which is of great importance for the general circulation of organic matter.

Most of the organic material produced in the photic surface layer sinks through the discontinuity layer and is deposited in the depths. Because of this, the balance of organic production would be critical, if the terrestrial discharges were not the main source of fertilizers for the Black Sea. After collecting 192 effluents, the Danube alone carries huge amounts of dissolved and suspended materials into the sea (Table 19.2). The values given vary from year to year; however, they indicate an increasing tendency during recent years.

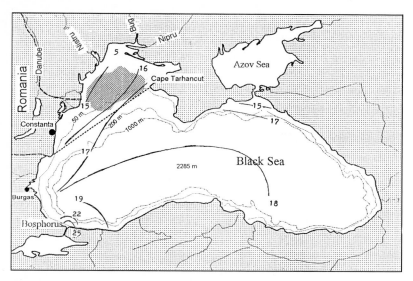

Fig. 19.1. Distribution of *"Phyllophora fields"* (*dotted area*) in the northwestern shelf area of the Black Sea

The general circulation of the water masses is primarily dependent on the morphology of the sea shore. In addition, it is influenced by the water exchange of water masses through the Bosphorus, as well as by the continental run-off.

The shoreline and subaquatic relief of the Black Sea is characterized by a great morphological, lithological and biogeographical variability, which has direct implications for the aquatic macrovegetation (Fig. 19.1; Bacescu et al. 1971). On the other hand, the sinuosity coefficient of the Black Sea shoreline is very reduced: the average is 1.77, and along the Romanian seashore values as low as 1.05–1.12 can be found (Trufas 1969). The stretched, open coastlines with only few capes, bays or peninsulas allow wind-induced water motion to persist.

The hydrodynamics of the open shoreline have implications not only for the attachment of macrophytes and stability of phytobenthic communities, but also for turbidity and transparency, i.e. the light climate of the coastal systems. As a result, the vertical distribution of the phytobenthos varies considerably, ranging from 0.5–12 m along the Romanian shores, and going down as deep as 45 m along the Caucasus and Bulgarian coasts.

The peculiarities of the physical environment create a variety of habitats, clearly distinguished from each other by their production potential. Among these, the halistatic areas show the lowest productivity, as for example the northwestern shelf of the Black Sea. On the other hand, this part of the Black Sea belongs to the areas most affected by human activities (Table 19.3). Because of this, and because of its geomorphological, hydrological, chemical and biological peculiarities (Zaitev 1979), eutrophication and pollution take place here.

The northwestern shelf area is bordered by part of the Bulgarian and Ukrainian, and by the entire Romanian coast, all in all nearly 1000 km long (Fig. 19.1). This shallow area with an average depth of only 30 m (maximum 100 m), a surface area of 63 900 km^2 and a volume of approximately 1910 km^3 receives about 226 km^3 land runoff annually, mainly as river discharges from the Danube, Nistru, Bug and Nipru.

High industrialization and urbanization of the bordering countries, particularly Romania and the Ukraine, and intensive agriculture based on extensive use of chemicals and irrigation with water from the tributary rivers, are the main sources of eutrophication and pollution. In fact, the northwestern shelf area of the Black Sea is the recipient of almost all anthropogenic waste from central and southeastern Europe, and this is reflected not only in significant changes of its hydrochemical pattern, but also in the structure and function the pelagic and benthic systems (Table 19.4).

19.2   Eutrophication and Pollution in Romanian Coastal Waters

In a synthesis of work on eutrophication and pollution in shallow waters, based on the analysis of the present state of the Black Sea coastal ecosystems, Gomoiu (1985a,b) distinguished between eutrophication and pollution as follows: (1) eutrophication: the improvement of the trophic situation on all the structural levels of the ecosystem; the improvement is the result of the augmentation of the quantities of nutrients up to the maximal critical level of the self-regulation capacity of the biochemical cycle; (2) pollution: the excess of nutrients beyond a critical maximum level that disturbs the cycling of matter and the flow of energy in the ecosystem.

The changes that occur with increasing eutrophication until it reaches the stage of pollution include several phases:

1. Increase in nutrient levels.
2 Emergence of planktonic algal blooms.

Table 19.3. Types of human activity and their impact on the marine coastal environment. (After Gomoiu 1985b)

Socio-economic activities	Morphology of the coastline	Dynamics and direction of the currents	Organic matter of sediments and seawater	Inorganic matter in the seawater	Type and morphology of the substratum	Thermal structure of the seawater	Optical structure of the seawater	Flora and fauna	Pattern of dissolved oxygen
Urban development	x	x	x	x			x	x	x
Entertainment, sport, tourism								x	
Balneology, thalassotherapy			x			x			x
Building of harbors, industrial platforms	x	x	x	x	x		x	x	x
Construction, consolidation of cliffs, beaches, and protective breakwaters	x	x	x	x	x		x	x	x
Harbor and industrial activities			x	x		x	x	x	x
Exploitation of marine and beach sand	x	x	x	x	x			x	
Marine drilling platforms, mineral oil and crude oil terminal lines			x	x	x			x	
Construction of navigable canals	x	x		x			x	x	
Commercial and sport fishing			x	x			x	x	
Agriculture, chemical industry and irrigation			x	x			x	x	x
Aquaculture				x			x	x	x

Table 19.4. Annual means of the main nutrient concentrations ($\mu g \ dm^{-3}$), organic matter (mg m^{-3})and of the phytoplanktonic biomass (mg m^{-3}) off the shallow water of Constanta. (Bodeanu 1989)

Period	1966–1970	1970–1975	1976–1980	1989
PO$_4$-P	10.50	177.50	197.50	137.90
NO$_3$-N	22.50	315.70	188.80	102.70
Organic matter	1.96	2.32	2.75	2.75
Phytoplankton biomass	495	719	2244	11967

3. Enrichment of dissolved and particulate organic matter both in sea and sediments.
4. Oxygen-consuming processes often followed by anoxia and increase in carbon dioxide with detrimental effects on plants and animals.
5. Decrease in zooplankton diversity compensated by the massive growth of only a few species, particularly those utilized by filter feeders.
6. Massive mortalities of flora and fauna, especially benthic components.
7. Decline of planktonic and benthal fish species.
8. Disappearance of Mammalia.

All these stages may also occur as a result of other harmful human activities as listed in Table 19.3. Therefore it is difficult, if not impossible, to separate the effects of eutrophication, pollution or other detrimental impacts from each other.

19.3   The Effects of Eutrophication on Phytobenthic Communities

19.3.1   The *Cystoseira* Belt

The brown algal belt of *Cystoseira*, which is common throughout the Mediterranean basin, is also characteristic of the shallow coastal waters of the Black Sea. *Cystoseira barbata* and *C. crinita* are the two physionomically most important species of the littoral along the coasts of Romania (between Cap Midia and Vama Veche), Russia, Bulgaria and Turkey. The two species and their ecotypes may form dense populations between 0.5–32 m depth. Depending on the area and season, the biomass ranges between 1.5–8.0 kg m^{-2} and may reach up to 21 kgm^{-2} wet weight (Kalughina-Gutnik 1975). A luxurious and diverse macrophytic flora develops together with *Cystoseira*, forming an association typical for the Black Sea (Muller and Bodeanu 1969). The quantitative estimates of *Cystoseira* standing stocks for the different subareas are given in Table 19.5.

The biomass of *Cystoseira* along the Romanian coast was first estimated in 1962. In the following three decades algal biomass declined rapidly to 5524 t wet weight in 1971, 4305 t in 1972, 775 t in 1973, 120 t in 1989, and only 89 t in 1991 (Vasiliu 1978; Vasiliu, unpubl. data).

One of the main reasons for the observed drastic decline is probably the deterioration of the light climate due to excessive growth of plankton–up to 807.6 × 10^6 cells per liter were reported (Bodeanu 1989)–and associated herewith increasing loads of dissolved and suspended organic materials. Since both *Cystoseira barbata* and *C. crinita* can be considered as "photophileous" algae, and Romanian coastal waters in general show a lower transparency compared to other Black Sea areas (Skolka and Vasiliu 1968), these seaweeds usually settled between 0.5 and 6.5 m depth. At present, however, *Cystoseira* can only be found between 1.5 and 3.5 m depth, except for the southern coast off Vama Veche where it still descends down to 5 m (Fig. 19.2).

Besides changes in light conditions other factors may contribute to the observed changes in biomass and distribution of *Cystoseira*. These include a general deterioration of the "chemical quality" of nearshore waters as a secondary effect of eutrophication (e.g. formation of certain "noxa" in connection with degradation of increased organic matter).

Another factor may be increased silting which can hinder settlement and attachment of spores or may be harmful to young *Cystoseira*. In addition, the main growth and reproduction periods March to May and September to October coincide with blooms and substantial mortalities

Table 19.5. Estimates of *Cystoseira* standing stocks in subareas of the northwestern Black Sea (10^3 t wet weight)

Area	Biomass
Southeastern littoral	74.8
North Caucasian coast	991.0
South crimean coast	353.6
Karkinit bay	112.0
Bulgarian Coast	331.2
Romanian coast	105.0

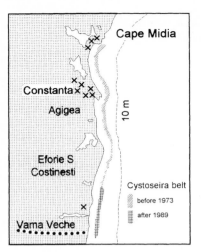

Fig. 19.2. The *Cystoseira* belt along the Romanian coast before 1973 and after 1989. The *crosses* indicate areas of main human activities

among macrophytes, molluscs, crustaceans, or fish, producing a huge mass of dead plant and animal remnants which are rolled and cast by the hydrodynamic play of waves and currents on the bottom inhabited by young *Cystoseira barbata* and *C. crinita*.

The above interpretation was supported by the results of earlier field experiments on the recruitment and restoration of the *Cystoseira* belt (Vasiliu 1978). In experimental areas with an initial density of 667 to 1850 *Cystoseira barbata* individuals per square meter, the development, growth and morphological characteristics of the plants were studied monthly. During the first 8 months under normal favorable conditions, the young plants developed normally and reached 6 cm height, on average. During this period, the above-mentioned eutrophication effects emerged in some test areas (Cape Midia, Agigea, and Tuzla) with massive deposits of animal and algal debris. Up to 5 kg m^{-2} of broken or intact shells, and 0.62 kg m^{-2} wet weight of algal debris (*Cladophora* sp., *Ceramium* sp., and especially *Enteromorpha linza*) were measured in the three test areas, respectively (Vasiliu 1980).

Annual monitoring of these processes from 1969 to the present shows that during the period from 1969 to 1975 off the Romanian coast, *Cystoseira barbata* and especially *C. crinita* declined rapidly, and a general "aging" of algal populations occurred.

In summary, we may say that both *Cystoseira* species have dramatically declined or disappeared from north to south along the Romanian coast between Cape Midia and Vama Veche. At present, the total remaining standing stock of *Cystoseira barbata* is not more than 89 t wet weight. These changes can clearly be related to human activities, and if no protective measures are taken, the complete extinction of the remaining *Cystoseira barbata* cannot be avoided.

The conspicuous recent decline of the *Cystoseira* biomass along the Romanian and other coasts of the Black Sea, will necessarily have consequences for the associated flora and fauna. The *Cystoseira* belt provides an ideal substrate and habitat for numerous photophileous and sciaphileous algal species. In the Black Sea the following four associations have been distinguished of which the first two have been described for the Romanian coasts:

1. The *Cystoseira barbata – Ulva rigida* association; polydominant, multiannual, mesosaprobic, optimum abundance depth 1 to 3 m, maximal 93 macroalgal species.
2. The *Cystoseira crinita – C. barbata – Cladostephus verticillatus – Corallina mediterranea* association; polydominant, multiannual, oligosaprobic, optimum depth between 1 – 10 m; up to 120 species.
3. The *Cystoseira barbata – C. crinita – Dilophus fasciola* association; oligodominant, multiannual, oligosaprobic, optimum depth 1 – 3 m; up to 60 species can be found in this association.
4. *Cystoseira barbata – Phyllophora nervosa – Cladophora dalmatica;* a polydominant, multiannual, oligosaprobic phytocoenosis; optimum growth-depth ranging between 12 – 15 m; there is a maximum of 68 species.

In recent years, the *Cystoseira – Ulva rigida* association (1) has been found only in some limited areas, far away from eutrophication or pollution (Fig. 19.2).

Among the most abundant species associated with *Cystoseira* were the red algae *Laurencia pinnatifida, L. coronopus, L. obtusa, Polysiphonia subulifera, P. opaca, Dasya pedicellata, Chondria tenuissima, Gracilaria verrucosa, Dermatolithon cystoseirae,* and the brown algae *Sphacelaria cirrhosa, Cladostephus verticillatus, Corynophlaea umbellata, Feldmannia irregularis, Stilophora rhizoides.*

The decreasing number of species recorded in Romanian waters is a warning signal for the deterioration of nearshore waters as a result of human impact Whereas Kalughina-Gutnik (1975) in a comprehensive monograph on macroscopic algae of the Black Sea listed 101 species for the Romanian coasts from papers published between 1908 – 1968, Skolka (1969) identified only 77 macroalgal species between Cape Midia and Vama Veche (Table 19.6). During a survey of the biomass and productivity in 1972 – 1981, only 69 seaweeds were found (Vasiliu 1984).

The substratum between 0.5 – 6.5 m depth, formerly inhabited by the *Cystoseira* associations, is now occupied by a diversity of other

Table 19.6. Number of macroalgal species along the Romanian Black Sea coasts between 1908–1981. (Skolka 1969; Kalughina-Gutnik 1975; Vasiliu 1984)

Phylum	Periods 1908–1968	1969	1981
Chlorophyta	36	23	29
Phaeophyta	20	13	9
Rhodophyta	45	41	31
Total	101	77	69

Table 19.7. The phytosaprobic structure of the macroscopic algae in the Romanian Black Sea littoral

Type	Green algae		Brown algae		Red algae	
	No. of species	(%)	No. of species	(%)	No. of species	(%)
Polysaprobic	11	31	0	0	14	17
Mesosaprobic	14	40	19	73	25	44
Oligosaprobic	10	29	7	27	19	39
Total	35	100	26	100	58	100

macroalgae which belong mainly to the meso- and polysaprobic type (Table 19.7).

In locations with little water movement such as harbors or small bays, *Enteromorpha, Cladophora, Ulothrix* can also develop on unstable substrate, building up monospecies populations (Celan and Vasiliu 1975).

The recolonization of denuded substratum took place under conditions ideal for mass development of certain algae, i.e. sufficient nutrients together with optimal temperature and light. Massive algal blooms, in particular of *Enteromorpha liza, Cladophora laetevirens* and *Ulva rigida*, have been observed along the Romanian coasts at Agigea, Eforie, Tuzla or Vama Veche in 1972–1979. The highest biomass (wet weight) measured was 4390 g m^{-2} for *Cladophora laetevirens* and 4860 g m^{-2} for *Enteromorpha linza*, the latter species reaching up to 80 cm length and 60 cm width. On a stretch of shore near Agigea of only 1.5 km length, for example, the algal biomass deposited on the beach in May–July 1972 was estimated at 90 t wet weight (Vasiliu 1980).

Similar observations have been reported for other areas in the Black Sea. On the Ukrainian coast near the bays of Odessa and Tarhankut, for example, the green algae biomass reached up to 2000 g m^{-2} wet weight at 4–5 m depth (Kalughina-Gutnik 1973).

Eutrophication and/or pollution are probably also the cause for the conspicuous decline of the marine phanerogams *Zostera marina* and *Z. noltii*, the latter being the predominant seagrass in Romanian waters (Bacescu et al. 1971). They have disappeared in 95% of the so-called enclaves (the term "enclave" is used by most Romanian researchers only for these marine phanerogams because both species have never formed large, compact fields as typical for the Ukrainian as well as for the Bulgarian coasts). Nowadays, along the Romanian coast between Cape Midia and Vama Veche, only few individual plants of *Zostera marina* can be found off Tuzla. *Zostera noltii* occurs only in small "clusters" of maximal 5–10 m² each, at 1–2 m depth off Tuzla and Vama Veche.

19.3.2 The *Phyllophora* Fields

An important element of the benthic vegetation of the Black Sea is the algal biomass pertaining to the genus *Phyllophora*, particularly in the so-called *Phyllophora* field, which later on became known under the name "Zernov's Field". The "field" is located in the sea area northeast of the Danube delta which is characterized by its very high productivity. It extends to a depth of 30 to 50 m, covers an area of more than 10 900 km², and is surrounded by a circular current (Fig.19.1). The first quantitative evaluation of the *Phyllophora* biomass approached 10 million tons wet weight (Kalughina-Gutnik 1975; Kaminer 1978). Since the beginning of the 1940s, these algae have been utilized by an agar-producing industry centered in Odessa. The biomass of the three species *Phyllophora nervosa*, *P. brodiaei* and *P. pseudoceranoides* exceeded by far all the other forms of marine macrovegetation, either seaweeds or spermatophytes, in the Pontic Basin (Vasiliu and Bodeanu 1972).

In 1960–1965, when ecological conditions were still considered normal by Russian experts, the stock of *Phyllophora* was estimated as not more than 5.5 million tons wet weight. We assume that the surprising difference between these two estimates is due either to an overestimation by Kalughina-Gutnik and by Kaminer or the result of uncontrolled overexploitation of the algal stock during years between the 1940s and 1960. In 1978 Kaminer estimated the *Phyllophora* biomass in the "Field" at 1.5 to 2.5 million tons wet weight.

Although there may be many causes for this alarming decline (cf. Table 19.3), we believe that eutrophication has played a major role. An indirect effect of eutrophication is possibly the deterioration of the light intensity due to the enormous increase in phytoplankton biomass from about 500 to 12 000 mg m³ during the past three decades (Table 19.4).

The formerly dominant diatom species have been replaced by dinoflagellates. Autotrophic species have partly been substituted by mixotrophic organisms, for example, *Exuviaella cordata* or *Goniaulax polygramma*. This change has probably supported the quantitative development of zooplankton (Bodeanu and Roban 1989).

Dinoflagellates may have noxious effects on the benthic communities by excretion of toxins, causing mass mortalities (Gomoiu 1985a). On this "ecological background" the formerly characteristic fauna of the *Phyllophora* community diminished significantly. Species of sponges, crustaceans, amphipods and isopods, fish, etc. disappeared (Zaitev 1979).

The described effects occurred particularly in the area bounded by the Chilia arm of the Danube and the Nistru bank. In August/September 1973, mass mortalities were observed at 6–23 m depth in an area covering 350 000 ha. An exception were the polychaetes which are known to be more resistant to toxins and to hypoxia. In the following years, the phenomenon was observed again and it extended even to areas towards the open sea, covering 1 million ha in 1978. Also, large areas of "Zernov's field" and of the bays of Karkinit, Tendra, and Egorlitk were affected (Zaitev 1979).

19.4   Summary

The observed changes in the benthic vegetation of the Black Sea, in particular along the Romanian coasts, are most probably the result of the specific hydrographic conditions. Industry and intensive agriculture along the coasts and in the drainage basin of the Black Sea, particularly in the north with the Danube as the main conveyer of nutrients and pollutants, are the main sources of eutrophication with the following effects:

1. Disappearance of the *Cystoseira* belt, with the complete extinction of *C. crinita* and a drastic decline in *C. barbata*, and a decline in biomass from 10 500 t wet weight to only 89 t wet weight today.
2. Decreasing species numbers, especially among red and brown algae; of the 101 species identified in 1968, only 69 can be found at present.
3. The last two species of marine phanerogams, *Zostera marina* and *Z. noltii* have disappeared or are disappearing; only individual plants or small patches can be found in Romanian waters.
4. The formerly qualitatively and quantitatively important *Phyllophora* communities ("Zernov's field") show alarming signs of destruction. The algal biomass has declined, and a great deal of the associated

flora and fauna (sponges, crustaceans, amphipods, isopods, etc.) has disappeared.

References

Bacescu M, Müller GJ, Gomoiu MT (1971) Ecologia marina – Cercetari de ecologie bentala in Marea Nearga, vol IV. Editura Academiei: Republic:: Socialiste Romania, Burcuresti, 358 pp

Bodeanu N (1989) Algal blooms and development of the main phytoplanktonic species at the Romanian Black Sea littoral under eutrophication conditions. Institutul Român de Cercetări Marine (IRCM) – Cercetari Mar 22: 107–126

Bodeanu N, Roban A (1989) Les développements massif du phytoplancton des eaux du littoral roumain de la mer Noire au cours de l'annèe 1989. Institul Român de Cercetări Marine (IRCM) – Cercetari Mar 22: 127–146

Celan N, Vasiliu F (1975) Nouvelles contributions à la connaissance de Enteromorpha du littoral roumain de la mer Noire. Institul Român de Cercetări Marine (IRCM) – Cercetari Mar 8: 83–89

Gomoiu MT (1985a) Problèmes concernant l'eutrophisation marine. Institul Român de Cercetări Marine (IRCM) – Cercetari Mar 18: 7–51

Gomoiu MT (1985b) Probleme ale reconstructiei ecologice in zonele marine costire de la litoralul romanesc. In: Ecologia si protectia ecosistemelor. Constanta 5: 68–72

Kalughina–Gutnik AA (1975) Fitobenthos Carnogo moria. IZd Naukova, Kiev, 246 pp

Kaminer KM (1978) Cernomorskaia filofora v usloviah anthropoghennogo vozdeistva. Tezisi docl II confer po biologhia selfa, Sevastopol, vol 1, pp 50–61

Müller GJ, Bodeanu N (1969) Date preliminare asupra populatiilor algale si animale asociate vegetatiei de Cystoseira barbata de la litoraluu romanesc al Marii Negre. Hidrobiologia 10: 279–289

Serbanescu O, Pecheanu I, Mihnea R, Cuingioglu E (1978) Le Danube comme ageant d' eutrophisation de la mer Noire. IV–es Journèes Etud Pollutions, Commission internationale pour l' exploration scientifique de la mer Méditerranée, Monaco, pp 457–459

Skolka H (1969) A propos de la répartition des algues macrophytes le long de la côte roumain de la mer Noire. Rev Roum Biol Bot 16, 4: 363–368

Skolka H, Vasiliu F (1968) Quelques observations sur le régime de la lumière dans les eaux de la partie ouestique de la mer Noire. Trav Mus Hist Nat grigore Antipa VIII: 285–289

Trufas V (1969) Hidrologia RSR I, Marea Neagra. Universitatea Buécuresti, 122 pp

Vasiliu F (1978) Donées sur l' ecologie et la productivité des espèces de Cystoseira de la mer Noire. Institul Român de Cercetări Marine (IRCM) – Cercetari Mar 13: 147–161

Vasiliu F (1980) La production des espèces d'Enteromorpha du littoral roumain de la mer Noire. Institul Român de Cercetări Marine (IRCM) – Cercetari Mar 13: 147–161

Vasiliu F (1984) Productia algelor macrofite la litoralul romanesc al Marii Negre. Ministerul Educaţiei şi Invatamantului Bucuresti, 23 pp

Vasiliu F, Bodeanu N (1972) Répartition et quantité d' algues rouges du genre Phyllophora sur le platforme continentale roumain de la mer Noire. Institul Român de Cercetări Marine (IRCM) – Cercetari Mar 3: 47–52

Zaitev I (1979) Problémes biologiques de la partie nord-ouest de la mer noire. Institul Român de Cercetări Marine (IRCM) – Cercetari Mar 12: 7–32

Conclusions

W. Schramm

The preceding chapters provide ample information on the response of different types of marine benthic vegetation to eutrophication along the European coasts under a wide range of different climatic and hydrographic conditions, and with different sources and levels of eutrophication. Despite qualitative or quantitative variations in the time course of the eutrophication process due to such local differences, a common pattern is obvious, as has already repeatedly been described earlier (e.g. Wallentinus 1981; Kemp et al. 1983; Nienhuis 1983, 1993; Sand-Jensen and Borum 1991; Vogt and Schramm 1991).

In general, we can observe a sequence of successive phases with increasing eutrophication as depicted in Fig. 1 in a schematic model where we have distinguished between four different phases of the eutrophication process.

Un-eutrophicated ("healthy") marine or brackish shallow coastal waters are characterized by a balanced nutrient regime, i.e. the nutrient load is low and nutrient concentrations may temporarily become limiting to primary production (phase I). In shallow nearshore systems, the dominant primary producer communities usually consist of perennial benthic macrophytes, such as seagrasses and other phanerogams on soft bottoms, or long-lived seaweeds on hard substrates, whereas seasonal macroalgae or phytoplankton play a lesser role in terms of biomass and production (see Fig. 4). Increasing nutrient loading, however, favours these seasonal forms.

A second phase (II) from slight to medium eutrophication is therefore characterized by increasing blooms of "nutrient opportunists", in particular fast-growing epiphytic macroalgae and bloom-forming phytoplankton taxa. In contrast, phanerogam and perennnial macroalgal communities gradually decline, usually combined with a change in their depth distribution limits, and finally disappear. During phase II, parallel to the blooms or mass development of seasonal fast growing seaweeds,

Institut für Meereskunde, Universität Kiel, Düsternbrooker Weg 20, 24105 Kiel, Germany

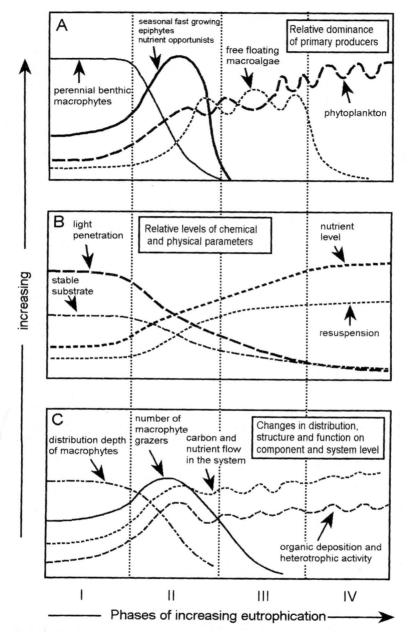

Fig. 1. Schematic presentation of the relative changes in the dominance of primary producers and some related structural and functional parameters during successive phases in the progress of eutrophication

often an increase in unattached or free-floating macroalgae can be observed.

With further increasing nutrient loads towards hypertrophic conditions in phase III, free-floating macroalgae, in particular "green tide"-forming taxa (e.g. *Ulva, Monostroma, Enteromorpha* spp.) alternating with heavy uncontrolled phytoplankton blooms, dominate and more or less completely replace the perennial and slow-growing benthic macrophytes. Finally, in phase IV under hypertrophic conditions with continuously high nutrient concentrations, phytoplankton constitute the dominant primary producers.

The first step in the eutrophication process, i.e. the increasing occurrence of seasonal fast-growing nutrient opportunists is without question a direct effect of increasing nutrient loads. It can be explained by the specific nutrient economy of these forms, i.e. by their ability to take up nutrients and to grow much faster than long-lived perennial macrophytes, when additional nutrients become available in the water column (Figs. 2, 3). In a nutrient-balanced system, under occasional nutrient

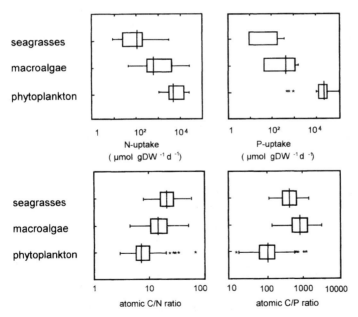

Fig. 2. Box plots showing the ranges of nitrogen and phosphorus uptake rates and of atomic ratios of carbon to nitrogen and phosphorus, respectively, for seagrasses, macroalgae, and phytoplankton. *Boxes* encompass the 25 and 75% quartiles of all the data for each plant group, the *central vertical line* represents the median, *bars* extend to the 95% confidence limits, and *asterisks* mark observations extending beyond the 95% confidence limits. (After Duarte 1995)

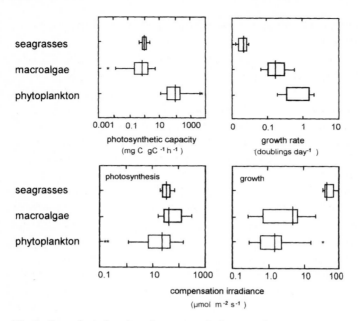

Fig. 3. Box plots showing the range of photosynthetic capacity, growth rates, and compensation irradiance for photosynthesis and growth of seagrasses, macroalgae and phytoplankton. *Boxes* encompass the 25 and 75% quartiles of all the data for each plant group, the *central vertical line* represents the median, *bars* extend to the 95% confidence limits, and *asterisks* mark observations extending beyond the 95% confidence limits. (After Duarte 1995)

limitation, the comparatively slow-growing perennnial macrophytes compete successfully with the faster growing seasonally occurring macro- or microalgae. One explanation is their significantly lower nutrient requirements as pointed out by Duarte (1995; Fig. 2). Another reason is that benthic plants can take advantage of the nutrient pools in the sediments. Rooting plants like seagrasses are able to acquire substantial proportions of their nutrient demands directly from the sediments in which nutrient concentrations can be higher by orders of magnitude compared to the overlying water column (e.g. Patriquin 1972). Macroalgae or the microphytobenthos living on the bottom can effectively scavenge the nutrients either released from the sediments or regenerated within the algal communities (e.g. Kautsky and Wallentinus 1980; Schramm et al. 1988). Another competitive advantage of seagrasses and perennial macroalgae over seasonal short-lived forms is their capability to accumulate and store nutrients when available (e.g. during winter) for later use when nutrients are depleted in the water (e.g. Chapman and Craigie 1977; Schramm et al. 1988). In addition, seagrasses and some macroalgae are able to translocate and store their internal nutrient resources for further use. In seagrasses, internal recycling or reabsorption of nutrients from the old leaves contributes nearly one fourth, or under oligotrophic conditions more than one-half of their annual nitrogen demands (e.g. Patriquin 1972; Borum et al. 1989; Hemminga et al. 1991; Pedersen and Borum 1992).

Field observations available so far, as well as numerous experimental works on the effects of nutrients on marine primary producers, suggest that the typical sequence of changes in benthic vegetation under eutrophication conditions cannot be attributed to increasing nutrient levels alone. Other indirect, direct or secondary effects and feedback mechanisms, triggered by the proliferation of fast growing nutrient opportunists as a direct response to increased nutrient levels, must become effective in the further progress of eutrophication (e.g. Harlin and Thorne-Miller 1981; Twilley et al. 1985; Vogt and Schramm 1991; Sand-Jensen and Borum 1991; Neundorfer and Kemp 1993). Such effects have recently been summarized and comprehensively discussed by Duarte (1995).

Among the indirect effects, a key role in the observed changes in benthic vegetation is obviously played by the deterioration of the light climate under eutrophication conditions. Enhanced growth of epiphytic algae and phytoplankton results in shading of the rooting or attached benthic plants confined to the bottom. A comparison of the light requirements of seagrasses, macroalgae and phytoplankton (Duarte 1995) shows that benthic plants, although competitively superior to phytoplankton under low nutrient supply, are outcompeted by epiphytes and by the free-floating phytoplankton or unattached macroalgae when light availability decreases. One reason is that since the latter group grows on the benthic plants or floats above them in the water column, they occupy positions more favourable for light harvesting. In addition, the photosynthetic capacity, efficiency, and consequently growth rates of phytoplankton and monostromeous or finely branched macroalgae are significantly higher. This is because of their higher proportion of photosynthetically active cells compared to seagrasses and perennial macrophytes (Fig. 3).

A result of the deterioration of the light climate under eutrophication is that the depth distribution limits of benthic macrophytes move upward, in response to their minimum light requirements for growth: macroalgae can grow down to depths receiving 0.12–1.5% of surface light, whereas seagrasses require at least 11% (Lüning 1985; Duarte 1991; Markager and Sand-Jensen 1992). The changes in depth distribution may considerably reduce the area inhabited by macrophyte communities, and thus their biomass and reproductive potential for recruitment. Along the

German coasts of Kiel Bay (western Baltic), for example, the formerly abundant perennial *Fucus* communities, still growing down to 6–8 m depth in the 1950s, had completely disappeared below 2 m depth by 1988/1989. The total biomass decreased by more than 90% during this period (Vogt and Schramm 1991).

Another indirect effect combined with feeback is that increasing blooms of fast-growing, short-lived phytoplankton and macroalgae and their subsequent breakdown are coupled with increasing input of organic materials to the bottom systems, enhancing oxygen-consuming heterotrophic activities. Resulting oxygen depletion, anoxia and finally hydrogen sulphide development in the sediments lead to further inpairment and mortality of the benthic communities. Reducing conditions under anoxia accelerate nutrient release from the sediments which in turn raises the eutrophication level.

The reduction of seagrasses, large seaweeds, but also of the benthic microalgal films, involves changes in the hydrodynamic conditions in the bottom layer and in the sediment stability. This results in enhanced resuspension of sediments and transfer of sediment nutrients into the water column which further increases turbidity and nutrient levels, respectively.

Other feedback effects, both negative and positive, are related to the activities of herbivores, which prefer to graze on microalgae and epiphytes (Fig. 4). The fact that epiphytic macroalgae at low nutrient loadings do not flourish and outcompete the slower growing perennials may be the result of grazing on epiphytes which under nutrient limitation cannot grow fast enough to compensate for grazing losses (Borum 1987; Sand-Jensen and Borum 1991; Neckles et al. 1993; Cebrián and Duarte 1994). Enhanced growth of epiphytic micro- and macroalgae or phytoplankton due to eutrophication, on the other hand, supports the development of associated grazer and filter-feeder populations which in turn may accelerate the eutrophication process through regeneration of nutrients (e.g. Kautsky and Wallentinus 1980).

In summary, the changes in benthic vegetation due to eutrophication are a reaction chain of direct and indirect effects, feedback mechanisms and self-accelerating processes, which are difficult to control once initiated. Duarte (1995) uses the term "domino effect", meaning that the eutrophication process once started is maintained and amplified until the final phase of eutrophication is reached, i.e. the extinction of the macrophytobenthic communities.

The importance of the eutrophication induced changes derives not only from changes in the primary producer compartment, but also from changes in the structure and function of the whole ecosystem. Duarte (1995, p. 99) discusses eutrophication effects on the ecosystem level and

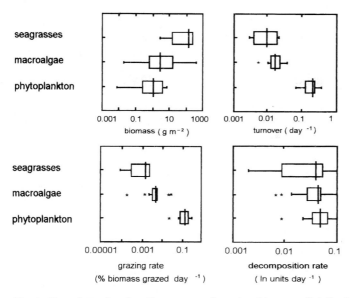

Fig. 4. Box plots showing the ranges of marine biomass distribution, biomass turnover rate, rates of grazing losses and decomposition for seagrasses, macroalgae, and phytoplankton. *Boxes* encompass the 25 and 75% quartiles of all the data for each plant group, the *central vertical line* represents the median, *bars* extend to the 95% confidence limits, and *asterisks* mark observations extending beyond the 95% confidence limits. (After Duarte 1995)

comes to the conclusion that the typical vegetation changes are to a lesser extent associated with quantitative alterations in primary production levels than they are with qualitative changes related to the different structural or functional roles of the different types of primary producers in the ecosystem. Changes in the structure of the primary producer component of the ecosystem from perennial communities to highly productive ephemeral forms as a result of additional nutrient input do not necessarily alter total primary production: seagrasses and long-lived macroalgae compensate for their slower growth rates by having a much higher biomass per unit area compared to plankton and short-lived macroalgae (Fig. 4). Probably more important is the fact that large macrophytes are essential components of the spatial structure of shallow-water systems: they provide a stable substrate for epiphytic communities, habitats for animals (physical support, shelter, spawning grounds), and strongly influence the hydrodynamic conditions by reducing turbulence and currents, resuspension and transport of sediments.

Another important ecosystem consequence of the change from long-lived to short-lived vegetation is the increased carbon and nutrient turnover rates and the associated heterotrophic activities (Fig. 4).

In his recent paper, Duarte (1995) emphasizes in his conclusions that present knowledge of the response of marine vegetation to eutrophication and the associated changes in the structure and function of the ecosystem is sufficient to make general qualitative predictions. However, he also concluded that we are still far away from predictive quantitative models, in particular from non-linear functional models which account for the high complexity of the eutrophication process due to feedback effects and self-acceleration. Such models which would enable us to define, for example, critical nutrient loadings at which the above-described reaction chain is triggered are urgently needed for coastal management and as a basis for environmental legislation for setting tolerable nutrient loadings.

Compared to our knowledge of the effects of increased nutrient loadings on marine vegetation, there is very little information available about the consequences of reduced nutrient loadings, or whether changes in benthic vegetation due to eutrophication are reversible. Comprehensive models that could predict the consequences of a reduction of nutrient loadings as planned, for example, on a larger scale for the Baltic or the North Sea (Helsinki and North Sea Convention), do not exist as far as we know. Only recently, Duarte (1995) presented a simple simulation model to show the time scales involved in the recovery of populations of different seagrasses in denuded areas. The model is based on the general observation that under eutrophication slow-growing plants are replaced by faster-growing ones in increasingly shorter recolonisation periods. This implies an inverse process under de-eutrophication, i.e. replacement of the fast-growing short-lived nutrient opportunists by slower-growing long-lived macrophytes. The length of the lag period between nutrient reduction and complete recovery of the system depends on the reproduction, recruitment and growth mode of the macrophytes. Re-colonization by rooting vascular plants, for example seagrasses, depends primarily on the supply of seeds from neighbouring fertile populations, formation of new patches by seedlings and subsequent patch growth through peripheral extension of the rhizome system (Duarte and Sand-Jensen 1990a, b). The few available reports on patch formation and patch growth reveal considerable differences between the seagrass species. For example, net patch formation ranges from 3×10^{-4} patches m^{-2} year^{-1} to 5×10^{-4} patches m^{-2} year^{-1} for *Posidonia ocenanica* and *Cymodocea nodosa*, respectively (Meinesz and Lefevre 1984; Duarte and Sand-Jensen 1990a), whereas rhizome elongation rates may range from 1 to over 500 cm year^{-1} in the largest and smallest seagrasses, respectively (Duarte 1991). Using these figures to simulate recovery for a wide combination of patch formation and patch growth, Duarte (1995) could show that the time scale of meadow formation (95% cover) would range from less than one year for fast growing seagrass species up to centuries for large, slow-growing forms. For example, the large Mediterranean seagrass *Posidonia oceanica* would need 280 years, the smaller, faster-growing *Cymodocea nodosa* about 6 years to re-colonize with 95% cover.

The few results from a rather simple model demonstrate impressively that coastal eutrophication and its effects may have severe long-lasting consequences, at least on rooting benthic vegetation and thus for coastal soft-bottom ecosystems. In hard-bottom systems, time scales involved in both the eutrophication and recovery processes are possibly different. Duarte (1995) points out that the recovery of seaweed populations may be considerably faster because of their greater reproduction and dispersion potential. Reproductive cells of macroalgae are shed in enormous numbers and can be transported over great distances because of their small size.

We also know little about the extent to which other functional or structural changes induced by nutrient loadings influence time course and time scales of both the eutrophication and the recovery processes. Among these other changes are secondary long-term effects associated with eutrophication, as for example altered hydrodynamic conditions after extinction of macrophyte communities, accumulated inorganic nutrients and organic materials in the sediments, or changed conditions for competition.

The development of more complex models that would allow quantitative prediciton of both eutrophication and recovery processes and that could provide measures for eutrophication control and management is a social demand and thus a challange for future research.

References

Borum J (1987) Dynamics of epiphyton on eelgrass (*Zostera marina* L.) leaves: relative roles of algal growth, herbivory, and substratum turnover. Limnol Oceanogr 32: 986–992

Borum J, Murray L, Kemp WM (1989) Aspects of nitrogen acquisition and conservation in eelgrass plants. Aquat Bot 35: 289–300

Cebrián J, Duarte CM (1994) Growth-dependent herbivory in natural plant communities. Funct Ecol 8: 518–525

Chapman ARO, Craigie JS (1977) Seasonal growth in *Laminaria longicruris*: relations with dissolved inorganic nutrients and internal reserves of nitrogen. Mar Biol 40: 197–205

Duarte CM (1991) Seagrass depth limits. Aquat Bot 40: 363–377

Duarte CM (1995) Submerged aquatic vegetation in relation to different nutrient regimes. Ophelia 41: 87–112

Duarte CM, Sand-Jensen K (1990a) Seagrass colonization: patch formation and patch growth in *Cymodocea nodosa*. Mar Ecol prog Ser 65: 183–191

Duarte CM, Sand-Jensen K (1990b) Seagrass colonization: biomass development and shoot demography in *Cymodocea nodosa* patches. Mar Ecol Prog Ser 67: 97–103

Harlin MM, Thorne-Miller B (1981) Nutrient enrichment of seagrass beds in a Rhode Island coastal lagoon. Mar Biol 65: 221–229

Hemminga MA, Harrison PG, van Lent F (1991) The balance between nutrient losses and gains in seagrass meadows. Mar Ecol Prog ser 71: 85–96

Kautsky N, Wallentinus I (1980) Nutrient release from a Baltic *Mytilus*-red algal community and its role in benthic and pelagic productivity. Ophelia (Suppl) 1: 17–30

Kemp WM, Twilley RT, Stevenson JC, Boynton WR, Means JC (1983) The decline of submerged vascular plants in upper Chesapeake Bay: summary of results concerning possible causes. Mar Technol Soc J 17: 78–89

Lüning K (1985) Meeresbotanik: Verbreitung, Ökophysiologie und Nutzung der marinen Meeresalgen. Thieme, Stuttgart, 374 pp

Markager S, Sand-Jensen K (1992) Light requirements and depth zonation of marine macroalgae. Mar Ecol Prog Ser 88: 83–92

Meinesz A, Lefevre JR (1984) Regeneration d'un herbiére de *Posidonia oceanica* quarante annes aprés sa destruction par une bombe dans la rade de Villefranche (Alpes-Maritimes. France). In: Boudouresque CF, Jeudy de Grissac A, Oliver J (eds) International workshop on *Posidonia oceanica* beds. GIS posidonie, Marseille, pp 39–44

Neckles HA, Wetzel RLK, Orth RJ (1993) Relative effects of nutrient enrichment and grazing on epiphytes-macrophyte (*Zostera marina* L.) dynamics. Oecologia 93: 285–295

Neundorfer JV, Kemp WM (1993) Nitrogen versus phosphorus enrichment of brackish waters: response of the submerged plant *Potamogeton perfoliatus* and its associated algal community. Mar Ecol Prog Ser 94: 71–82

Nienhuis PH (1983) Temporal and spatial patterns of eelgrass (*Zostera marina* L.) in a former estuary in the Netherlands, dominated by human activities. Mar Technol Soc J 17: 69–77

Nienhuis PH (1992) Ecology of coastal lagoons in the Netherlands (Veerse Meer and Grevelingen). Vie Milieu 42(2): 59–72

Patriquin DG (1972) The origin of nitrogen and phosphorus for growth of the marine angiosperm *Thalassia testudinum*. Mar Biol 15: 35–46

Pedersen MF, Borum J (1992) Nitrogen dynamics of eelgrass *Zostera marina* during late summer period of high growth and low nutrient availability. Mar Ecol Prog Ser 80: 65–73

Sand-Jensen K, Borum J (1991) Interactions among phytoplankton, periphyton, and macrophytes in temperate freshwaters and estuaries. Aquat Bot 41: 137–175

Schramm W, Abele D, Breuer G (1988) Nitrogen and phosphorus nutrition of two community forming seaweeds (*Fucus vesiculosus, Phycodrys rubens*) from the western Baltic (Kiel Bight) in the light of eutrophication processes. Kiel Meeresforsch Sonderh 6: 221–241

Twilley RR, Kemo WM, Staver KW, Stevenson JC, Boynton WR (1985) Nutrient enrichment of estuarine vascular plant communities. I. Algal growth and effects on production of plants and associated communities. Mar Ecol Prog Ser 23: 178–191

Vogt H, Schramm W (1991) Conspicuous decline of *Fucus* in Kiel Bay (western Baltic): what are the causes? Mar Ecol Prog Ser 69: 189–194

Wallentinus I (1981) Phytobenthos. In: Melvasalo T, Pawlak J, Grasshoff K, Thorell L, Tsiban A (eds) Assessment of the effects of pollution on the natural resources of the Baltic Sea, 1980. Baltic Environment Protection Commission, Helsinki, pp 322–342

Species Index

Subject Index

Ecological Studies
Volumes published since 1990

Ecological Studies
Volumes published since 1990

Springer-Verlag
and the Environment

We at Springer-Verlag firmly believe that an international science publisher has a special obligation to the environment, and our corporate policies consistently reflect this conviction.

We also expect our business partners – paper mills, printers, packaging manufacturers, etc. – to commit themselves to using environmentally friendly materials and production processes.

The paper in this book is made from low- or no-chlorine pulp and is acid free, in conformance with international standards for paper permanency.

Printing: Saladruck, Berlin
Binding: Buchbinderei Lüderitz & Bauer, Berlin